Contents

APPLIED CONTINUUM MECHANICS

T. J. CHUNG
University of Alabama in Huntsville

CAMBRIDGE
UNIVERSITY PRESS

Published by the Press Syndicate of the University of Cambridge
The Pitt Building, Trumpington Street, Cambridge CB2 1RP
40 West 20th Street, New York, NY 10011-4211, USA
10 Stamford Road, Oakleigh, Melbourne 3166, Australia

First published 1996

Printed in the United States of America

Library of Congress Cataloguing-in-Publication Data

Chung, T. J., 1929–
Applied continuum mechanics / T. J. Chung.
p. cm.
Includes bibliographical references (p. –) and index.
ISBN 0-521-48297-6
1. Continuum mechanics. I. Title.
QA808.2.C547 1996
620. 1 – dc20 95–10969
CIP

A catalogue record for this book is available from the British Library.

ISBN 0-521-48297-6 hardback

To my family

Preface

Continuum mechanics is one of the most important interdisciplinary subjects in engineering. Often, physicists and applied mathematicians are deeply involved in continuum mechanics in their pursuit of the physical and mathematical aspects of universal laws. Continuum mechanics is regarded as a mathematical process of engineering problems in general. For this reason, this book provides the first-year graduate student with a bridge to a destination – mechanical, aerospace, civil, or chemical engineering, whichever it may be.

At the University of Alabama in Huntsville, continuum mechanics is a required course for many engineering students, especially those who wish to pursue advanced degrees in the areas of solid mechanics, fluid mechanics, and heat transfer. This book begins with a review of vectors and tensors and their applications to domain and boundary integrals. This is followed by a study of kinematics that addresses deformations of a body – the concept of a strain tensor and the rate of deformation tensor. A clear view of Lagrangian coordinates and Eulerian coordinates is constructed, as they relate to the formation of the foundations for solid mechanics and fluid mechanics. This idea is then extended to kinetics – the concept of stress. Various approaches toward definitions of stress are demonstrated in small and large strain conditions. The kinematics and kinetics then constitute the basic building blocks toward solid mechanics, thermodynamics, fluid mechanics, and heat transfer; thus, the role of continuum mechanics becomes that of the mediator, thereby uniting all engineering disciplines under one roof.

In accomplishing this goal, a common language used is that of the tensor. This is a powerful tool that makes continuum mechanics a powerful subject. The basic concept of the tensor is introduced to the uninitiated reader at the beginning, and its full utilization with significant consequences is demonstrated in later chapters.

This volume is a revised version of *Continuum Mechanics*, published by Prentice-Hall in 1988. To make the present volume more suitable for a three credit hour semester course, some advanced topics such as inelastic materials and non-Newtonian fluids have been excluded, whereas introductory subjects have been expanded considerably.

In retrospect, two well-known treatises, "The Classical Field Theories" by Truesdell and Toupin, Vol. III/1, pp. 226–793 (1960) and "The Nonlinear Field Theories of Mechanics" by Truesdell and Noll, Vol. IV/3 (1965), both in the *Encyclopedia of Physics* (Springer-Verlag), have been important contributions to continuum mechanics with an emphasis on applied mathematics and physics. Such a focus, however, puts them beyond the scope of the traditional engineering curriculum. On the other hand, continuum mechanics textbooks designed specifically for engineering have been appearing since the 1960s, with varying degrees of mathematical rigor and with emphasis on different engineering topics. This text is an attempt to provide a reasonable balance between mathematical rigor and practical applications, solids and fluids, and linear and nonlinear theories. With these considerations in mind, this book is dedicated primarily to the practitioner in engineering.

I thank the reviewers of the original manuscript for many suggestions for improvement. A number of graduate students have also contributed to this book; I owe them a debt of gratitude.

My thanks are also due to Florence Padgett, Physical Sciences Editor at Cambridge University Press, who has most effectively managed the publication process of this book.

The manuscript was typed by Barbara Moore. To her also I am truly thankful.

1
Introduction

1.1 General

Continuum mechanics encompasses all scientific disciplines that describe the global behavior of gases, liquids, or solids under the influence of external disturbances. All real materials, as is well known, when studied at sufficiently large magnifications (that is, on a microscopic scale), display a molecular or atomic structure. In continuum mechanics, however, we adopt a macroscopic viewpoint: We ignore all the fine details of the molecular or atomic structure and, for the purpose of study, we replace the discontinuous microscopic medium with a *hypothetical continuum*. According to the resulting model, the field quantities, such as displacements, velocities, and stresses, are piecewise continuous functions of time and appropriate spatial coordinate systems.

The concept of a continuous medium makes the powerful methods of calculus available for the study of nonuniform distributions of physical variables and provides an easily visualized physical model that closely approximates everyday observations of matter in the large. Thus, in continuum mechanics, the mathematical analysis is often guided by intuition.

To distinguish the continuum or macroscopic model from a microscopic one, we may list a number of criteria. It is well known that an atom is an extremely small, electrically neutral particle that has a core, or nucleus, and one or more electrons outside its nucleus that are in constant motion around the nucleus. An uncharged particle resulting from the union of two or more atoms is called a molecule. This is the smallest particle of nonionic matter that has the characteristics of the substance. A concept of fundamental importance here is that of the *mean free path*, which can be defined as the average distance a molecule travels between successive collisions with other molecules. The ratio of the mean free path λ to the characteristic length S of the physical boundaries of interest, called the *Knudsen number Kn*, may be used to determine the dividing line between the macroscopic and microscopic models:

$$Kn = \frac{\lambda}{S} < 1 \quad \textit{macroscopic} \tag{1.1.1}$$

$$Kn = \frac{\lambda}{S} \geqslant 1 \quad \textit{microscopic}, \tag{1.1.2}$$

where $\lambda \simeq 10^{-7}$ cm for solids and liquids and $\lambda \simeq 10^{-6}$ cm for gases.[1]

For the macroscopic model, mass m is defined as a continuous function of volume Ω, such that density ρ is determined by the relation

$$\rho = \frac{\mathrm{d}m}{\mathrm{d}\Omega}, \tag{1.1.3}$$

whereas in the microscopic model, applicable mainly to rarefied gases,

$$\rho = \sum_{i=1}^{N} \rho_i = \sum_{i=1}^{N} m_i n_i, \tag{1.1.4}$$

where n_i denotes the number density of molecules per unit volume of a gas composed of a chemical species i.

The definitions of pressure for the macroscopic and microscopic models involve many variables. Temperatures, velocities, and densities are associated with the pressure according to different formulas, depending on the types of solids, liquids, and gases under consideration. We pursue these topics in the chapters that follow. In general, however, it may safely be assumed that, in a macroscopic model, the maximum pressure gradient $|\nabla p|_{\max}$ is bounded as:

$$|\nabla p|_{\max} \ll \frac{\Delta p}{\Delta L}, \tag{1.1.5}$$

where Δp denotes the pressure difference within an infinitesimal volume $\Delta\Omega$ and ΔL is a corresponding linear dimension of $\Delta\Omega$, taken as approximately 10^4 times the mean free path. For rarefied gases, in

[1] The mean free path in air at standard temperature and pressure is of the order of 6×10^{-6} cm. A sound wave with a frequency equal to 20,000 Hz would have, under the same conditions, a wavelength equal to 1.7 cm, so that the Knudsen number would be $Kn = 3.5 \times 10^{-6}$. Lower frequencies would correspond to proportionately lower Knudsen numbers. Thus, audible acoustic waves fall well within the limits of the continuum model. However, for extremely high frequency waves, the wavelength may be comparable to the mean free path. In such a case, a molecular model for matter would have to be used.

which ΔL is very large, the requirement expressed by Eq. (1.1.5) cannot be satisfied. The continuum model is then invalid, and we must employ the microscopic model, as required by Eqs. (1.1.2) and (1.1.4). The microscopic model is based on statistical mechanics, quantum mechanics, or the kinetic theory of gases, in any of which the details of atomic or molecular structure must be considered. Because this book is not concerned with the microscopic viewpoint we shall dispense with this subject in what follows.

Once the constitution of the medium is determined in a continuum model, it is then possible to predict the responses that will be produced by external agents, which may take the form of contact forces; heat; electrical, chemical, or mechanical energy; or any other type of disturbance. These external agents are commonly known as *boundary* and/or *initial conditions*.

The unified approach to the study of the global behavior of materials consists of, first, a thorough study of the basic principles common to all media and, second, a clear demonstration of the properties specific to the medium under consideration. The basic principles include the conservation of mass, the balance of linear and angular momentum, the conservation of energy, and the principle of entropy. The underlying assumption of the unified theory is that these principles are valid for all materials irrespective of their constitution. Thus, to account for the nature of different materials – the various types of solids, liquids, or gases – we require additional equations to describe the basic characteristics of the body and its response to the external agent under consideration.

The concept of a continuous medium enables us to develop so-called continuum theories based on continuity, homogeneity, and isotropy. Reasonable approximations, however, can extend the usefulness of continuum theories to include such conditions of discontinuity, inhomogeneity, and anisotropy as occur in fracture mechanics and anisotropic composite materials. Similarly, turbulence fluctuations, shock discontinuities, and reacting flows may be resolved by an extension of the continuum model through proper approximations without resorting to microscopic models.

In the course of our study, we will encounter the notion of *invariance* and various methods of mathematical analysis for dealing with it. *Tensor analysis* plays a major role throughout the text. We begin with the basic operations of the tensor in Section 1.2 and continue to work with this indispensable tool for the rest of the book.

Chapters 2 and 3 develop the concepts of *strain* (kinematics) and *stress* (kinetics), which will become the foundation for all other topics, as characterized by Lagrangian and Eulerian coordinates. Throughout this text, Cartesian and curvilinear tensors are emphasized equally. Properties of elastic materials are discussed in Chapter 4. We present modern, nonlinear theories, as well as classical, linear theories, with regard to the constitutive equations and the energy principles, along with the thermodynamics of solids based on the first and second laws of thermodynamics. Applications include thermomechanically coupled equations of motion and heat conduction, finite elasticity, torsion, and anisotropic composite materials. Chapter 5 deals with Newtonian fluids. The basic idea used in elastic solids, the first and second laws of thermodynamics, will be applied to derive the most general form of the Navier–Stokes system of equations for compressible viscous fluids. We then demonstrate how this general form can be simplified or modified to produce the governing equations for ideal flow, rotational flow, boundary-layer flow, turbulence, high-speed flow, acoustics, and reacting flow.

1.2 Vectors and Tensors

A vector is determined in a given reference frame by a set of components. If a new coordinate system is introduced, the same vector is determined by another set of components, and these new components are related, in a definite way, to the old ones. The law of transformation of components of a vector is the essence of the vector representation.

Tensors are founded on a notion similar to that of vectors, but are much broader in conception. Tensor analysis is concerned with the study of *abstract objects*, called *tensors*, whose properties are independent of the reference frames used to describe the objects. A tensor is represented in a particular reference frame by a set of functions, termed its *components*, just as a vector is determined by a set of components. Tensor analysis deals with entities and properties that are independent of the choice of reference frames. Thus, it forms an ideal tool for the study of natural laws because tensor equations are *invariant* with respect to a given category of coordinate transformations. Tensors are capable of delineating a variety of objects, ranging from scalars to multiple components of matter encountered in various physical phenomena. To discuss this subject further, however, it is necessary to introduce some notation and rules which will be applied to tensors and also to other topics in continuum mechanics.

Index Notation

A vector is denoted by a boldfaced letter symbol. A vector may be written in terms of its components by using indices. For example, consider a vector in a right-handed rectangular Cartesian coordinate system (Fig. 1.2.1),

$$\mathbf{A} = A_1\mathbf{i}_1 + A_2\mathbf{i}_2 + A_3\mathbf{i}_3,$$

where each $\mathbf{i}_i$ denotes one of the unit vectors and the indices $i = 1, 2, 3$ have a range of 3. This expression may be written as:

$$\mathbf{A} = \sum_{i=1}^{3} A_i\mathbf{i}_i. \tag{1.2.1}$$

A_i is known as the first-order system indicating the components of the vector $\mathbf{A}$. For simplicity, we may write Eq. (1.2.1) in the form

$$\mathbf{A} = A_i\mathbf{i}_i \quad (i = 1, 2, 3).$$

Note that *repeated indices* (sometimes called *dummy indices*) imply summing over the range of the index. Consider

$$x_i' = a_{ij}x_j = a_{i1}x_1 + a_{i2}x_2 + a_{i3}x_3 \quad (i, j = 1, 2, 3). \tag{1.2.2}$$

Here, j is the dummy index and i must change independently of x_j to give x_1', x_2', x_3', which indicates that Eq. (1.2.2) represents three equations. The index i, which is not repeated here, is called a *free index*.

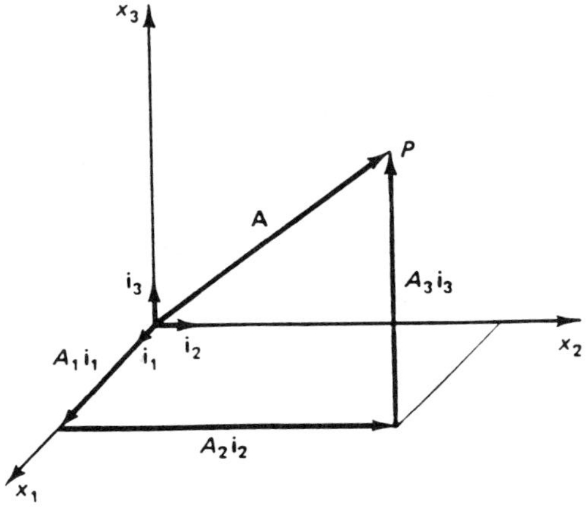

Figure 1.2.1 Three-dimensional representation of the vector **A** and its components.

The Kronecker delta is defined as:

$$\delta_{ij} = \begin{Bmatrix} 1 & if\ i = j \\ 0 & if\ i \neq j \end{Bmatrix}. \tag{1.2.3}$$

The permutation symbol is given by

$$\epsilon_{ijk} = \begin{Bmatrix} 1 & \text{for even permutations of } ijk \ (123, 231, 312) \\ -1 & \text{for odd permutations of } ijk \ (132, 213, 321) \\ 0 & \text{for two or more equal indices } (112, 111, \text{etc.}) \end{Bmatrix}. \tag{1.2.4}$$

Vector Multiplication

A dot product of any two vectors in index notation reads

$$\mathbf{A} \cdot \mathbf{B} = A_i\mathbf{i_i} \cdot B_j\mathbf{i_j} = A_iB_j\delta_{ij} = A_iB_i = A_jB_j = \lambda. \tag{1.2.5}$$

Because of the orthogonal or orthonormal coordinate system, a dot product of the unit vectors produces a Kronecker delta:

$$\mathbf{i}_i \cdot \mathbf{i}_j = \delta_{ij}. \tag{1.2.6}$$

The role of the Kronecker delta is to interchange the index of a component of a vector, as demonstrated in Eq. (1.2.5). Furthermore, the dot product of two vectors results in a scalar, λ, which does not have a free index.

On the other hand, a cross product of any two vectors in index notation reads:

$$\mathbf{A} \times \mathbf{B} = A_i\mathbf{i}_i \times B_j\mathbf{i}_j = A_iB_j\epsilon_{ijk}\mathbf{i}_k,$$

where the cross product of unit vectors produces a permutation symbol ϵ_{ijk} such that

$$\mathbf{i}_i \times \mathbf{i}_j = \epsilon_{ijk}\mathbf{i}_k. \tag{1.2.7}$$

Tensors

The quantities that appear in the foregoing paragraphs may be identified as tensors, because they satisfy the basic properties set forth at the beginning of this section. To provide a specific example, let us examine Eq. (1.2.2). Let $\mathbf{r}$ be a position vector in Fig. 1.2.2:

$$\mathbf{r} = x_i\mathbf{i}_i \quad (i = 1, 2, 3). \tag{1.2.8a}$$

If the old coordinates x_i are rotated by an angle θ about the x_3 axis to a set of new coordinates x_i', then the position vector can be represented by

$$\mathbf{r} = x_i'\mathbf{i}_i' \quad (i = 1, 2, 3). \tag{1.2.8b}$$

Figure 1.2.2 shows that a set of old coordinates x_i are rotated about the x_3 axis counterclockwise by an angle θ to a set of new coordinates x_i' given by

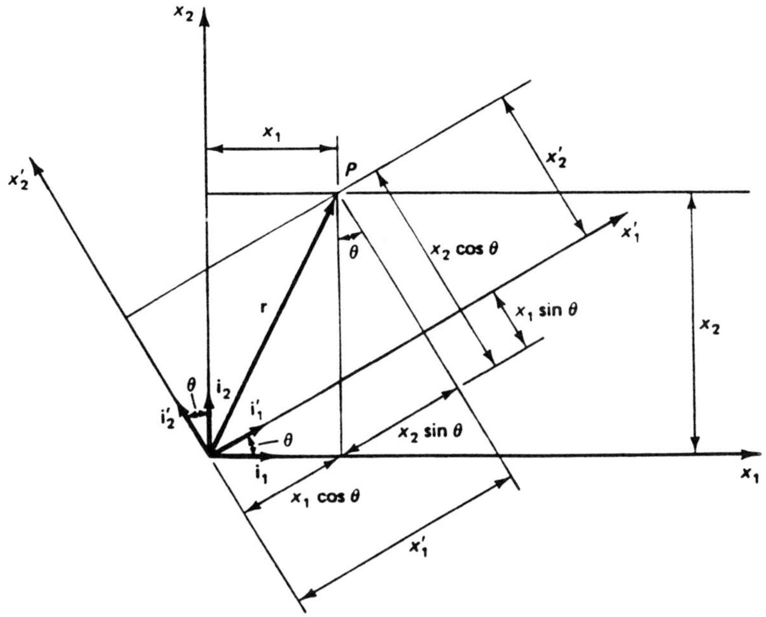

Figure 1.2.2 Coordinate transformation. Rotation between x_i and x_i' coordinates by an angle θ.

$$\begin{aligned} x_1' &= (\cos\theta)x_1 + (\sin\theta)x_2, \\ x_2' &= -(\sin\theta)x_1 + (\cos\theta)x_2, \\ x_3' &= x_3. \end{aligned} \tag{1.2.9}$$

These three equations may be combined to read (Fig. 1.2.3)

$$x_i' = a_{ij}x_j \quad (i, j = 1, 2, 3) \tag{1.2.10}$$

with

$$\begin{aligned} a_{11} &= \cos\theta_{11} = \cos\theta, \\ a_{12} &= \cos\theta_{12} = \cos\left(\frac{\pi}{2} - \theta\right) = \sin\theta, \\ a_{13} &= \cos\frac{\pi}{2} = 0, \\ a_{21} &= \cos\theta_{21} = \cos\left(\frac{\pi}{2} + \theta\right) = -\sin\theta, \\ a_{22} &= \cos\theta_{22} = \cos\theta, \\ a_{23} &= \cos\frac{\pi}{2} = 0, \end{aligned}$$

$$a_{31} = \cos\theta_{31} = \cos\frac{\pi}{2} = 0,$$

$$a_{32} = \cos\theta_{32} = \cos\frac{\pi}{2} = 0,$$

$$a_{33} = \cos\theta_{33} = \cos 0 = 1, \tag{1.2.11}$$

or, in matrix notation,

$$\begin{bmatrix} x'_1 \\ x'_2 \\ x'_3 \end{bmatrix} = \begin{bmatrix} \cos\theta & \sin\theta & 0 \\ -\sin\theta & \cos\theta & 0 \\ 0 & 0 & 1 \end{bmatrix} \begin{bmatrix} x_1 \\ x_2 \\ x_3 \end{bmatrix}. \tag{1.2.12}$$

Notice that the old coordinates x_j are transformed into the new coordinates x'_i by means of the quantities called the *transformation matrix* a_{ij}, whose components are the cosines of angles measured from the new coordinates x'_i to the old coordinates x_j. We have seen that the position vector **r** remains invariant through coordinate transformations. The quantities x_j, x'_i, and a_{ij} are the abstract objects whose properties remain invariant with the coordinate transformations. Therefore, they are all tensors, as defined earlier.

Similarly, we may consider a vector $\mathbf{A} = A_i\mathbf{i}_i$ oriented at θ_1, θ_2, and θ_3 from the respective rectangular Cartesian coordinate axes (Fig. 1.2.3a) so that

$$A_i = An_i \quad (i = 1, 2, 3),$$

where

$$A = \sqrt{A_1^2 + A_2^2 + A_3^2}$$

$$n_1 = \frac{A_1}{A} \quad n_2 = \frac{A_2}{A} \quad n_3 = \frac{A_3}{A}.$$

Here $n_i(n_1, n_2, n_3)$ are the direction cosines whose properties must satisfy $\mathbf{n}\cdot\mathbf{n} = n_in_i = n_1^2 + n_2^2 + n_2^2 = 1$. Note also in Eq. (1.2.12),

$$\mathbf{n}^{(1)}\cdot\mathbf{n}^{(1)} = a_{11}^2 + a_{12}^2 + a_{13}^2 = \cos^2\theta + \sin^2\theta = 1$$
$$\mathbf{n}^{(2)}\cdot\mathbf{n}^{(2)} = a_{21}^2 + a_{22}^2 + a_{23}^2 = (-\sin\theta)^2 + \cos^2\theta = 1$$
$$\mathbf{n}^{(3)}\cdot\mathbf{n}^{(3)} = a_{31}^2 + a_{32}^2 + a_{33}^2 = 1.$$

Proceeding in a similar manner, we may write (Fig. 1.2.3b)

$$A'_i = a_{ij}A_j,$$

where A'_i refers to a set of new coordinates so that the vector **A** is now measured in terms of the new coordinates, and a_{ij} acts as a set of direction cosines with i and j indicating the new and old coordinates, respectively. Again, the vector components A'_i, A_j, as well as a_{ij}, are

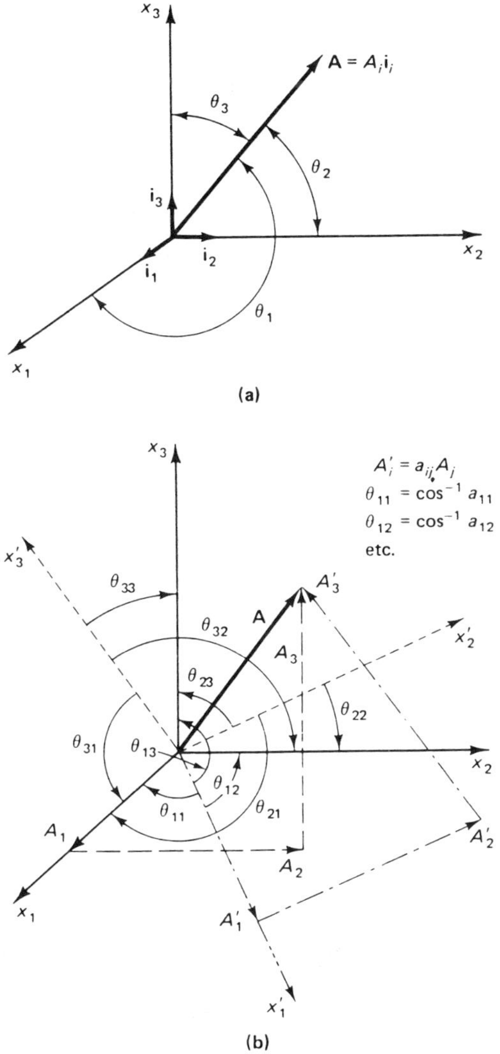

Figure 1.2.3 Coordinate transformations. (a) Representation of vector **A**. (b) Representation of a vector **A** in two sets of right-handed Cartesian axes with different orientation.

tensors. Once a quantity is determined to be a tensor, then the number of indices determines the order of the tensor. For example, A_i and a_{ij} are first- and second-order tensors, respectively. Thus we define, in general,

$$A = \text{zero-order tensor},$$

$$A_i = \text{first-order tensor},$$
$$A_{ij} = \text{second-order tensor},$$
$$A_{ijk} = \text{third-order tensor},$$
$$A_{ijkl} = \text{fourth-order tensor}.$$

Example 1.1 Consider a set of new axes x_i' obtained by rotating the old axes x_i through a 60° angle counterclockwise about the x_2 axis. What are the components of a vector **A** in the new coordinates if A_i in the old coordinates is (2, 1, 3)?

Solution. The direction cosines a_{ij} between the old and new axes are

$$a_{ij} = \begin{bmatrix} \cos 60° & 0 & -\sin 60° \\ 0 & 1 & 0 \\ \sin 60° & 0 & \cos 60° \end{bmatrix}.$$

From the transformation law

$$A_i' = a_{ij}A_j$$

we obtain

$$A_1' = a_{11}A_1 + a_{12}A_2 + a_{13}A_3 = \frac{2 - 3\sqrt{3}}{2}$$
$$A_2' = a_{21}A_1 + a_{22}A_2 + a_{23}A_3 = 1$$
$$A_3' = a_{31}A_1 + a_{32}A_2 + a_{33}A_3 = \frac{2\sqrt{3} + 3}{2},$$

which demonstrates that the vector **A** remains invariant, as shown below.

$$\text{Old coordinates, } |\mathbf{A}| = \sqrt{A_1^2 + A_2^2 + A_3^2} = \sqrt{14}$$
$$\text{New coordinates, } |\mathbf{A}| = \sqrt{A_1'^2 + A_2'^2 + A_3'^2} = \sqrt{14}.$$

Example 1.2 Let the coordinate transformations be carried out successively by first rotating about the x_1 axis counterclockwise with $\theta_x = 30°$, then about the x_2' axis clockwise with $\theta_y = 46°$, and finally, about the x_3'' axis clockwise with $\theta_z = 60°$, as shown in Fig. 1.2.4. Determine the resulting direction cosines and calculate the components of **A** (2, 1, 3) in terms of these rotations.

Solution. First, rotation about the x_1 axis gives

$$A_i' = a_{ij}'A_j.$$

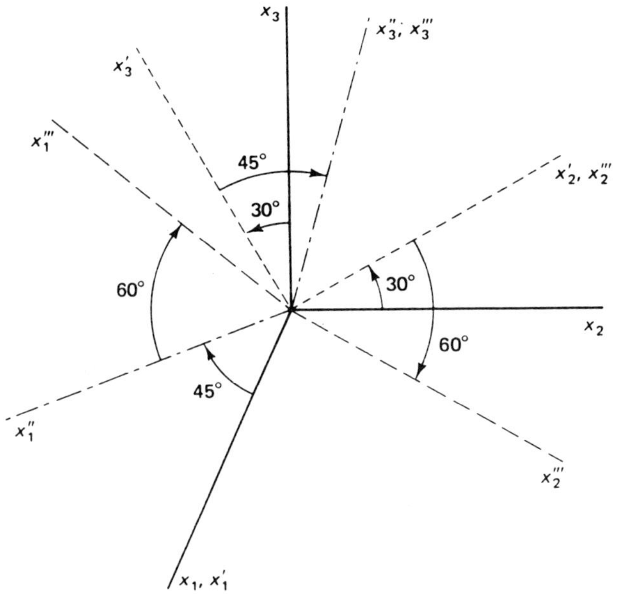

Figure 1.2.4 Successive rotations in the right-handed system first about x_1 counterclockwise ($\theta_x = 30°$), and then about x_2' clockwise ($\theta_y = 45°$), and finally about x_3'' clockwise ($\theta_x = 60°$) (Example 1.2).

Second, following the rotation about the x_2' axis,

$$A_i'' = a_{ij}''A_j' = a_{ij}''a_{jk}'A_k.$$

The third rotation about the x_3'' axis gives

$$A_i''' = a_{ij}'''A_j'' = a_{ij}'''a_{jk}''a_{km}'A_m = a_{ij}A_j.$$

Note that the summing of the repeated indices is equivalent to the matrix multiplications:[2]

$$a_{ij} = \begin{bmatrix} c_z & -s_z & 0 \\ s_z & c_z & 0 \\ 0 & 0 & 1 \end{bmatrix} \begin{bmatrix} c_y & 0 & s_y \\ 0 & 1 & 0 \\ -s_y & 0 & c_y \end{bmatrix} \begin{bmatrix} 1 & 0 & 0 \\ 0 & c_x & s_x \\ 0 & -s_x & c_x \end{bmatrix}$$
$$= \begin{bmatrix} c_z\,c_y & -c_z\,s_y\,s_x - s_z\,c_x & c_z\,s_y\,c_x - s_z\,s_x \\ s_z\,c_y & -s_z\,s_y\,s_x + c_z\,c_x & s_z\,s_y\,c_x + c_z\,s_x \\ -s_y & -c_y\,s_x & c_y\,c_x \end{bmatrix}$$

[2] Matrix multiplications for the second-order tensors are often possible only if there are repeated indices between two second-order tensors. Depending on the position of repeated indices, the summing operation may require the transpose of matrices to be multiplied. To avoid any confusion, the reader is advised not to attempt matrix multiplication if the repeated indices occur irregularly, but rather to perform tensorial summing faithfully, as dictated by the repeated indices.

with $s_x = \sin\theta_x$, $s_y = \sin\theta_y$, $s_z = \sin\theta_z$, $c_x = \cos\theta_x$, $c_y = \cos\theta_y$, $c_z = \cos\theta_z$

$$= \frac{1}{\sqrt{2}}\begin{bmatrix} \frac{1}{2} & \frac{-1-3\sqrt{2}}{4} & \frac{\sqrt{3}-\sqrt{6}}{4} \\ \frac{\sqrt{3}}{2} & \frac{-\sqrt{3}+\sqrt{6}}{4} & \frac{3+\sqrt{2}}{4} \\ -1 & -\frac{1}{2} & \frac{\sqrt{3}}{2} \end{bmatrix}.$$

Thus, the components of **A** based on the new coordinates x_i''' are given by

$$A_i''' = a_{ij}A_j,$$
$$A_1''' = \frac{3}{4\sqrt{2}}(1 - \sqrt{2} + \sqrt{3} - \sqrt{6}),$$
$$A_2''' = \frac{1}{4\sqrt{2}}(3\sqrt{3} + \sqrt{6} + 9 + 3\sqrt{2}),$$
$$A_3''' = \frac{1}{2\sqrt{2}}(-5 + 3\sqrt{3}).$$

Once again, the vector **A** remains invariant with the coordinate transformation ($|\mathbf{A}| = \sqrt{14}$).

It is interesting to note that the old coordinates x_j may be solved from Eq. (1.2.12) in terms of the new coordinates x_i' in the form

$$x_j = a_{ij}x_i', \tag{1.2.13a}$$

or

$$x_i = a_{ji}x_j'. \tag{1.2.13b}$$

Substitute Eq. (1.2.13a) into Eq. (1.2.12) to give

$$x_i' = a_{ij}a_{kj}x_k', \tag{1.2.14}$$

which requires

$$a_{ij}a_{kj} = \delta_{ik}. \tag{1.2.15a}$$

Substitute this into Eq. (1.2.14) to obtain

$$x_i' = \delta_{ik}x_k' = x_i'.$$

In matrix notation Eq. (1.2.15a) can be written as

$$[a]\,[a]^T = [I], \tag{1.2.15b}$$

where I is the unit matrix. The transformation matrix of this kind is called the *proper orthogonal matrix* and has the property

$$[a]^T = [a]^{-1} \tag{1.2.16a}$$

and

$$\det[a] = 1. \tag{1.2.16b}$$

Notice that the transpose arises in Eq. (1.2.13a, b) from the repeated index occurring with the first index in a_{ij} rather than the second index, as in Eq. (1.2.10). The matrix form of a_{ij} in Eq. (1.2.12) is the proper orthogonal matrix because it satisfies the property given by Eq. (1.2.16a).

A tensor that has the same components in all orthogonal coordinate systems is referred to as an *isotropic tensor*. The Kronecker delta is an isotropic tensor of order 2:

$$a_{ik}a_{jn}\delta_{ij} = a_{jk}a_{jn} = \delta_{kn}. \tag{1.2.17}$$

A scalar such as λ in Eq. (1.2.5) is an isotropic tensor of order 0. There is no first-order isotropic tensor. The permutation symbol in orthogonal coordinate systems is the isotropic tensor of order 3. Isotropic tensors of a higher order arise in the descriptions of material constants, which will be discussed in Chapter 4 (see Eq. [4.1.23]).

The permutation symbol is related to the Kronecker delta as:

$$\begin{aligned}\epsilon_{ijk} &= \epsilon_{rst}\delta_{ir}\delta_{js}\delta_{kt}\\ &= \delta_{i1}\delta_{j2}\delta_{k3} - \delta_{i1}\delta_{j3}\delta_{k2} + \delta_{i2}\delta_{j3}\delta_{k1} - \delta_{i2}\delta_{j1}\delta_{k3} + \delta_{i3}\delta_{j1}\delta_{k2}\\ &\quad - \delta_{i3}\delta_{j2}\delta_{k1}.\end{aligned}$$

Thus, ϵ_{ijk} becomes the determinant of the form

$$\epsilon_{ijk} = \begin{vmatrix} \delta_{i1} & \delta_{j1} & \delta_{k1} \\ \delta_{i2} & \delta_{j2} & \delta_{k2} \\ \delta_{i3} & \delta_{j3} & \delta_{k3} \end{vmatrix}. \tag{1.2.18}$$

A product $\epsilon_{ijk}\epsilon_{mnp}$ can then be computed in this manner, noting that the determinant of a matrix is equal to the determinant of its transposed matrix so that $|A||B| = |A^{\mathrm{T}}||B| = |[A]^{\mathrm{T}}[B]|$, with $[A]$ and $[B]$ being square matrices. See Problem 1.5(b) at the end of this chapter.

Derivatives of Vectors

A spatial rate of change of tensor field is obtained by the *del operator*, $\boldsymbol{\nabla}$, defined as:

$$\boldsymbol{\nabla} = \mathbf{i}_i\frac{\partial}{\partial x_i} = \mathbf{i}_1\frac{\partial}{\partial x_1} + \mathbf{i}_2\frac{\partial}{\partial x_2} + \mathbf{i}_3\frac{\partial}{\partial x_3}. \tag{1.2.19}$$

For a scalar field, ψ, we write the del of ψ as

$$\nabla\psi = \mathbf{i}_i \frac{\partial \psi}{\partial x_i} = \psi_{,i}\mathbf{i}_i. \tag{1.2.20}$$

Here, the partial derivative of ψ with respect to x_i is written in a compact form

$$\frac{\partial \psi}{\partial x_i} = \psi_{,i},$$

where the comma denotes a partial derivative with respect to an independent variable x_i. We use this notation for simplicity unless confusion is likely to arise (that is, this practice must be avoided if the derivative is with respect to a dependent variable).

The dot product of $\nabla\psi$ and a vector normal to a surface may be written in the form

$$\mathbf{n}\cdot\nabla\psi = n_i\mathbf{i}_i \cdot \mathbf{i}_j \frac{\partial \psi}{\partial x_j} = \psi_{,j} n_i \delta_{ij} = \psi_{,i} n_i = \frac{\partial \psi}{\partial n},$$

where the alternate notation $\partial/\partial n = n_i(\partial/\partial x_i)$ is used.

The divergence of a vector $\mathbf{V}$ is given by the notation

$$\nabla\cdot\mathbf{V} = \mathbf{i}_i \frac{\partial}{\partial x_i} \cdot V_j \mathbf{i}_j = \frac{\partial V_j}{\partial x_i}\delta_{ij} = \frac{\partial V_i}{\partial x_i} = V_{i,i}. \tag{1.2.21}$$

Likewise, the curl of a vector $\mathbf{V}$ takes the form

$$\nabla\times\mathbf{V} = \mathbf{i}_i \frac{\partial}{\partial x_i} \times V_j \mathbf{i}_j = \frac{\partial V_j}{\partial x_i}\epsilon_{ijk}\mathbf{i}_k = V_{j,i}\epsilon_{ijk}\mathbf{i}_k. \tag{1.2.22}$$

Similarly, the curl of the cross product of two vectors $\mathbf{V}$ and $\mathbf{W}$ reads

$$\begin{aligned}
\nabla\times(\mathbf{V}\times\mathbf{W}) &= \mathbf{i}_j \frac{\partial}{\partial x_j} \times (V_k\mathbf{i}_k \times W_n\mathbf{i}_n) \\
&= \mathbf{i}_j \frac{\partial}{\partial x_j} \times (V_k W_n \epsilon_{knm}\mathbf{i}_m) = (V_k W_n)_{,j}\epsilon_{knm}\epsilon_{jmi}\mathbf{i}_i \\
&= (V_{k,j}W_n + V_k W_{n,j})(\delta_{ki}\delta_{nj} - \delta_{kj}\delta_{ni})\mathbf{i}_i \\
&= (V_{i,j}W_j + V_i W_{j,j} - V_{j,j}W_i - V_j W_{i,j})\mathbf{i}_i \\
&= (\mathbf{W}\cdot\nabla)\mathbf{V} + \mathbf{V}(\nabla\cdot\mathbf{W}) - \mathbf{W}(\nabla\cdot\mathbf{V}) - (\mathbf{V}\cdot\nabla)\mathbf{W} \\
&= A_1\mathbf{i}_1 + A_2\mathbf{i}_2 + A_3\mathbf{i}_3,
\end{aligned} \tag{1.2.23}$$

where

$$\begin{aligned}
A_1 = {} & V_{1,2}W_2 + V_{1,3}W_3 + V_1 W_{2,2} + V_1 W_{3,3} - V_{2,2}W_1 - V_{3,3}W_1 \\
& - V_2 W_{1,2} - V_3 W_{1,3},
\end{aligned}$$

$$A_2 = V_{2,1}W_1 + V_{2,3}W_3 + V_2W_{1,1} + V_2W_{3,3} - V_{1,1}W_2 - V_{3,3}W_2 - V_1W_{2,1} - V_3W_{2,3},$$

$$A_3 = V_{3,1}W_1 + V_{3,2}W_2 + V_3W_{1,1} + V_3W_{2,2} - V_{1,1}W_3 - V_{2,2}W_3 - V_1W_{3,1} - V_2W_{3,2}.$$

It is important to realize that in the preceding example one of the powerful features of tensor analysis has been demonstrated; that is, we have shown the vector identity

$$\nabla \times (\mathbf{V} \times \mathbf{W}) = (\mathbf{W} \cdot \nabla)\mathbf{V} + \mathbf{V}(\nabla \cdot \mathbf{W}) - \mathbf{W}(\nabla \cdot \mathbf{V}) - (\mathbf{V} \cdot \nabla)\mathbf{W}.$$

Note that this identity is a consequence of the routine tensorial manipulations in the summing of repeated indices. The reader is encouraged to verify the components A_1, A_2, and A_3 in Eq. (1.2.23) using the standard determinant calculations.

Additional discussions of vectors and tensors will appear in later chapters; it will be seen that they play an indispensable role in continuum mechanics.

1.3 The Green–Gauss Theorem and Boundary Surfaces

The Green–Gauss theorem is the relation that connects the surface integral for a vector $\mathbf{V}$ in the direction normal to an inclined surface with the volume integral of the divergence of $\mathbf{V}$:

$$\int_\Gamma \mathbf{V} \cdot \mathbf{n}\, d\Gamma = \int_\Omega \nabla \cdot \mathbf{V}\, d\Omega, \tag{1.3.1}$$

where Γ and Ω denote the boundary surface and the domain, respectively, and $\mathbf{n}$ is the vector normal to the surface (Fig. 1.3.1). In terms of index notation, the Green–Gauss theorem is written

$$\int_\Gamma V_i n_i\, d\Gamma = \int_\Omega V_{i,i}\, d\Omega. \tag{1.3.2}$$

To prove a more general case of Eq. (1.3.1) let us consider

$$\int_\Gamma \alpha\mathbf{V} \cdot \mathbf{n}\, d\Gamma = \int_\Omega \nabla \cdot (\alpha\mathbf{V})\, d\Omega = \int_\Omega \alpha\nabla \cdot \mathbf{V}\, d\Omega + \int_\Omega (\mathbf{V} \cdot \nabla)\alpha\, d\Omega, \tag{1.3.3a}$$

or

$$\int_\Gamma \alpha V_i n_i\, d\Gamma = \int_\Omega \alpha V_{i,i}\, d\Omega + \int_\Omega V_i \alpha_{,i}\, d\Omega. \tag{1.3.3b}$$

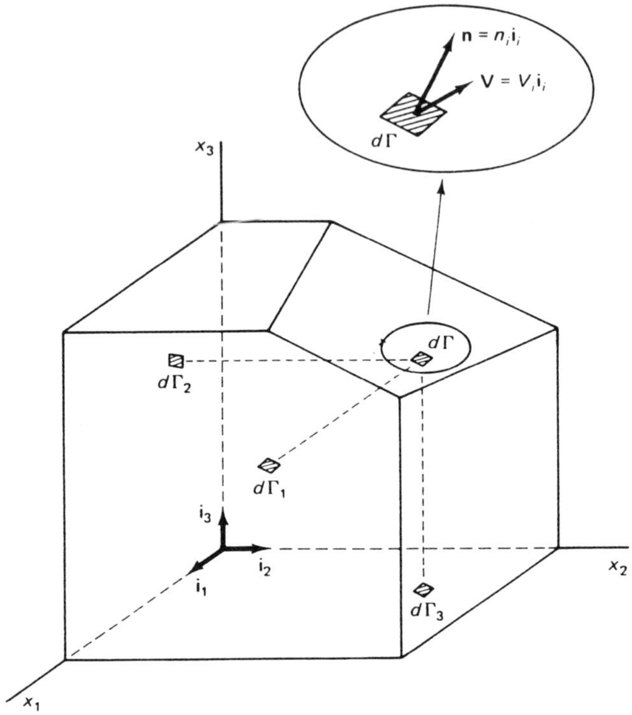

Figure 1.3.1 Projections of an inclined boundary surface area $\mathrm{d}\Gamma$ on each of the three orthogonal planes of the three-dimensional body are $\mathrm{d}\Gamma_1$, $\mathrm{d}\Gamma_2$, and $\mathrm{d}\Gamma_3$, which may be regarded as shadows or images of $\mathrm{d}\Gamma$ on the x_1, x_2, and x_3 planes, respectively.

To see that αV_i acting on $\mathrm{d}\Gamma$ on the inclined surface may be broken into three components on the plane surfaces of three orthogonal walls (Fig. 1.3.1), the first term of the right-hand side of Eq. (1.3.3b) will be integrated by parts,

$$\begin{aligned}\int_\Omega \alpha\left(\frac{\partial V_1}{\partial x_1} + \frac{\partial V_2}{\partial x_2} + \frac{\partial V_3}{\partial x_3}\right) \mathrm{d}x_1\,\mathrm{d}x_2\,\mathrm{d}x_3 \\ = \int_\Gamma \alpha V_1\,\mathrm{d}x_2\,\mathrm{d}x_3 + \int_\Gamma \alpha V_2\,\mathrm{d}x_1\,\mathrm{d}x_3 + \int_\Gamma \alpha V_3\,\mathrm{d}x_1\,\mathrm{d}x_2 \qquad (1.3.4) \\ - \int_\Omega \left(V_1\frac{\partial \alpha}{\partial x_1} + V_2\frac{\partial \alpha}{\partial x_2} + V_3\frac{\partial \alpha}{\partial x_3}\right) \mathrm{d}x_1\,\mathrm{d}x_2\,\mathrm{d}x_3.\end{aligned}$$

Notice that $\mathrm{d}x_2\,\mathrm{d}x_3$, $\mathrm{d}x_1\,\mathrm{d}x_3$, and $\mathrm{d}x_1\,\mathrm{d}x_2$ are the infinitesimal areas of images or shadows of $\mathrm{d}\Gamma$ projected upon plane surfaces of three orthogonal walls, normal to the directions x_1, x_2, and x_3, respectively.

$$dx_2\,dx_3 = d\Gamma_1 = n_1\,d\Gamma, \tag{1.3.5a}$$

$$dx_1\,dx_3 = d\Gamma_2 = n_2\,d\Gamma, \tag{1.3.5b}$$

$$dx_1\,dx_2 = d\Gamma_3 = n_3\,d\Gamma, \tag{1.3.5c}$$

where n_i (n_1, n_2, n_3) are the direction cosines denoting the components of the unit vector normal to the surface, as shown in Fig. 1.3.1. Therefore, Eq. (1.3.4) may be written:

$$\int_\Omega \alpha(V_{1,1} + V_{2,2} + V_{3,3})\,d\Omega = \int_\Gamma \alpha(V_1 n_1 + V_2 n_2 + V_3 n_3)\,d\Gamma - \int_\Omega (V_1\alpha_{,1} + V_2\alpha_{,2} + V_3\alpha_{,3})\,d\Omega. \tag{1.3.6}$$

This is identical to Eq. (1.3.3b) expanded into components. We conclude that integration by parts of an equation in the form given by Eq. (1.3.3b) is equivalent to an application of the Green–Gauss theorem, as presented in Eq. (1.3.3a). Although this demonstration appears to be a trivial exercise, Eq. (1.3.5) shows how we construct the appropriate direction cosines for inclined surfaces for use in numerical models of boundary-value problems with irregular multidimensional boundary geometries.

To demonstrate a further application, we may consider the second-order differential equation

$$\boldsymbol{\nabla}^2\psi = 0 \tag{1.3.7}$$

or

$$(\boldsymbol{\nabla}\cdot\boldsymbol{\nabla})\psi = \left(\mathbf{i}_i\frac{\partial}{\partial x_i}\cdot\mathbf{i}_j\frac{\partial}{\partial x_j}\right)\psi = \frac{\partial^2}{\partial x_i\partial x_j}\delta_{ij}\psi = \frac{\partial^2\psi}{\partial x_i\partial x_i} = \psi_{,ii} = 0. \tag{1.3.8}$$

Let us now take the inner product of Eq. (1.3.8) and a scalar ϕ and twice integrate by parts:

$$(\nabla^2\psi, \phi) \equiv \int_\Omega \psi_{,ii}\phi\,d\Omega = \int_\Gamma \psi_{,i}n_i\phi\,d\Gamma - \int_\Omega \psi_{,i}\phi_{,i}\,d\Omega = \int_\Gamma (\psi_{,i}n_i\phi - \psi n_i\phi_{,i})\,d\Gamma + \int_\Omega \psi\phi_{,ii}\,d\Omega. \tag{1.3.9}$$

Here, $\psi_{,i}n_i$ and ψ, which appear in the integrand of the surface integral, represent the gradients of ψ and the value of ψ normal to the surface, respectively. They are the boundary data, called *boundary conditions*, to be specified on the boundary surfaces. This implies that the solution of governing equation (1.3.7) is subject to the boundary conditions consisting of derivatives of all orders lower than the highest derivatives in the governing differential equation. Here, the first-order derivative $\psi_{,i}n_i$ is called the Neumann or natural boundary condition, whereas the zero derivative ψ is known as the Dirichlet or essential boundary condition.

As an additional example, consider the inner product of a biharmonic equation and a scalar ϕ

$$(\nabla^4\psi, \phi) \equiv \int_\Omega (\nabla^4\psi)\phi \, d\Omega = \int_\Omega \left(\frac{\partial^4\psi}{\partial x^4} + 2\frac{\partial^4\psi}{\partial x^2 \partial y^2} + \frac{\partial^4\psi}{\partial y^4} \right) \phi \, dx \, dy. \tag{1.3.10}$$

This may be written in index notation and integrated by parts four consecutive times:

$$\int_\Omega \psi_{,iijj}\phi \, d\Omega = \int_\Gamma (\psi_{,iij}n_j\phi - \psi_{,ii}n_j\phi_{,j} + \psi_{,i}n_i\phi_{,jj} - \psi n_i\phi_{,jji}) \, d\Gamma + \int_\Omega \psi\phi_{,jjii} \, d\Omega \quad (i, j = 1, 2) \tag{1.3.11}$$

in which integrations by parts have been carried out four times with one set of boundary conditions emerging at each integration. Notice that integration by parts can be carried out much more efficiently using the index notation; otherwise, the mixed derivative term in Eq. (1.3.10) must be split into two terms:

$$2\frac{\partial^4\psi}{\partial x^2 \partial y^2} = \frac{\partial^4\psi}{\partial x^2 \partial y^2} + \frac{\partial^4\psi}{\partial y^2 \partial x^2}. \tag{1.3.12}$$

Integrations by parts will be carried out for the first term on the right-hand side of Eq. (1.3.12), twice with respect to x and then twice with respect to y. This process is reversed for the second term on the right-hand side. Then the fourth-order equation produces boundary conditions with third, second, first, and zero derivatives of ψ. They are important considerations for the bending of plates and stream function descriptions of Navier–Stokes equations. To distinguish between the Neumann and Dirichlet boundary conditions, in general, we may consider a $2m$ order differential equation, where m is an integer (i.e., $m = 1$ and $m = 2$ correspond to the second- and fourth-order differential equations, respectively). The Neumann boundary conditions then consist of derivatives of the order $2m - 1, 2m - 2, \ldots, m$, whereas the Dirichlet boundary conditions consist of derivatives of the order $m - 1, \ldots, 0$. Thus, the Neumann boundary conditions in Eq.

(1.3.11) are $\psi_{,iij}n_j$ and $\psi_{,ii}n_j$, which represent the boundary shears and moments for the bending of plates and boundary forces and velocity gradients in incompressible viscous flows, respectively. The Dirichlet boundary conditions are given by $\psi_{,i}n_i$ and ψn_i, which represent, respectively, the boundary slopes and displacements for the plate bending and the boundary velocity and stream function in incompressible viscous flows.

If $\psi = \phi$ in Eq. (1.3.10) represents the transverse displacement of a plate in bending, then $(\nabla^4\psi)\phi$ is the energy or work mobilized in undergoing the transverse displacement. Thus, each boundary term in Eq. (1.3.11) implies the energy exerted at the boundaries due to various kinds of boundary conditions in balance with the work done in the domain of the plate. Obviously, for free boundaries, all boundary terms must vanish and the left-hand side is equal to the right-hand side for $\phi = \psi$.

In general, for the governing equations in engineering, the highest derivative is of an even order so that the above criteria for boundary conditions always hold. However, for the equation characterized by Eq. (1.3.3),

$$\int_\Omega \boldsymbol{\nabla} \cdot (\rho \mathbf{V})\, \mathrm{d}\Omega = \int_\Gamma \rho \mathbf{V} \cdot \mathbf{n}\, \mathrm{d}\Gamma, \tag{1.3.13}$$

in which α is now set equal to the density ρ and $\mathbf{V}$ is the velocity, we have $2m = 1$ and $m = 1/2$. In this case, the Neumann boundary condition occurs at the order $2m - 1 = 0$ and the Dirichlet boundary condition also occurs at the order $m - m = 0$. Physically, the quantity $\rho\mathbf{V} \cdot \mathbf{n}$ is the mass flow normal to the boundary surface, identified here as both Neumann and Dirichlet boundary conditions. The Neumann and Dirichlet boundary conditions coincide in this case.

In summary, the integration of derivatives in partial differential equations produces various boundary conditions. To this end, tensor algebra is a convenient tool to perform integration by parts. Further discussions of boundary conditions will be given in Chapters 4 and 5.

Remarks

This chapter has introduced a short summary of the definitions for continuum mechanics and tensor analysis. We recommend Sokolnikoff (1958) and Aris (1962) for additional reading. Before embarking on the remaining chapters, the reader is reminded that this book is intended as a reasonable compromise between the highly theoretical treatise and the practical engineering approach. Eringen (1962) or

Gurtin (1981), among others, represents the former, and Frederick and Chang (1965) or Malvern (1969), among others, belongs to the latter. The reader may wish to examine them all in light of this text. At the journey's end, we hope that the wisdom of balance between mathematical rigor and practical applications, and that the title 'applied' continuum mechanics be recognized and appreciated.

Problems

1.1 Let $\mathbf{A} = 3\mathbf{i}_1 + 2\mathbf{i}_2 + 4\mathbf{i}_3$. Compute the direction cosines with respect to the (x_1, x_2, x_3) axes of the line containing the vector $\mathbf{A}$. Calculate the corresponding angles between $\mathbf{A}$ and each component line (A_1, A_2, A_3).

1.2 Let x_i be rotated 30° counterclockwise about the x_1 axis to x_i' through transformations a_{ij} in Problem 1.1 such that

$$A_i' = a_{ij}A_j \quad (i, j = 1, 2, 3).$$

Determine the new components A_i' of this vector.

1.3 Let x_i be rotated 60° clockwise successively about the x_2, x_1', and x_3'' axes. Calculate the components of a vector $\mathbf{A}$ in terms of the final set of coordinates x_i''' if the components of $\mathbf{A}$ based on the original coordinates x_i are (1, 1, 1).

1.4 Calculate the component of a vector $\mathbf{A}$ (2, 3, 1) in a direction normal to the plane $x_1 = x_2$.

1.5 Prove:

(a) $\delta_{ii} = 3,$

(b) $\epsilon_{inm}\epsilon_{jpq} = \delta_{ij}\delta_{np}\delta_{mq} + \delta_{ip}\delta_{nq}\delta_{mj} + \delta_{iq}\delta_{nj}\delta_{mp} - \delta_{ij}\delta_{nq}\delta_{mp} - \delta_{ip}\delta_{nj}\delta_{mq} - \delta_{iq}\delta_{np}\delta_{mj},$

(c) $\epsilon_{inm}\epsilon_{ipq} = \delta_{np}\delta_{mq} - \delta_{nq}\delta_{mp},$

(d) $\epsilon_{inm}\epsilon_{inq} = 2\delta_{mq},$

(e) $\epsilon_{inm}\epsilon_{inm} = 6.$

1.6 Prove Eq. (1.2.23).

1.7 Using index notation, prove the relation

$$(\mathbf{V} \cdot \nabla)\mathbf{V} = \tfrac{1}{2}\nabla(\mathbf{V} \cdot \mathbf{V}) - \mathbf{V} \times (\nabla \times \mathbf{V}).$$

1.8 Show that

(a) $(\mathbf{V} \cdot \nabla)\phi = V_i\phi_{,i},$

(b) $\nabla(\nabla \cdot \mathbf{V}) = V_{j,ji}\mathbf{i}_i,$

and expand completely for $i, j = 1, 2, 3$.

1.9 Consider an inclined boundary surface identified by the coordinates at A $(4, 4, 3)$, B $(0, 4, 2)$, and C $(4, 2, 4)$. Determine the unit vector normal to the surface ABC satisfying $\mathbf{n} \cdot \mathbf{n} = 1$.

1.10 Let an angle θ be measured from the x_1 axis counterclockwise to a line drawn normal to the boundary surface in a two-dimensional domain ($x_1 x_2$ plane). What are the direction cosines n_i (that is, the components of a vector normal to the surface) according to Eq. (1.3.5)? Hint: Draw a sketch for a two-dimensional domain with the boundary $d\Gamma$ and show that $\mathbf{n} = \cos\theta \mathbf{i}_1 + \sin\theta \mathbf{i}_2$, with $dx_3 = 1$ and $n_3 = 0$ in Eq. (1.3.5).

1.11 Prove Eq. (1.3.11).

1.12 Carry out integration by parts, using the form of the right-hand side of Eq. (1.3.10), and compare the results by expanding Eq. (1.3.11). Identify the boundary data.

2

Deformations and Kinematics

2.1 General

One of the basic variables in continuum mechanics is displacement, the geometrical change of a point in the continuum. In a uniaxial tension or compression test of a solid bar, strain is defined as the change in length or displacement per unit of initial length. Thus, if a bar of 10 cm in length is elongated to 10.1 cm, then the strain is 0.01 or 1% in tension. If, instead, this bar is shortened to 9.9 cm, the strain is -0.01 or 1% in compression. But, if the bar is elongated to 10.2 cm and it is subsequently compressed to 10.1 cm, a statement that the final strain is still 1% in tension does not adequately describe what has happened. If this deformed state is the same as in the case of the specimen which was merely strained to a length of 10.1 cm, such deformation is referred to as *elastic*. On the other hand, some materials exhibit *inelastic* rearrangements of particles under applied loads and the final states of deformation in these two cases are different. Definitions of strain also become complicated when displacements are very large and when they are caused by sustained loading, cyclic loading, torsional loading, or flexural (bending) loading.

In viscous fluids, on the other hand, we cannot visualize the strain as defined in solids. We prefer to measure the velocity of a fluid particle that passes through a given point in space rather than to keep track of the distance of a particle traveling downstream at high speed. Thus, in fluid mechanics, we are concerned with the rates of change of velocity, known as *deformation rate*, which may be regarded as the *time rate of change of strain*.

The mechanics of deformations, strains, deformation rates, and accelerations is referred to as *kinematics*. In this chapter, we discuss the concept of strain, the deformation rate, and related topics, including the various coordinate systems, derivatives of variables on these coordinates, strain-displacement relations, coordinate transformations, dilatational and deviatoric strains, and compatibility equations. Studies of these subjects will prepare for the introduction of the concept of equilibrium and kinetics, which will be discussed in Chapter 3.

2.2 Coordinate Systems

2.2.1 *Lagrangian Coordinates*

If a body undergoes geometric changes, we need, in addition to the reference Cartesian coordinates, a coordinate system that follows the deformed shape. Such a coordinate system is called the *Lagrangian* or material coordinates, and is commonly used in solid mechanics. On the other hand, in fluid mechanics, we are interested in measuring the velocity of fluid particles as they pass through a fixed point in space. The coordinate system required for this purpose is referred to as the spatial description or the *Eulerian* coordinate system.

The Lagrangian coordinate system (Fig. 2.2.1) refers to a particle motion given by

$$\mathbf{R} = \mathbf{R}(\mathbf{r}, t), \tag{2.2.1}$$

where $\mathbf{r}$ denotes the position vector from the origin of the rectangular Cartesian coordinates to a point P_0 at time $t = t_0$ in the undeformed state, whereas $\mathbf{R}$ is the position vector to the point P in the deformed state at time $t = t$. Note that P_0 is located at x_i $(i = 1, 2, 3)$, where the x_i's are called the *material labels*. As deformation takes place, P_0 moves to P and the coordinate lines initially given in rectangular Cartesian coordinates at P_0 may be convected arbitrarily in nonrectangular, or curvilinear, form. For this reason, such a coordinate system is sometimes referred to as the *convective coordinate* system.

The position vectors $\mathbf{r}$ and $\mathbf{R}$ are defined as:

$$\mathbf{r} = x_i \mathbf{i}_i, \tag{2.2.2}$$

$$\mathbf{R} = z_i \mathbf{i}_i. \tag{2.2.3}$$

Let us consider the infinitesimal change in $\mathbf{R}$ such that

$$d\mathbf{R} = \frac{\partial \mathbf{R}}{\partial x_i} dx_i = \frac{\partial z_k}{\partial x_i} \mathbf{i}_k \, dx_i = \mathbf{G}_i \, dx_i, \tag{2.2.4}$$

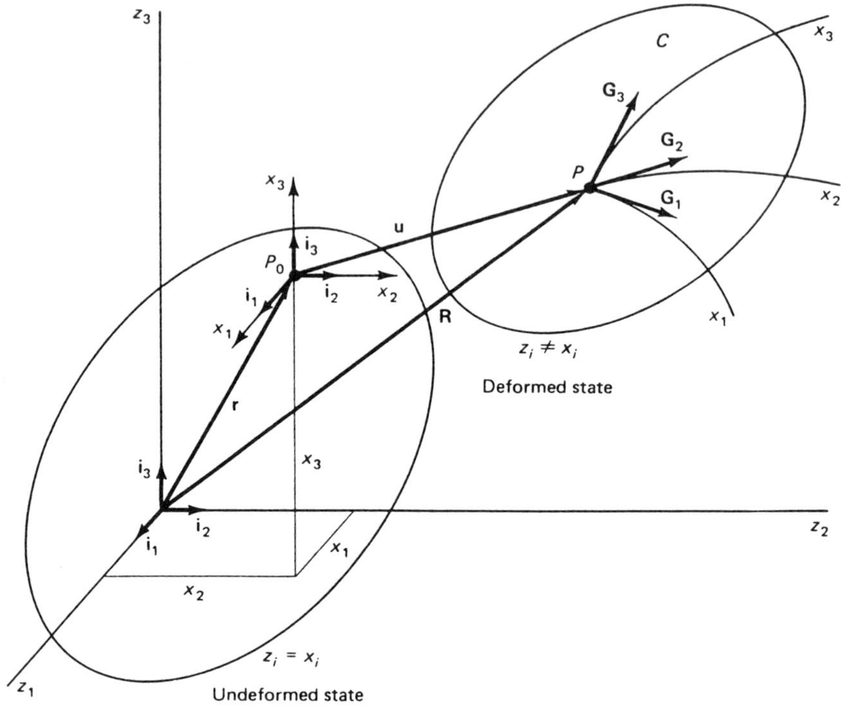

Figure 2.2.1 Lagrangian coordinates. The undeformed state is defined by rectangular Cartesian coordinates, and the deformed state by arbitrary curvilinear (convected) coordinates.

where

$$\mathbf{G}_i = \frac{\partial z_k}{\partial x_i}\mathbf{i}_k = z_{k,i}\mathbf{i}_k \tag{2.2.5}$$

is known as the base vector or tangent vector, in which $\partial z_k/\partial x_i$ is called the *deformation gradient*. Notice commas are used to represent partial derivatives only when differentiation is performed with respect to the independent variables x_i. Otherwise, to avoid confusion, this short notation should not be used. For example, $\partial x_i/\partial z_j$ should not be written as $x_{i,j}$. Note that $\partial z_k/\partial x_i$ in Eq. (2.2.5) transforms the unit vectors into tangent vectors, and these need no longer be unit orthonormal vectors. To distinguish the deformed coordinate lines from undeformed coordinate lines x_i, the deformed coordinate lines may be written as x^i, with the index i given as a superscript instead of a subscript, implying that x^i is non-Cartesian, so that we may write

$$d\mathbf{R} = \mathbf{G}_i\, dx^i. \tag{2.2.6}$$

Although Eq. (2.2.6) is consistent with a mathematical argument, we show that this is unacceptable from the physical reason for space conservation (see Eq. [2.2.17]).

Figure 2.2.1 shows that the displacement vector $\mathbf{u}$ is given by the relation

$$\mathbf{u} = \mathbf{R} - \mathbf{r}.$$

Thus, from Eq. (2.2.4)

$$\mathrm{d}\mathbf{R} = \frac{\partial}{\partial x_i}(\mathbf{r} + \mathbf{u})\,\mathrm{d}x_i = \left(\frac{\partial x_k}{\partial x_i}\mathbf{i}_k + \frac{\partial u_k}{\partial x_i}\mathbf{i}_k\right)\mathrm{d}x_i.$$

Note that $\partial x_k/\partial x_i = \delta_{ki}$, which is the Kronecker delta, so it follows that

$$\mathrm{d}\mathbf{R} = \left(\delta_{ki} + \frac{\partial u_k}{\partial x_i}\right)\mathbf{i}_k\,\mathrm{d}x_i. \tag{2.2.7}$$

Equate Eqs. (2.2.4) and (2.2.7) to yield

$$\mathbf{G}_i = \left(\delta_{ki} + \frac{\partial u_k}{\partial x_i}\right)\mathbf{i}_k \tag{2.2.8}$$

or

$$\frac{\partial z_k}{\partial x_i} = \delta_{ki} + \frac{\partial u_k}{\partial x_i}. \tag{2.2.9}$$

In the Lagrangian coordinate system, we regard $\mathbf{r}$ and t as independent variables, and $\mathbf{R}(\mathbf{r}, t)$ and $\mathbf{u}$ as dependent variables. The particle velocity $\mathbf{v}(\mathbf{r}, t)$ is the Lagrangian velocity, defined as:

$$\mathbf{v}(\mathbf{r}, t) = \frac{\partial \mathbf{R}}{\partial t} = \frac{\partial}{\partial t}[\mathbf{r} + \mathbf{u}(\mathbf{r}, t)] = \frac{\partial \mathbf{u}(\mathbf{r}, t)}{\partial t} = \dot{\mathbf{u}}. \tag{2.2.10}$$

Note here that $\mathbf{r}$ is independent of time and the superposed dot denotes the time derivative. Likewise, the particle acceleration or Lagrangian acceleration is

$$\mathbf{a}(\mathbf{r}, t) = \dot{\mathbf{v}}(\mathbf{r}, t) = \frac{\partial^2 \mathbf{R}}{\partial t^2} = \frac{\partial^2 \mathbf{u}}{\partial t^2} = \ddot{\mathbf{u}}. \tag{2.2.11}$$

A fundamental requirement in mechanics is the conservation of mass. In Lagrangian coordinates, the mass conservation principle may be established by examining the undeformed and the deformed infinitesimal bodies, as depicted in Fig. 2.2.2. First, consider the undeformed cube with its volume given by

$$\mathrm{d}\Omega_0 = (\mathrm{d}x_1\mathbf{i}_1 \times \mathrm{d}x_2\mathbf{i}_2)\cdot \mathrm{d}x_3\mathbf{i}_3 = \mathrm{d}x_1\,\mathrm{d}x_2\,\mathrm{d}x_3. \tag{2.2.12}$$

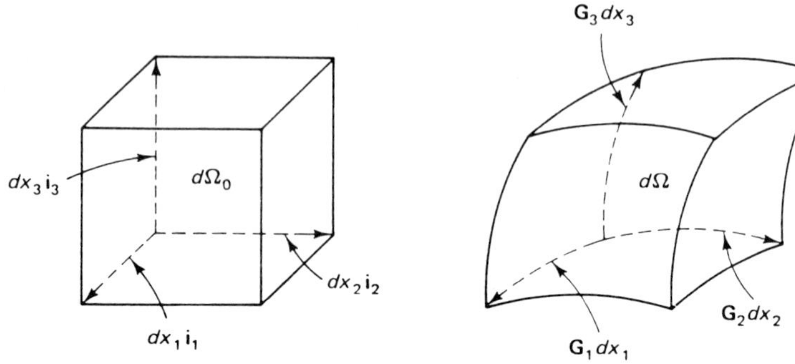

Figure 2.2.2 Infinitesimal volumes in the undeformed state ($d\Omega_0$) and deformed state ($d\Omega$).

The deformed volume is then calculated as

$$\begin{aligned} d\Omega &= (\mathbf{G}_1\, dx_1 \times \mathbf{G}_2\, dx_2) \cdot \mathbf{G}_3\, dx_3 \\ &= (\mathbf{G}_1 \times \mathbf{G}_2 \cdot \mathbf{G}_3)\, dx_1\, dx_2\, dx_3. \end{aligned} \tag{2.2.13}$$

Write the tangent vectors in the form

$$\begin{aligned} \mathbf{G}_1 &= \frac{\partial z_m}{\partial x_1}\mathbf{i}_m = \frac{\partial z_1}{\partial x_1}\mathbf{i}_1 + \frac{\partial z_2}{\partial x_1}\mathbf{i}_2 + \frac{\partial z_3}{\partial x_1}\mathbf{i}_3, \\ \mathbf{G}_2 &= \frac{\partial z_m}{\partial x_2}\mathbf{i}_m = \frac{\partial z_1}{\partial x_2}\mathbf{i}_1 + \frac{\partial z_2}{\partial x_2}\mathbf{i}_2 + \frac{\partial z_3}{\partial x_2}\mathbf{i}_3, \\ \mathbf{G}_3 &= \frac{\partial z_m}{\partial x_3}\mathbf{i}_m = \frac{\partial z_1}{\partial x_3}\mathbf{i}_1 + \frac{\partial z_2}{\partial x_3}\mathbf{i}_2 + \frac{\partial z_3}{\partial x_3}\mathbf{i}_3, \end{aligned} \tag{2.2.14}$$

and substitute into Eq. (2.2.13) to obtain

$$d\Omega = J\, d\Omega_0, \tag{2.2.15}$$

where J is called the Jacobian, which is the determinant of the deformation gradient $\partial z_i/\partial x_j$

$$J = \begin{vmatrix} \dfrac{\partial z_1}{\partial x_1} & \dfrac{\partial z_1}{\partial x_2} & \dfrac{\partial z_1}{\partial x_3} \\[2ex] \dfrac{\partial z_2}{\partial x_1} & \dfrac{\partial z_2}{\partial x_2} & \dfrac{\partial z_2}{\partial x_3} \\[2ex] \dfrac{\partial z_3}{\partial x_1} & \dfrac{\partial z_3}{\partial x_2} & \dfrac{\partial z_3}{\partial x_3} \end{vmatrix} \tag{2.2.16}$$

Note that the determinant of $\partial z_i/\partial x_j$ is equal to the determinant of its transpose $\partial z_j/\partial x_i$. To prove the relation given by Eq. (2.2.15), write Eq. (2.2.13) in the form

$$
\begin{aligned}
d\Omega &= \frac{\partial z_m}{\partial x_1}\frac{\partial z_n}{\partial x_2}\frac{\partial z_p}{\partial x_3}\mathbf{i}_m \times \mathbf{i}_n \cdot \mathbf{i}_p \, dx_1\, dx_2\, dx_3 \\
&= \epsilon_{mnp}\frac{\partial z_m}{\partial x_1}\frac{\partial z_n}{\partial x_2}\frac{\partial z_p}{\partial x_3}\, dx_1\, dx_2\, dx_3 \\
&= \left|\frac{\partial z_i}{\partial x_j}\right| dx_1\, dx_2\, dx_3 = \sqrt{\left|\frac{\partial z_k}{\partial x_i}\frac{\partial z_k}{\partial x_j}\right|}\, dx_1\, dx_2\, dx_3 \\
&= \sqrt{|G_{ij}|}\, dx_1\, dx_2\, dx_3 = \sqrt{G}\, dx_1\, dx_2\, dx_3 = J\, dx_1\, dx_2\, dx_3 \\
&= J\, d\Omega_0
\end{aligned}
\tag{2.2.17}
$$

where

$$J = \sqrt{G} = \sqrt{|G_{ij}|} \tag{2.2.18}$$

and

$$G_{ij} = \mathbf{G}_i \cdot \mathbf{G}_j = \frac{\partial \mathbf{R}}{\partial x_i} \cdot \frac{\partial \mathbf{R}}{\partial x_j} = \frac{\partial z_m}{\partial x_i}\frac{\partial z_m}{\partial x_j}. \tag{2.2.19}$$

Here, G_{ij} is called the *metric tensor*, *Green's deformation tensor*, or simply the *deformation tensor*; it will be treated further in Section 2.4. Had we used the notation given in Eq. (2.2.6) we would then have a physical inconsistency, $d\Omega = J\, dx^1\, dx^2\, dx^3$, whereas $d\Omega_0 = dx_1\, dx_2\, dx_3$. Thus, although the notation $\mathbf{G}_i\, dx^i$ is mathematically consistent for convective or curvilinear coordinates, as shown in most textbooks, it is physically incorrect, as demonstrated here, for the volume change between the undeformed and deformed states, $dx^1\, dx^2\, dx^3 = dx_1\, dx_2\, dx_3$.

The physical significance of the existence of the Jacobian is that the mass is conserved during the transition from $d\Omega_0$ to $d\Omega$, as signified by the fact that the motion

$$z_i = z_i(x_i, t)$$

is continuously differentiable. We also note that

$$d\Omega_0 = J^{-1}\, d\Omega, \tag{2.2.20}$$

which implies that the matrix in Eq. (2.2.16) is nonsingular.

We now point out that a continuous density function ρ exists:

$$\rho = \lim_{\Delta\Omega \to 0} \frac{\Delta m}{\Delta\Omega} = \frac{dm}{d\Omega}, \tag{2.2.21}$$

where m is the mass. By the law of conservation of mass, we find

$$m = \int_{\Omega} \rho \, d\Omega = \int_{\Omega_0} \rho_0 \, d\Omega_0. \tag{2.2.22}$$

Substitute Eq. (2.2.15) into Eq. (2.2.22) to yield

$$\int_{\Omega_0} (\rho J - \rho_0) \, d\Omega_0 = 0. \tag{2.2.23}$$

For this equation to be valid for all arbitrary volumes $d\Omega_0$, it is required that the integrand vanish. Thus,

$$\rho_0 = J\rho. \tag{2.2.24}$$

Equation (2.2.24) is the formula for conservation of mass as it applies to motion in solid mechanics. Once again, we emphasize that to assure mass conservation, the coordinate lines should be written as x_i, not x^i, for the deformed coordinates.

2.2.2 Eulerian Coordinates

A fluid particle may travel with such velocity that a human eye is unable to trace it downstream. The Lagrangian coordinates described in subsection 2.2.1 are obviously inadequate because our aim is to determine the velocity of fluid particles at any point in space rather than to calculate the displacement of fluid particles. For this purpose, we introduce Eulerian coordinates (see Fig. 2.2.3), in which we consider the velocity vector $\mathbf{V}(\mathbf{R}, t)$ and density ρ to be dependent variables and $\mathbf{R}$ and t to be independent variables. The velocity is measured at the current position P located at $\mathbf{R}$. Thus, the initial position vector $\mathbf{r}$, located at P_0, used in Lagrangian coordinates, is irrelevant in Eulerian coordinates. Since the dependent variables change with respect to both time t and spatial coordinates z_i in the Eulerian coordinates, we introduce a special form of a derivative given by

$$\frac{D}{Dt} = \frac{\partial}{\partial t} + \frac{\partial}{\partial z_i}\frac{\partial z_i}{\partial t} = \frac{\partial}{\partial t} + V_i \frac{\partial}{\partial z_i},$$

where $\partial z_i/\partial t$ is defined as the Eulerian velocity V_i. The above expression may be written in vector notation

$$\frac{D}{Dt} = \frac{\partial}{\partial t} + \mathbf{V} \cdot \boldsymbol{\nabla}. \tag{2.2.25}$$

Here, the symbol D/Dt is called the *substantial derivative* or the *material derivative*, with $\mathbf{V}(\mathbf{R}, t) = \partial \mathbf{R}/\partial t$. Therefore, the Eulerian

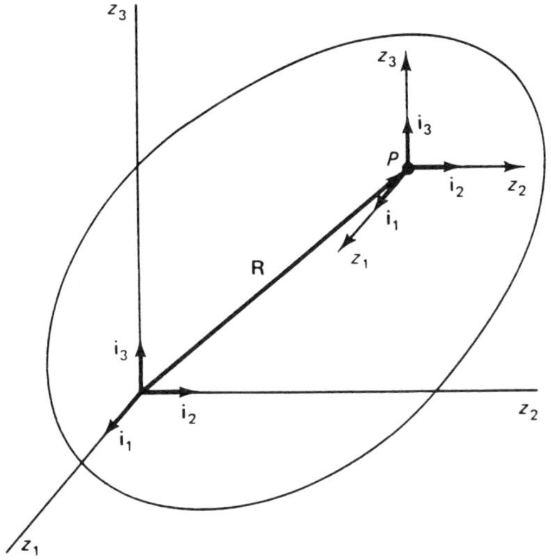

Figure 2.2.3 Eulerian coordinates—Cartesian representation. Velocity $V_i = \partial z_i/\partial t$ is measured at P.

acceleration $\mathbf{A}(\mathbf{R}, t)$ is given by

$$\mathbf{A}(\mathbf{R}, t) = \frac{D\mathbf{V}(\mathbf{R}, t)}{Dt} = \frac{\partial \mathbf{V}}{\partial t} + \frac{\partial \mathbf{V}}{\partial z_j} V_j = \frac{\partial \mathbf{V}}{\partial t} + (\mathbf{V} \cdot \boldsymbol{\nabla})\mathbf{V}$$
$$= \left(\frac{\partial \mathbf{V}_i}{\partial t} + V_{i,j} V_j\right)\mathbf{i}_i.$$

Note that z_i is the dependent variable with respect to time, but is an independent variable with respect to the velocity. Therefore, we can use the comma to represent the derivative of the velocity with respect to z_i, contrary to the case of Lagrangian coordinates.

Let us now consider a domain Ω with an infinitesimal boundary surface $\mathrm{d}\Gamma$ with a unit vector $\mathbf{n}$ normal to $\mathrm{d}\Gamma$ (see Fig. 2.2.4). The mass rate of flow out of $\mathrm{d}\Gamma$ is

$$\frac{\mathrm{d}m}{\mathrm{d}t} = \int_\Gamma \rho \mathbf{V} \cdot \mathbf{n}\, \mathrm{d}\Gamma = \int_\Gamma \rho V_i n_i\, \mathrm{d}\Gamma. \tag{2.2.26a}$$

On the other hand, the mass rate of flow in Ω is

$$\frac{\mathrm{d}m}{\mathrm{d}t} = -\int_\Omega \frac{\partial \rho}{\partial t}\, \mathrm{d}\Omega. \tag{2.2.26b}$$

Since it is necessary that the mass be conserved, Eqs. (2.2.26a) and (2.2.26b) must be equal:

$$-\int_\Omega \frac{\partial \rho}{\partial t}\, \mathrm{d}\Omega = \int_\Gamma \rho \mathbf{V} \cdot \mathbf{n}\, \mathrm{d}\Gamma.$$

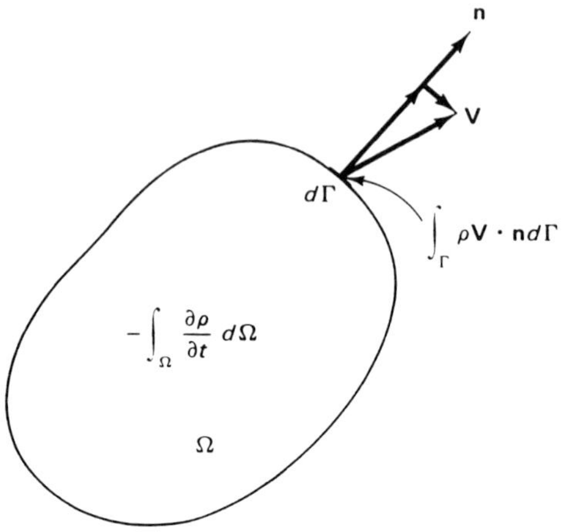

Figure 2.2.4 Conservation of mass in Eulerian coordinates.

Use the Green–Gauss theorem to obtain

$$\int_\Gamma \rho \mathbf{V} \cdot \mathbf{n} \, \mathrm{d}\Gamma = \int_\Omega \boldsymbol{\nabla} \cdot (\rho \mathbf{V}) \, \mathrm{d}\Omega. \tag{2.2.27}$$

It follows that

$$\int_\Omega \left[\frac{\partial \rho}{\partial t} + \boldsymbol{\nabla} \cdot (\rho \mathbf{V}) \right] \mathrm{d}\Omega = 0. \tag{2.2.28a}$$

For the above integral equation to remain valid for all arbitrary volumes $\mathrm{d}\Omega$, it is necessary that the integrand vanish. Therefore,

$$\frac{\partial \rho}{\partial t} + \boldsymbol{\nabla} \cdot (\rho \mathbf{V}) = 0 \tag{2.2.28b}$$

or

$$\frac{\partial \rho}{\partial t} + (\rho V_i)_{,i} = 0. \tag{2.2.28c}$$

Once again, we must realize that

$$(\rho V_i)_{,i} = \frac{\partial}{\partial z_i} (\rho V_i),$$

which implies that the comma represents partial derivatives with respect to z_i instead of x_i, contrary to the case of Lagrangian coordinates. Equation (2.2.28) represents conservation of mass in Eulerian coordinates or for fluid mechanics.

The conservation of mass for fluids can be obtained by taking a substantial derivative of Eq. (2.2.23), which represents a transformation from Lagrangian coordinates to Eulerian coordinates as:

$$\begin{aligned}
\int_\Omega \frac{D}{Dt}(\rho J - \rho_0)\,\mathrm{d}\Omega_0 &= \int_\Omega \frac{D}{Dt}(\rho\,\mathrm{d}\Omega) = \int_\Omega\left(\frac{D\rho}{Dt}\,\mathrm{d}\Omega + \rho\frac{D\,\mathrm{d}\Omega}{Dt}\right)\\
&= \int_\Omega\left(\frac{D\rho}{Dt}\,\mathrm{d}\Omega + \rho\frac{DJ}{Dt}\,\mathrm{d}\Omega_0\right)\\
&= \int_\Omega\left(\frac{D\rho}{Dt} + \rho\boldsymbol{\nabla}\cdot\mathbf{V}\right)\mathrm{d}\Omega\\
&= \int_\Omega\left(\frac{\partial\rho}{\partial t} + \boldsymbol{\nabla}\cdot(\rho\mathbf{V})\right)\mathrm{d}\Omega = 0,
\end{aligned} \tag{2.2.29}$$

where

$$\begin{aligned}
\frac{DJ}{Dt} &= \frac{\partial J}{\partial t} + \frac{\partial J}{\partial z_j}\frac{\partial z_j}{\partial t}\\
&= \frac{\partial}{\partial t}\left(\epsilon_{mnp}\frac{\partial z_m}{\partial x_1}\frac{\partial z_n}{\partial x_2}\frac{\partial z_p}{\partial x_3}\right) + V_j\frac{\partial}{\partial z_j}\left(\epsilon_{\mathrm{mnp}}\frac{\partial z_m}{\partial x_1}\frac{\partial z_n}{\partial x_2}\frac{\partial z_p}{\partial x_3}\right)\\
&= \epsilon_{mnp}\left(\frac{\partial V_m}{\partial z_k}\frac{\partial z_k}{\partial x_1}\frac{\partial z_n}{\partial x_2}\frac{\partial z_p}{\partial x_3} + \frac{\partial z_m}{\partial x_1}\frac{\partial V_n}{\partial z_k}\frac{\partial z_k}{\partial x_2}\frac{\partial z_p}{\partial x_3}\right.\\
&\qquad \left. + \frac{\partial z_m}{\partial x_1}\frac{\partial z_n}{\partial x_2}\frac{\partial V_p}{\partial z_k}\frac{\partial z_k}{\partial x_3}\right)\\
&\quad + V_j\epsilon_{mnp}\left[\frac{\partial}{\partial x_1}(\delta_{mj})\frac{\partial z_n}{\partial x_2}\frac{\partial z_p}{\partial x_3} + \frac{\partial z_m}{\partial x_1}\frac{\partial}{\partial x_2}(\delta_{nj})\frac{\partial z_p}{\partial x_3}\right.\\
&\qquad \left. + \frac{\partial z_m}{\partial x_1}\frac{\partial z_n}{\partial x_2}\frac{\partial}{\partial x_3}(\delta_{pj})\right]\\
&= \left|\frac{\partial z_i}{\partial x_j}\right|\frac{\partial V_k}{\partial z_k} = J\frac{\partial V_k}{\partial z_k} = J\boldsymbol{\nabla}\cdot\mathbf{V}.
\end{aligned} \tag{2.2.30}$$

Note that 54 terms arise from the temporal variation of J, whereas the spatial variation of J vanishes because

$$\frac{\partial}{\partial z_j}\left(\frac{\partial z_m}{\partial x_1}\right) = \frac{\partial}{\partial x_1}\left(\frac{\partial z_m}{\partial z_j}\right) = \frac{\partial}{\partial x_1}(\delta_{mj}) = 0, \quad \text{etc.}$$

In view of these results, we have reconfirmed the conservation of mass shown in Eq. (2.2.28a) as a consequence of transforming the conservation of mass in solids from Lagrangian to Eulerian coordinates. It is important to recognize that $\partial z_i/\partial x_j$ is an expression that is valid only in Lagrangian coordinates, whereas $\partial V_i/\partial z_j$ can be defined only in Eulerian coordinates, thus requiring Eq. (2.2.30) to participate in both Lagrangian and Eulerian coordinates in this transformation process.

Multiplying Eq. (2.2.30) by $d\Omega_0$ and integrating over the domain, we obtain

$$\frac{\partial}{\partial t}\int_{\Omega} J \, d\Omega_0 = \int_{\Omega} \nabla \cdot \mathbf{V} J \, d\Omega_0 = \int_{\Omega} \nabla \cdot \mathbf{V} \, d\Omega$$

or

$$\frac{\partial}{\partial t}\int_{\Omega} d\Omega = \int_{\Gamma} \mathbf{V} \cdot \mathbf{n} \, d\Gamma, \tag{2.2.31}$$

which represents an important process in temporal moving boundary problems, known as the *space conservation law*. This equation is also the basis for the so-called *volume of fluid* (VOF) computations in multiphase flow.

2.2.3 Relationships between Lagrangian and Eulerian Coordinates

In continuum mechanics the use of material description (Lagrangian coordinates) or spatial description (Eulerian coordinates) is dictated by the nature of the continuum. Although these coordinate systems are distinctly different from each other, they are related in such a way that properties in one coordinate system can be expressed in an alternative way in the other. If $\mathbf{R}(\mathbf{r}, t)$ denotes particle paths, then $\mathbf{R}$ is the solution of the ordinary differential equation

$$\frac{d\mathbf{R}}{dt} = \mathbf{v}(\mathbf{r}, t).$$

However, particle paths, in general, are unimportant in fluid mechanics. Rather, the streamlines, which are the lines drawn parallel to the velocity field, are defined by

$$d\mathbf{R} = \mathbf{V}(\mathbf{R}, t) \, d\lambda,$$

where $d\lambda$ is a time parameter. This can be written in the Pfaffian form (Sneddon 1957):

$$\frac{dz_1}{V_1(\mathbf{R}, t)} = \frac{dz_2}{V_2(\mathbf{R}, t)} = \frac{dz_3}{V_3(\mathbf{R}, t)} = d\lambda. \tag{2.2.32}$$

For a steady-state condition with the streamlines remaining constant, we may set $\lambda = t$. For steady flows, we have $\mathbf{V}(\mathbf{R}, t) = \mathbf{V}(\mathbf{R}) = \partial \mathbf{R}(\mathbf{R}, t)/\partial t = \partial z_i \mathbf{i}_i/\partial t$, and the streamlines and the particle paths coincide. Throughout this book, we designate $\mathbf{v}$ (lower case) and $\mathbf{V}$ (upper case) as the Lagrangian and Eulerian velocities, respectively.

To illustrate the relationship between Lagrangian and Eulerian coordinates, let us assume that the motion is given by

$$z_1 = \frac{x_1}{1 + tx_1} \quad z_2 = x_2 \quad z_3 = x_3. \tag{2.2.33}$$

The Lagrangian velocities are determined as

$$\mathbf{v}(\mathbf{r}, t) = \frac{\partial \mathbf{R}(\mathbf{r}, t)}{\partial t},$$

or

$$v_1 = \frac{dz_1}{dt} = -\frac{x_1^2}{(1 + tx_1)^2} \tag{2.2.34}$$
$$v_2 = v_3 = 0.$$

The Eulerian velocities are calculated by first solving for x_1 from the Lagrangian motion Eq. (2.2.33)

$$x_1 = \frac{z_1}{1 - tz_1} \tag{2.2.35}$$

and then by substituting Eq. (2.2.35) into Eq. (2.2.34),

$$\mathbf{V}(\mathbf{R}, t) = \mathbf{v}[\mathbf{r}(\mathbf{R}, t)]$$

or

$$V_1 = -\frac{x_1^2}{(1 + tx_1)^2} = \frac{-[z_1/(1 - tz_1)]^2}{[1 + tz_1/(1 - tz_1)]^2} = -z_1^2 \tag{2.2.36}$$
$$V_2 = V_3 = 0.$$

The Lagrangian velocity v_1 and the Eulerian velocity V_1 are shown in Figs. 2.2.5a and b, respectively. The Eulerian velocities, Eq. (2.2.36), may be transformed back into the Lagrangian velocities by solving for z_1 from the differential equation

$$\frac{dz_1}{dt} + z_1^2 = 0, \tag{2.2.37}$$

where $V_1 = dz_1/dt$ is substituted into Eq. (2.2.36), subject to the initial condition $z_1 = x_1$ at $t = 0$. The solution of Eq. (2.2.37) gives Eq. (2.2.33) and, finally, the Lagrangian velocities are obtained by differentiating the Lagrangian motion Eq. (2.2.33), as shown in Eq. (2.2.34). Problem 2.3 at the end of Chapter 2 illustrates a reverse process in which Eulerian velocities are given initially and Lagrangian velocities are to be calculated.

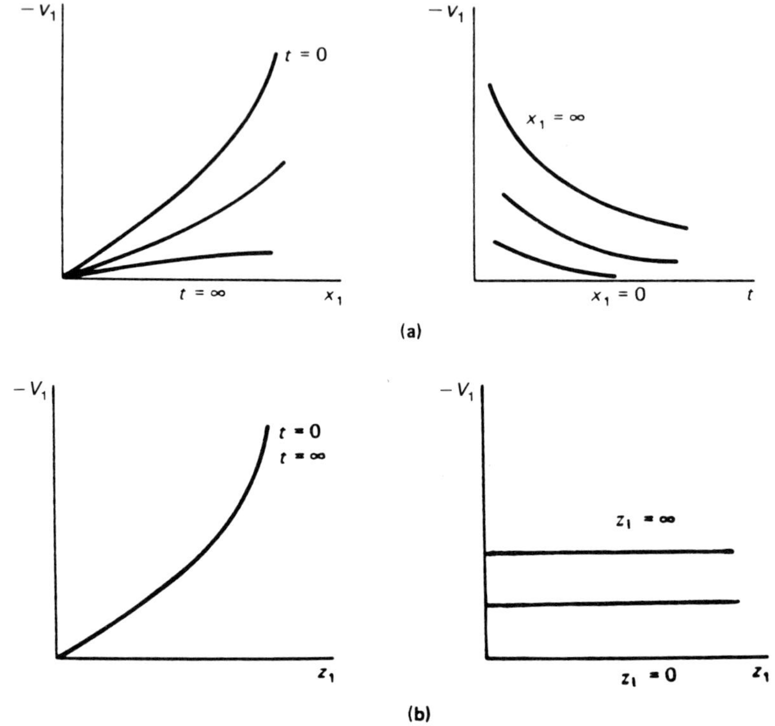

Figure 2.2.5 Relationship between Lagrangian and Eulerian velocities for the motion, $z_1 = x_1/(1 + tx_1)$, $z_2 = x_2$, $z_3 = x_3$. (a) Velocity distributions in material coordinates and time (Lagrangian coordinates). (b) Velocity distributions in space, independent of time (Eulerian coordinates).

Remarks

Even though we draw a clear line by saying that Lagrangian descriptions are applicable to solids and Eulerian descriptions to fluids, engineering applications often call for a reversal of that rule or a combined use of both systems in a given analysis. For example, in hypervelocity impact problems in which phase changes are involved between solids and fluids, it is advantageous to invoke both Lagrangian and Eulerian descriptions. Another example is the particle tracking of spray combustion, in which, for convenience, both coordinate systems are used. Furthermore, convective terms in fluid mechanics (see Chapter 5) may be eliminated by coordinate transformation from Eulerian to Lagrangian descriptions, which results in numerical expediency.

2.3 Non-Cartesian Coordinates

2.3.1 Basic Properties

We have discussed tangent vectors and metric tensors associated with undeformed and deformed states. Cartesian coordinates were utilized to describe the undeformed state of three-dimensional solids. In the analysis of cylinders or spheres, however, the undeformed state itself can be non-Cartesian, although orthogonal. Figure 2.3.1 shows the use of a general Lagrangian non-Cartesian or curvilinear coordinate system for the undeformed state and its convected coordinates for the deformed state. The base vectors or tangent vectors $\mathbf{g}_i$ are drawn tangent to the initial undeformed curvilinear coordinates ξ_i. These tangent vectors $(\mathbf{g}_1, \mathbf{g}_2, \mathbf{g}_3)$ may not necessarily be orthogonal to each other. Here again, the coordinate lines ξ_i are subscripted rather than superscripted (ξ^i) for the same reason stated in connection with Eq. (2.2.17), except for the case of mathematical consistency such as occurs in Eq. (2.4.10).

The tangent (base) vectors $\mathbf{g}_i$ on an undeformed state represent the

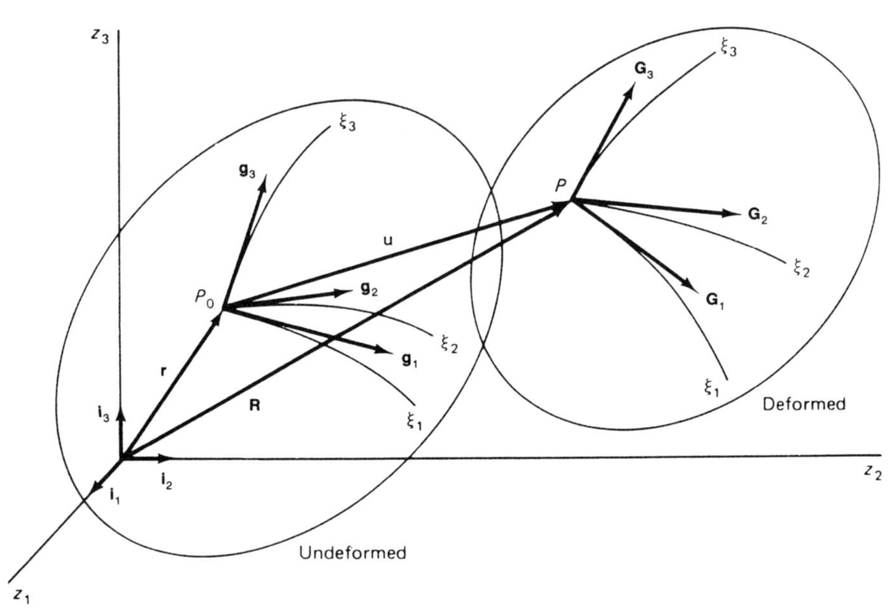

Figure 2.3.1 Lagrangian curvilinear coordinates.

rate of change of the position vector $\mathbf{r}$ with respect to the curvilinear coordinates ξ_i,

$$\mathbf{g}_i = \frac{\partial \mathbf{r}}{\partial \xi_i} = \frac{\partial z_m}{\partial \xi_i}\mathbf{i}_m, \tag{2.3.1}$$

whereas the *reciprocal tangent* vectors are defined as

$$\mathbf{g}^i = \frac{\partial \xi_i}{\partial z_m}\mathbf{i}_m. \tag{2.3.2}$$

These tangent vectors are related by

$$\mathbf{g}_i \cdot \mathbf{g}^j = \frac{\partial z_m}{\partial \xi_i}\mathbf{i}_m \cdot \frac{\partial \xi_j}{\partial z_n}\mathbf{i}_n = \frac{\partial z_m}{\partial \xi_i}\frac{\partial \xi_j}{\partial z_n}\delta_{mn} = \frac{\partial z_m}{\partial \xi_i}\frac{\partial \xi_j}{\partial z_m}, \tag{2.3.3a}$$

where $\mathbf{g}_i$ and $\mathbf{g}^j$ are referred to as the *covariant* and *contravariant* tangent (base) vectors, respectively. The contravariant (reciprocal) tangent vectors $\mathbf{g}^1$, $\mathbf{g}^2$, $\mathbf{g}^3$ are orthogonal to the planes constructed by $\mathbf{g}_2$ and $\mathbf{g}_3$, $\mathbf{g}_3$ and $\mathbf{g}_1$, and $\mathbf{g}_1$ and $\mathbf{g}_2$, respectively (see Fig. 2.3.2). This definition requires that $\mathbf{g}_i \cdot \mathbf{g}^j = \delta_{ij}$ and

$$\frac{\partial z_m}{\partial \xi_i}\frac{\partial \xi_j}{\partial z_m} = \delta_i^j = \delta_{ij}, \tag{2.3.3b}$$

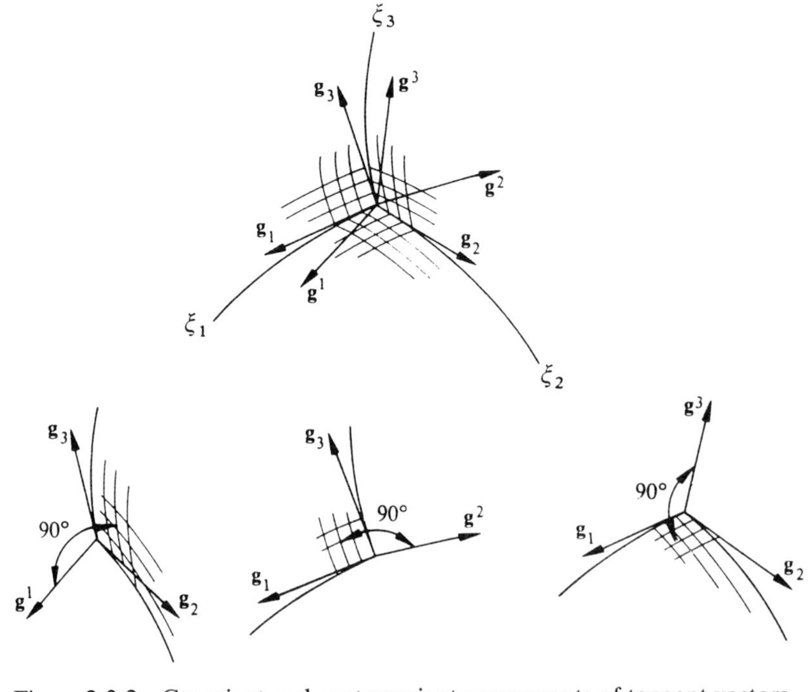

Figure 2.3.2 Covariant and contravariant components of tangent vectors.

where the Kronecker delta δ_i^j performs the same way as δ_{ij}. We also define g_{ij} and g^{ij} as the covariant and contravariant metric tensors, respectively:

$$g_{ij} = \mathbf{g}_i \cdot \mathbf{g}_j = \frac{\partial z_m}{\partial \xi_i} \frac{\partial z_m}{\partial \xi_j}, \tag{2.3.4a}$$

$$g^{ij} = \mathbf{g}^i \cdot \mathbf{g}^j = \frac{\partial \xi_i}{\partial z_m} \frac{\partial \xi_j}{\partial z_m}. \tag{2.3.4b}$$

Note also the additional properties:

$$\mathbf{g}_i = g_{ij}\mathbf{g}^j, \quad \mathbf{g}^i = g^{ij}\mathbf{g}_j, \quad g_{ik}g^{jk} = \delta_i^j = \delta_{ij}, \tag{2.3.5a}$$

$$\begin{aligned} \mathbf{i}_p &= \frac{\partial \xi_i}{\partial z_p}\mathbf{g}_i = \frac{\partial \xi_i}{\partial z_p} g_{ik}\mathbf{g}^k = \frac{\partial \xi_i}{\partial z_p} \frac{\partial z_m}{\partial \xi_i} \frac{\partial z_m}{\partial \xi_k}\mathbf{g}^k \\ &= \delta_{mp} \frac{\partial z_m}{\partial \xi_k}\mathbf{g}^k = \frac{\partial z_p}{\partial \xi_k}\mathbf{g}^k. \end{aligned} \tag{2.3.5b}$$

The cross product of the covariant tangent vectors takes the form

$$\mathbf{g}_i \times \mathbf{g}_j = \sqrt{g}\,\epsilon_{ijk}\mathbf{g}^k, \tag{2.3.6}$$

where

$$g = |g_{ij}|$$

$$\sqrt{g} = \frac{\partial z_m}{\partial \xi_1} \frac{\partial z_n}{\partial \xi_2} \frac{\partial z_p}{\partial \xi_3}\epsilon_{mnp} \equiv z_{m,1} z_{n,2} z_{p,3} \epsilon_{mnp}$$

(see the similar expression for $\sqrt{G}$ in Eq. (2.2.17)). For curvilinear coordinates, all repeated indices are aligned diagonally, one is superscripted and the other is subscripted, with exceptions involving x_i and ξ_i, as noted earlier in subsection 2.2.1. To prove Eq. (2.3.6) we may proceed by:

$$\begin{aligned} \mathbf{g}_i \times \mathbf{g}_j &= z_{m,i} z_{n,j} \epsilon_{mnp} \mathbf{i}_p = z_{m,i} z_{n,j} z_{p,k} \epsilon_{mnp} \mathbf{g}^k \\ &= z_{m,r} z_{n,s} z_{p,t} \delta_{ir} \delta_{js} \delta_{kt} \epsilon_{mnp} \mathbf{g}^k \\ &= [(z_{1,1} z_{2,1} z_{3,1} - z_{1,1} z_{3,1} z_{2,1} + \ldots)\delta_{i1}\delta_{j1}\delta_{k1} + \ldots \\ &\quad \ldots + (\ldots + z_{3,3} z_{1,3} z_{2,3} - z_{3,3} z_{2,3} z_{1,3})\delta_{i3}\delta_{j3}\delta_{k3}]\mathbf{g}^k. \end{aligned}$$

The complete expansion results in 27 (sum on r, s, t) $\times$ 6 (sum on m, n, p) $=$ 162 terms. However, only those terms with $r, s, t = 1, 2, 3$; $1, 3, 2$; $2, 1, 3$; $2, 3, 1$; $3, 1, 2$; and $3, 2, 1$ survive. Thus, using (1.2.18), we obtain

$$\begin{aligned} \mathbf{g}_i \times \mathbf{g}_j &= z_{m,1} z_{n,2} z_{p,3} (\delta_{i1}\delta_{j2}\delta_{k3} - \delta_{i1}\delta_{j3}\delta_{k2} + \delta_{i2}\delta_{j3}\delta_{k1} \\ &\quad - \delta_{i2}\delta_{j1}\delta_{k3} + \delta_{i3}\delta_{j1}\delta_{k2} - \delta_{i3}\delta_{j2}\delta_{k1})\epsilon_{mnp}\mathbf{g}^k \\ &= z_{m,1} z_{n,2} z_{p,3} \epsilon_{ijk} \epsilon_{mnp} \mathbf{g}^k = \sqrt{g}\,\epsilon_{ijk}\mathbf{g}^k. \end{aligned}$$

Again, a caution here is that the commas representing the partial derivative are used only when differentiations are performed with respect to independent variables. When confusion is likely to occur, as in Eq. (2.3.5b), the full derivative notation should be used.

Using a similar approach, it can be shown that

$$\mathbf{g}^i \times \mathbf{g}^j = \frac{1}{\sqrt{g}}\epsilon^{ijk}\mathbf{g}_k. \tag{2.3.7}$$

The permutation symbol ϵ^{ijk} performs identically as ϵ_{ijk}. The superscripts are used merely for convenience so that the repeated indices can be placed diagonally, consistent with the basic rule for curvilinear coordinates.

A vector $\mathbf{V}$ can be written in terms of covariant components V_i or contravariant components V^i,

$$\mathbf{V} = V_i\mathbf{g}^i = V^i\mathbf{g}_i$$

or

$$V^i = \mathbf{V}\cdot\mathbf{g}^i, \quad V_i = \mathbf{V}\cdot\mathbf{g}_i$$

in contrast with the Cartesian coordinates, in which we write

$$V_i = \mathbf{V}\cdot\mathbf{i}_i.$$

Example 2.1 Tangent Vectors We consider the functions z_i:

$$z_1 = 3\xi_1 - \xi_2$$
$$z_2 = 2\xi_1 + 2\xi_2$$
$$z_3 = \xi_3.$$

The covariant tangent vectors are

$$\mathbf{g}_i = \frac{\partial z_m}{\partial \xi_i}\mathbf{i}_m$$
$$\mathbf{g}_1 = 3\mathbf{i}_1 + 2\mathbf{i}_2, \quad \mathbf{g}_2 = -\mathbf{i}_1 + 2\mathbf{i}_2, \quad \mathbf{g}_3 = \mathbf{i}_3.$$

The metric tensors are

$$g_{ij} = \mathbf{g}_i\cdot\mathbf{g}_j = \begin{bmatrix} 13 & 1 & 0 \\ 1 & 5 & 0 \\ 0 & 0 & 1 \end{bmatrix}.$$

From

$$\xi_1 = \tfrac{1}{4}z_1 + \tfrac{1}{8}z_2$$
$$\xi_2 = -\tfrac{1}{4}z_1 + \tfrac{3}{8}z_2$$
$$\xi_3 = z_3$$

we calculate the contravariant tangent vectors:

$$\mathbf{g}^1 = \tfrac{1}{4}\mathbf{i}_1 + \tfrac{1}{8}\mathbf{i}_2, \quad \mathbf{g}^2 = -\tfrac{1}{4}\mathbf{i}_1 + \tfrac{3}{8}\mathbf{i}_2, \quad \mathbf{g}^3 = \mathbf{i}_3.$$

We now have the check:

$$\mathbf{g}_i \cdot \mathbf{g}^j = \begin{bmatrix} 1 & 0 & 0 \\ 0 & 1 & 0 \\ 0 & 0 & 1 \end{bmatrix} = \delta_{ij}.$$

Example 2.2. Covariant and Contravariant Components Given $\mathbf{g}_1 = 3\mathbf{i}_1$, $\mathbf{g}_2 = 6\mathbf{i}_1 + 8\mathbf{i}_2$, $\mathbf{g}_3 = \mathbf{i}_3$, and $\mathbf{V} = 4\mathbf{i}_1 + 3\mathbf{i}_2$ (see Fig. 2.3.3), calculate the metric tensors and covariant and contravariant components of $\mathbf{V}$.

First, we determine

$$g_{ij} = \begin{bmatrix} 9 & 18 & 0 \\ 18 & 100 & 0 \\ 0 & 0 & 1 \end{bmatrix}.$$

It follows from Eq. (2.3.5a) that

$$g^{ij} = [g_{ij}]^{-1} = \frac{1}{576}\begin{bmatrix} 100 & -18 & 0 \\ -18 & 9 & 0 \\ 0 & 0 & 576 \end{bmatrix}.$$

From

$$\mathbf{g}^i = g^{ij}\mathbf{g}_j$$

we obtain

$$\mathbf{g}^1 = \tfrac{1}{3}\mathbf{i}_1 - \tfrac{1}{4}\mathbf{i}_2, \quad \mathbf{g}^2 = \tfrac{1}{8}\mathbf{i}_2, \quad \mathbf{g}^3 = \mathbf{i}_3.$$

The covariant components of $\mathbf{V}$ take the form

$$\begin{aligned} V_i &= \mathbf{V} \cdot \mathbf{g}_i, \\ V_1 &= (4\mathbf{i}_1 + 3\mathbf{i}_2) \cdot 3\mathbf{i}_1 = 12, \\ V_2 &= (4\mathbf{i}_i + 3\mathbf{i}_2) \cdot (6\mathbf{i}_1 + 8\mathbf{i}_2) = 48, \\ V_3 &= 0. \end{aligned}$$

Note that this gives the check:

$$\mathbf{V} = V_i\mathbf{g}^i = 12\mathbf{g}^1 + 48\mathbf{g}^2 = 4\mathbf{i}_1 + 3\mathbf{i}_2.$$

Similarly, the contravariant components of $\mathbf{V}$ are

$$\begin{aligned} V^i &= \mathbf{V} \cdot \mathbf{g}^i, \\ V^1 &= (4\mathbf{i}_1 + 3\mathbf{i}_2) \cdot (\tfrac{1}{3}\mathbf{i}_1 - \tfrac{1}{4}\mathbf{i}_2) = \tfrac{7}{12}, \\ V^2 &= (4\mathbf{i}_1 + 3\mathbf{i}_2) \cdot (\tfrac{1}{8}\mathbf{i}_2) = \tfrac{3}{8}, \\ V^3 &= 0. \end{aligned}$$

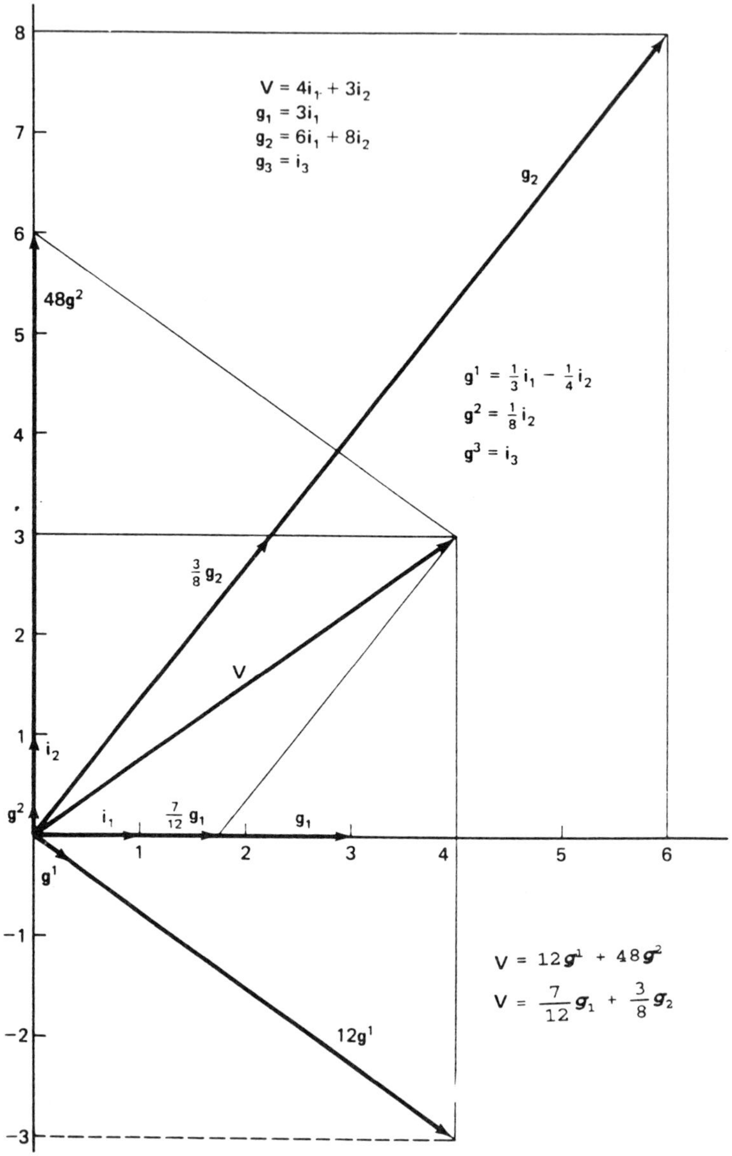

Figure 2.3.3 Relationships between covariant and contravariant components of a vector.

Once again, we have a check:

$$\mathbf{V} = V^i\mathbf{g}_i = \tfrac{7}{12}\mathbf{g}_1 + \tfrac{3}{8}\mathbf{g}_2 = 4\mathbf{i}_1 + 3\mathbf{i}_2.$$

Now consider the derivatives of the covariant tangent vectors:

$$\mathbf{g}_{i,j} = \frac{\partial \mathbf{g}_i}{\partial \xi_j} = \frac{\partial}{\partial \xi_j}\left(\frac{\partial z_m}{\partial \xi_i}\mathbf{i}_m\right) = \frac{\partial^2 z_m}{\partial \xi_j \partial \xi_i}\frac{\partial \xi_n}{\partial z_m}\mathbf{g}_n = \Gamma^n_{ij}\mathbf{g}_n, \quad (2.3.8)$$

where Γ^n_{ij} is called the Christoffel symbol of the second kind, defined as:

$$\Gamma^n_{ij} = \frac{\partial^2 z_m}{\partial \xi_j \partial \xi_i}\frac{\partial \xi_n}{\partial z_m} = z_{m,ij}\frac{\partial \xi_n}{\partial z_m}. \quad (2.3.9)$$

Note also that

$$\frac{\partial}{\partial \xi_i}(\mathbf{g}^j \cdot \mathbf{g}_k) = \mathbf{g}_k \cdot \mathbf{g}^j_{,i} + \mathbf{g}^j \cdot \mathbf{g}_{k,i} = 0$$

or

$$\mathbf{g}_k \cdot \mathbf{g}^j_{,i} = -\mathbf{g}^j \cdot \mathbf{g}_{k,i} = -\mathbf{g}^j \cdot \mathbf{g}_n \Gamma^n_{ki} = -\Gamma^j_{ki}.$$

Thus, the derivative of contravariant tangent vectors is of the form

$$\frac{\partial \mathbf{g}^j}{\partial \xi_i} = \mathbf{g}^j_{,i} = -\Gamma^j_{ik}\mathbf{g}^k. \quad (2.3.10)$$

The relation shown in Eq. (2.3.8) may be written in the alternate form

$$\frac{\partial \mathbf{g}_i}{\partial \xi_j} = \frac{\partial^2 z_m}{\partial \xi_i \partial \xi_j}\frac{\partial z_m}{\partial \xi_k}\mathbf{g}^k = \Gamma_{ijk}\mathbf{g}^k = \Gamma^k_{ij}\mathbf{g}_k, \quad (2.3.11)$$

where Γ_{ijk} is known as the Christoffel symbol of the first kind,

$$\Gamma_{ijk} = \frac{\partial^2 z_m}{\partial \xi_i \partial \xi_j}\frac{\partial z_m}{\partial \xi_k}, \quad (2.3.12)$$

which is related to the Christoffel symbol of the second kind as:

$$\Gamma^k_{ij} = \Gamma_{ijm} g^{mk}. \quad (2.3.13)$$

Here the contravariant metric tensor transforms the Christoffel symbol of the first kind into that of the second kind.

With the definitions stated previously, the Christoffel symbols are related to the derivatives of metric tensors as:

$$\Gamma_{ijk} = \tfrac{1}{2}(g_{ik,j} + g_{jk,i} - g_{ij,k}),$$
$$\Gamma^k_{ij} = \tfrac{1}{2}g^{km}(g_{mi,j} + g_{mj,i} - g_{ij,m}),$$
$$\Gamma^i_{ij} = \tfrac{1}{2}g^{im}g_{im,j}.$$

The usefulness of the Christoffel symbols will be demonstrated in the examples of cylindrical and spherical coordinates given in subsection 2.3.3.

2.3.2 Covariant Derivatives

We define the del operator in curvilinear coordinates as:

$$\nabla = \mathbf{g}^i \frac{\partial}{\partial \xi_i}. \tag{2.3.14}$$

The divergence of a vector is given by

$$\begin{aligned} \nabla \cdot \mathbf{V} &= \mathbf{g}^i \frac{\partial}{\partial \xi_i} \cdot (V^j \mathbf{g}_j) = \mathbf{g}^i \cdot \mathbf{g}_j V^j_{,i} + \mathbf{g}^i \cdot \Gamma^k_{ji} \mathbf{g}_k V^j \\ &= V^i_{,i} + \Gamma^i_{ij} V^j = V^i_{|i}, \end{aligned} \tag{2.3.15a}$$

where the subscripted vertical stroke stands for what is known as a *covariant derivative*, which indicates that the term contains the partial derivative plus the change of the tangent vector, as given by the Christoffel symbol.

Note that the contravariant metric tensor g^{ij} is the inverse of g_{ij} or the adjoint of g_{ij} divided by the determinant of g_{ij} (Cramer's rule):

$$g^{ij} = \frac{1}{g} \frac{\partial g}{\partial g_{ij}},$$

so we may write

$$\nabla \cdot \mathbf{V} = V^i_{|i} = \frac{1}{\sqrt{g}} (\sqrt{g}\, V^i)_{,i}. \tag{2.3.15b}$$

The curl of a vector takes the form

$$\begin{aligned} \nabla \times \mathbf{V} &= \mathbf{g}^i \frac{\partial}{\partial \xi_i} \times V_j \mathbf{g}^j = \mathbf{g}^i \times \mathbf{g}^j V_{j,i} + \mathbf{g}^i \times \mathbf{g}^j_{,i} V_j \\ &= \mathbf{g}^i \times \mathbf{g}^j (V_{j,i} - \Gamma^r_{ij} V_r) = \frac{1}{\sqrt{g}} \epsilon^{ijk} \mathbf{g}_k V_{j|i} \end{aligned} \tag{2.3.16}$$

with

$$V_{j|i} = V_{j,i} - \Gamma^r_{ij} V_r.$$

The second derivative of a scalar or a vector can be carried out in a similar manner. For example, let us consider the Laplace equation

$$\nabla^2 \phi = 0. \tag{2.3.17}$$

Since $\nabla^2\phi$ is the divergence of the gradient of ϕ we obtain

$$\begin{aligned}
\nabla^2\phi = (\nabla\cdot\nabla)\phi &= \left(\mathbf{g}^j\frac{\partial}{\partial\xi_j}\cdot\mathbf{g}^i\frac{\partial}{\partial\xi_i}\right)\phi \\
&= \mathbf{g}^j\cdot\mathbf{g}^i_{,j}\phi_{,i} + \mathbf{g}^j\cdot\mathbf{g}^i\phi_{,ij} \\
&= \mathbf{g}^j\cdot(g^{ik}\mathbf{g}_k)_{,j}\phi_{,i} + g^{ij}\phi_{,ij} \qquad (2.3.18)\\
&= g^{ij}_{,j}\phi_{,i} + g^{ij}\Gamma^k_{ki}\phi_{,j} + g^{ij}\phi_{,ij} \\
&= g^{ij}_{,i}\phi_{,j} + \phi_{|ji}g^{ij} = \frac{1}{\sqrt{g}}(\sqrt{g}\,g^{ij}\phi_{,j})_{,i},
\end{aligned}$$

where $g^{ij}_{,i} = 0$ (see Problem 2.8) and $\phi_{|ji}$ is defined as the second-order covariant derivative of ϕ:

$$\phi_{|ji} = \phi_{,ji} + \Gamma^k_{ki}\phi_{,j} = \frac{1}{\sqrt{g}}(\sqrt{g}\,\phi_{,j})_{,i}. \qquad (2.3.19)$$

If a tensor is of higher order or mixed covariant contravariant form, then the covariant derivatives are more complicated. For example, consider a covariant derivative of A_{ij}. First, write a dot product of two vectors

$$\mathbf{V}\cdot\mathbf{W} = V_i\mathbf{g}^i\cdot W_j\mathbf{g}^j = A_{ij}\mathbf{g}^i\cdot\mathbf{g}^j$$

and take the partial derivative of the above

$$\begin{aligned}
\frac{\partial}{\partial\xi_s}(\mathbf{V}\cdot\mathbf{W}) &= A_{ij,s}g^{ij} - A_{ij}\Gamma^i_{sr}g^{rj} - A_{ij}\Gamma^j_{sr}g^{ri} \\
&= (A_{ij,s} - \Gamma^r_{is}A_{rj} - \Gamma^r_{js}A_{ir})g^{ij} = A_{ij|s}g^{ij},
\end{aligned}$$

where

$$A_{ij|s} = A_{ij,s} - \Gamma^r_{is}A_{rj} - \Gamma^r_{js}A_{ir}. \qquad (2.3.20a)$$

Similarly, set $\mathbf{V}\cdot\mathbf{W} = V^i\mathbf{g}_i\cdot W^j\mathbf{g}_j$ to obtain

$$\frac{\partial}{\partial\xi_s}(\mathbf{V}\cdot\mathbf{W}) = A^{ij}_{|s}g_{ij},$$

where

$$A^{ij}_{|s} = A^{ij}_{,s} + \Gamma^i_{rs}A^{rj} + \Gamma^j_{rs}A^{ir}. \qquad (2.3.20b)$$

The covariant derivative of the mixed tensor may be determined by setting $\mathbf{V}\cdot\mathbf{W} = V^i\mathbf{g}_i\cdot W_j\mathbf{g}^j$. This gives

$$\frac{\partial}{\partial\xi_s}(\mathbf{V}\cdot\mathbf{W}) = A^i_{j|s}\delta^j_i,$$

where

$$A^i_{j|s} = A^i_{j,s} - \Gamma^r_{js}A^i_r + \Gamma^i_{rs}A^r_j, \qquad (2.3.20c)$$

where A^i_j is called the *mixed tensor*.

It follows from these observations that covariant derivatives of multiple-order mixed tensors can be written in the form

$$
\begin{aligned}
A^{j_1 \ldots j_m}_{i_1 \ldots i_n}|_s = {} & A^{j_1 \ldots j_m}_{i_1 \ldots i_n, s} - \Gamma^{r}_{i_1} A^{j_1 \ldots j_m}_{r i_2 \ldots i_n} - \Gamma^{r}_{i_2 s} A^{j_1 \ldots j_m}_{i_1 r i_3 \ldots i_n} \\
& - \ldots - \Gamma^{r}_{i_n s} A^{j_1 \ldots j_m}_{i_1 \ldots r} + \Gamma^{j_1}_{rs} A^{r j_2 \ldots j_m}_{i_1 \ldots i_n} \\
& + \Gamma^{j_2}_{rs} A^{j_1 r \ldots j_m}_{i_1 \ldots i_n} + \ldots + \Gamma^{j_m}_{rs} A^{j_1 \ldots r}_{i_1 \ldots i_n}.
\end{aligned}
\tag{2.3.21}
$$

Covariant derivatives are useful in obtaining differential equations in curvilinear coordinates (cylindrical or spherical geometries), and also for automatic grid generation in finite difference methods of computational mechanics.

2.3.3 Physical Components of Tensors

It has been shown that the tangent vectors are not unit vectors; therefore, all tensor components must be re-evaluated in terms of unit vectors to obtain physical components. To this end, we first consider the tangent vectors $\mathbf{g}_i$ or $\mathbf{g}^i$. Denoting $\bar{\mathbf{g}}_i$ and $\bar{\mathbf{g}}^i$ as unity, we write

$$
\begin{aligned}
\bar{\mathbf{g}}_i &= \frac{\mathbf{g}_i}{|\mathbf{g}_i|} = \frac{\mathbf{g}_i}{\sqrt{\mathbf{g}_i \cdot \mathbf{g}_i}} = \frac{\mathbf{g}_i}{\sqrt{g_{(ii)}}} \\
\mathbf{g}_i &= \sqrt{g_{(ii)}}\, \bar{\mathbf{g}}_i
\end{aligned}
\tag{2.3.22a}
$$

$$
\begin{aligned}
\bar{\mathbf{g}}^i &= \frac{\mathbf{g}^i}{|\mathbf{g}^i|} = \frac{\mathbf{g}^i}{\sqrt{\mathbf{g}^i \cdot \mathbf{g}^i}} = \frac{\mathbf{g}^i}{\sqrt{g^{(ii)}}} \\
\mathbf{g}^i &= \sqrt{g^{(ii)}}\, \bar{\mathbf{g}}^i,
\end{aligned}
\tag{2.3.22b}
$$

where the index inside the parentheses does not imply a summing. Therefore, any vector $\mathbf{V}$ can be written

$$
\mathbf{V} = V_i \mathbf{g}^i = \bar{V}_i \bar{\mathbf{g}}^i = \bar{V}_i \frac{\mathbf{g}^i}{\sqrt{g^{(ii)}}}.
$$

This gives

$$
V_i = \frac{\bar{V}_i}{\sqrt{g^{(ii)}}},
$$

which implies that the physical components $\bar{V}_i$ of the tensor components V_i of the vector $\mathbf{V}$ are given by

$$
\bar{V}_i = \sqrt{g^{(ii)}}\, V_i. \tag{2.3.23}
$$

Similarly,

$$
\bar{V}^i = \sqrt{g_{(ii)}}\, V^i. \tag{2.3.24}
$$

It is easy to see that for orthogonal coordinates in which all off-diagonal terms are zero, $g_{(ii)}$ can be written as

$$g_{(ii)} = \frac{1}{g^{(ii)}}.$$

Note that with the conversion of tensor components into physical components, all units and dimensions are consistent throughout the differential equations, as will be seen in the examples of cylindrical and spherical coordinates (see Examples 2.3 and 2.4).

Physical components of second-order tensors will be discussed in the later chapters.

Example 2.3 Write the Laplace equation in cylindrical coordinates. Consider the cylindrical coordinates $\xi_1 = r$, $\xi_2 = \theta$, and $\xi_3 = z_3$, $z_1 = r\cos\theta$, $z_2 = r\sin\theta$, and $z_3 = z$ (see Fig. 2.3.4a). This gives

$$\begin{aligned}
\mathbf{r} &= z_i\mathbf{i}_i = r(\cos\theta)\mathbf{i}_1 + r(\sin\theta)\mathbf{i}_2 + z\mathbf{i}_3,\\
\mathbf{g}_1 &= \mathbf{r}_{,1} = (\cos\theta)\mathbf{i}_1 + (\sin\theta)\mathbf{i}_2,\\
\mathbf{g}_2 &= \mathbf{r}_{,2} = -r(\sin\theta)\mathbf{i}_1 + r(\cos\theta)\mathbf{i}_2,\\
\mathbf{g}_3 &= \mathbf{r}_{,3} = \mathbf{i}_3.
\end{aligned}$$

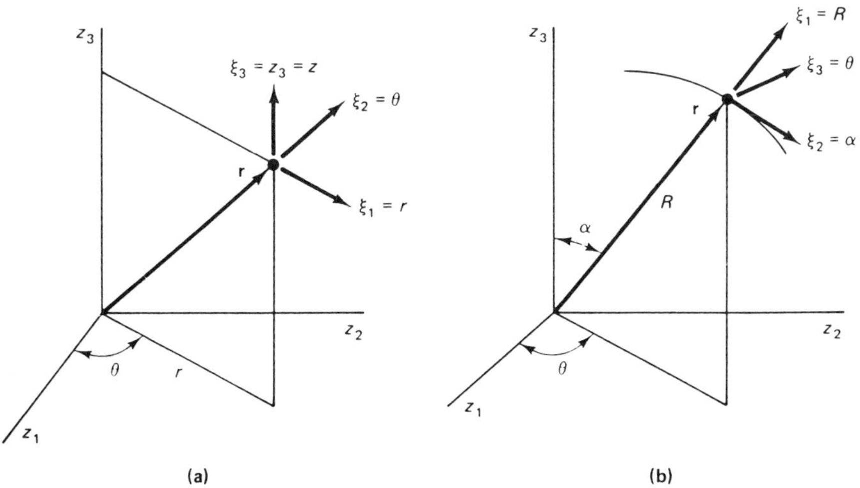

Figure 2.3.4 Curvilinear coordinates. (a) Cylindrical coordinates: r = radius of cylinder, θ = circumferential angle, z = axial coordinate. (b) Spherical coordinates: R = radius of sphere, α = meridian angle, θ = circumferential angle.

To calculate the Christoffel symbols of the second kind, we proceed as follows: The covariant metric tensors are computed by

$$
\begin{aligned}
g_{11} &= \frac{\partial z_1}{\partial \xi_1}\frac{\partial z_1}{\partial \xi_1} + \frac{\partial z_2}{\partial \xi_1}\frac{\partial z_2}{\partial \xi_1} + \frac{\partial z_3}{\partial \xi_1}\frac{\partial z_3}{\partial \xi_1} = \cos^2\theta + \sin^2\theta = 1, \\
g_{22} &= r^2\sin^2\theta + r^2\cos^2\theta = r^2, \\
g_{33} &= 1,
\end{aligned}
$$

with all other $g_{ij} = 0$.

Thus

$$
g_{ij} = \begin{bmatrix} 1 & 0 & 0 \\ 0 & r^2 & 0 \\ 0 & 0 & 1 \end{bmatrix}.
$$

The contravariant metric tensors are given by the reciprocal of g_{ij} as

$$
g^{ij} = \begin{bmatrix} 1 & 0 & 0 \\ 0 & \dfrac{1}{r^2} & 0 \\ 0 & 0 & 1 \end{bmatrix}.
$$

Recall that the Christoffel symbols of the first and second kinds are:

$$
\begin{aligned}
\Gamma_{ijs} &= \tfrac{1}{2}(g_{si,j} + g_{sj,i} - g_{ij,s}), \\
\Gamma^r_{ij} &= g^{rs}\Gamma_{ijs}.
\end{aligned}
$$

The Christoffel symbols of the first kind are determined from derivatives of covariant metric tensors:

$$
\Gamma_{122} = \tfrac{1}{2}(g_{21,2} + g_{22,1} - g_{12,2}) = \tfrac{1}{2}(0 + 2r - 0) = r.
$$

In summary, the Christoffel symbols of the first kind are:

$$
\Gamma_{122} = \Gamma_{212} = r, \quad \Gamma_{221} = -r,
$$

and all other $\Gamma_{ijk} = 0$. From these we calculate the Christoffel symbols of the second kind:

$$
\begin{aligned}
\Gamma^1_{22} &= g^{11}\Gamma_{221} + g^{12}\Gamma_{222} = -r, \\
\Gamma^2_{21} &= \Gamma^2_{12} = \frac{1}{r},
\end{aligned}
$$

and all other $\Gamma^i_{jk} = 0$.

Note that in Eq. (2.3.18)

$$
g^{ij}_{,i}\phi_{,j} = 0,
$$

so we can expand the Laplace equation in the form

$$\begin{aligned}
\nabla^2\phi &= \phi_{|ji}g^{ij} \\
&= \phi_{|11} + \phi_{|22}\frac{1}{r^2} + \phi_{|33} \\
&= \phi_{,11} + \Gamma^1_{11}\phi_{,1} + \Gamma^2_{21}\phi_{,1} + (\phi_{,22} + \Gamma^1_{12}\phi_{,2} + \Gamma^2_{22}\phi_{,2})\frac{1}{r^2} + \phi_{,33} \\
&\quad + \Gamma^1_{13}\phi_{,3} + \Gamma^2_{23}\phi_{,3} = 0
\end{aligned}$$

or

$$\nabla^2\phi = \frac{\partial^2\phi}{\partial r^2} + \frac{1}{r}\frac{\partial\phi}{\partial r} + \frac{1}{r^2}\frac{\partial^2\phi}{\partial\theta^2} + \frac{\partial^2\phi}{\partial z^2} = 0.$$

Example 2.4 Repeat Example 2.3 for spherical coordinates. Consider the spherical coordinates (see Fig. 2.3.4b) $\xi_1 = R$, $\xi_2 = \alpha$, $\xi_3 = \theta$, $z_1 = R\sin\alpha\cos\theta$, $z_2 = R\sin\alpha\sin\theta$, and $z_3 = R\cos\alpha$. The position vector takes the form

$$\mathbf{r} = R(\sin\alpha\cos\theta)\mathbf{i}_1 + R(\sin\alpha\sin\theta)\mathbf{i}_2 + R(\cos\alpha)\mathbf{i}_3.$$

Thus, the tangent vectors are

$$\begin{aligned}
\mathbf{g}_1 &= \mathbf{r}_{,1} = (\sin\alpha\cos\theta)\mathbf{i}_1 + (\sin\alpha\sin\theta)\mathbf{i}_2 + (\cos\alpha)\mathbf{i}_3, \\
\mathbf{g}_2 &= \mathbf{r}_{,2} = R(\cos\alpha\cos\theta)\mathbf{i}_1 + R(\cos\alpha\sin\theta)\mathbf{i}_2 - R(\sin\alpha)\mathbf{i}_3, \\
\mathbf{g}_3 &= \mathbf{r}_{,3} = -R(\sin\alpha\sin\theta)\mathbf{i}_1 + R(\sin\alpha\cos\theta)\mathbf{i}_2.
\end{aligned}$$

From these the metric tensors are calculated as:

$$g_{ij} = \begin{bmatrix} 1 & 0 & 0 \\ 0 & R^2 & 0 \\ 0 & 0 & R^2\sin^2\alpha \end{bmatrix}.$$

Again, g^{ij} is the reciprocal of g_{ij}. The Christoffel symbols are

$$\begin{aligned}
\Gamma^1_{22} &= -R, & \Gamma^1_{33} &= -R\sin^2\alpha, \\
\Gamma^2_{12} &= \Gamma^2_{21} = \frac{1}{R}, & \Gamma^2_{33} &= -\sin\alpha\cos\alpha, \\
\Gamma^3_{13} &= \Gamma^3_{31} = \frac{1}{R}, & \Gamma^3_{32} &= \Gamma^3_{23} = \cot\alpha,
\end{aligned}$$

with all other $\Gamma^i_{jk} = 0$. Therefore, we obtain

$$\nabla^2\phi = \frac{\partial^2\phi}{\partial R^2} + \frac{2}{R}\frac{\partial\phi}{\partial R} + \frac{1}{R^2}\frac{\partial^2\phi}{\partial\alpha^2} + \frac{\cot\alpha}{R^2}\frac{\partial\phi}{\partial\alpha} + \frac{1}{R^2\sin^2\alpha}\frac{\partial^2\phi}{\partial\theta^2}.$$

Example 2.5 Divergence and Curl of Vectors with Physical Components Physical components of first-order tensors are computed as:

$$u_i = \sqrt{g_{(ii)}}\,\bar{u}_i, \quad u^i = \sqrt{g^{(ii)}}\,\bar{u}^i.$$

Cylindrical Coordinates

$$\begin{aligned}
&u_1 = \sqrt{g_{11}}\bar{u}_1 = \bar{u}_1 = u_r, \quad u_2 = \sqrt{g_{22}}\bar{u}_2 = r\bar{u}_2 = ru_\theta,\\
&u_3 = \sqrt{g_{33}}\bar{u}_3 = \bar{u}_3 = u_z,\\
&u^1 = \sqrt{g^{11}}\bar{u}^1 = \bar{u}^1 = u_r, \quad u^2 = \sqrt{g^{22}}\bar{u}^2 = \frac{1}{r}\bar{u}^2 = \frac{1}{r}u_\theta,\\
&u^3 = \sqrt{g^{33}}\bar{u}^3 = \bar{u}^3 = u_z.
\end{aligned}$$

Spherical Coordinates

$$\begin{aligned}
&u_1 = \bar{u}_1 = u_R, \quad u_2 = R\bar{u}_2 = Ru_\alpha, \quad u_3 = R(\sin\alpha)\bar{u}_3 = R(\sin\alpha)u_\theta,\\
&u^1 = \bar{u}^1 = u_R, \quad u^2 = \frac{\bar{u}^2}{R} = \frac{u_\alpha}{R}, \quad u^3 = \frac{\bar{u}^3}{R\sin\alpha} = \frac{u_\theta}{R\sin\alpha}.
\end{aligned}$$

Thus, the divergence and the curl of a vector $\mathbf{u}$ in cylindrical coordinates can be written as:

$$\begin{aligned}
\nabla\cdot\mathbf{u} &= u^i_{|i}\\
&= u^i_{,i} + \Gamma^i_{ij}u^j\\
&= u^1_{,1} + u^2_{,2} + u^3_{,3} + \Gamma^1_{11}u^1 + \Gamma^1_{12}u^2 + \Gamma^1_{13}u^3 + \Gamma^2_{21}u^1 + \Gamma^2_{22}u^2\\
&\quad + \Gamma^2_{23}u^3 + \Gamma^3_{31}u^1 + \Gamma^3_{32}u^2 + \Gamma^3_{33}u^3\\
&= u^1_{,1} + u^2_{,2} + u^3_{,3} + \Gamma^2_{21}u^1\\
&= (\sqrt{g^{11}}\bar{u}^1)_{,1} + (\sqrt{g^{22}}\bar{u}^2)_{,2} + (\sqrt{g^{33}}\bar{u}^3)_{,3} + \frac{1}{r}\sqrt{g^{11}}\bar{u}^1\\
&= \frac{\partial u_r}{\partial r} + \frac{1}{r}\frac{\partial u_\theta}{\partial\theta} + \frac{\partial u_z}{\partial z} + \frac{u_r}{r}.
\end{aligned}$$

Similarly,

$$\begin{aligned}
\nabla\times\mathbf{u} &= \frac{1}{\sqrt{g}}\epsilon^{ijk}(u_{j,i} - \Gamma^r_{ji}u_r)\mathbf{g}_k\\
&= \frac{1}{\sqrt{g}}(u_{2,1} - \Gamma^2_{21}u_2)\mathbf{g}_3 + \frac{1}{\sqrt{g}}(-u_{1,2} + \Gamma^2_{12}u_2)\mathbf{g}_3\\
&\quad + \frac{1}{\sqrt{g}}(u_{1,3} - u_{3,1})\mathbf{g}_2 + \frac{1}{\sqrt{g}}(u_{3,2} - u_{2,3})\mathbf{g}_1
\end{aligned}$$

$$= \left(\frac{1}{r}\frac{\partial u_z}{\partial \theta} - \frac{\partial u_\theta}{\partial z}\right)\mathbf{e}_r + \left(\frac{\partial u_r}{\partial r} - \frac{\partial u_z}{\partial r}\right)\mathbf{e}_\theta$$
$$+ \left(\frac{\partial u_\theta}{\partial r} - \frac{1}{r}\frac{\partial u_r}{\partial \theta} + \frac{u_\theta}{r}\right)\mathbf{e}_z,$$

where

$$\sqrt{g} = \sqrt{|g_{ij}|} = r,$$
$$\mathbf{g}_1 = \sqrt{g_{11}}\bar{\mathbf{g}}_1 = \bar{\mathbf{g}}_1 = \mathbf{e}_r,$$
$$\mathbf{g}_2 = \sqrt{g_{22}}\bar{\mathbf{g}}_2 = r\bar{\mathbf{g}}_2 = r\mathbf{e}_\theta,$$
$$\mathbf{g}_3 = \sqrt{g_{33}}\bar{\mathbf{g}}_3 = \bar{\mathbf{g}}_3 = \mathbf{e}_z.$$

For spherical coordinates, we proceed in a similar manner and find:

$$\boldsymbol{\nabla}\cdot\mathbf{u} = \frac{\partial u_R}{\partial R} + \frac{1}{R}\frac{\partial u_\alpha}{\partial \alpha} + \frac{1}{R\sin\alpha}\frac{\partial u_\theta}{\partial \theta} + \frac{2u_R}{R} + \frac{u_\alpha \cot\alpha}{R},$$
$$\boldsymbol{\nabla}\times\mathbf{u} = \left(\frac{1}{R}\frac{\partial u_\theta}{\partial \alpha} - \frac{1}{R\sin\alpha}\frac{\partial u_\alpha}{\partial \theta} + \frac{(\cot\alpha)u_\theta}{R}\right)\mathbf{e}_R$$
$$+ \left(\frac{1}{R\sin\alpha}\frac{\partial u_R}{\partial \theta} - \frac{u_\theta}{R} - \frac{\partial u_\theta}{\partial R}\right)\mathbf{e}_\alpha$$
$$+ \left(\frac{\partial u_\alpha}{\partial R} - \frac{1}{R}\frac{\partial u_R}{\partial \alpha} + \frac{u_\alpha}{R}\right)\mathbf{e}_\theta,$$

where

$$\mathbf{g}_1 = \bar{\mathbf{g}}_1 = \mathbf{e}_R$$
$$\mathbf{g}_2 = R\bar{\mathbf{g}}_2 = R\mathbf{e}_\alpha$$
$$\mathbf{g}_3 = R\sin\alpha\bar{\mathbf{g}}_3 = R\sin\alpha\mathbf{e}_\theta.$$

Note that the material presented in this section is important information that is applicable to structured grid generations by means of the solution of elliptic partial differential equations in computational mechanics between the physical and the computational domains.

Discussions of the curvilinear coordinates that involve the deformed state are not rigorously pursued in this chapter, although strain-displacement relations associated with a deformed state (nonlinear strains) will be discussed briefly in the next section. A complete treatment of this subject would require theories of curved surfaces or differential geometries. See Chapter 6 in Chung (1988) for this subject.

Further exercises for curvilinear coordinates that govern equations in solid and fluid mechanics problems appear in Chapters 3, 4, and 5.

2.4 Strain Tensor

2.4.1 Cartesian Coordinates

It was shown in subsection 2.2.1 that the Jacobian is considered a measure of deformation:

$$J = \left| \frac{\partial z_i}{\partial x_j} \right| = |F_{ij}|,$$

where

$$F_{ij} = \frac{\partial z_i}{\partial x_j} \tag{2.4.1}$$

which is the deformation gradient, as defined in Eq. (2.2.5). However, it is more convenient to define the deformation specifically through strain-displacement relationships. Toward this end, we consider squares of infinitesimal line segments on the undeformed and deformed surfaces (see Fig. 2.2.1):

$$ds_0^2 = \mathbf{dr} \cdot \mathbf{dr} = dx_i \, dx_i,$$

$$ds^2 = \mathbf{dR} \cdot \mathbf{dR} = \frac{\partial \mathbf{R}}{\partial x_i} \cdot \frac{\partial \mathbf{R}}{\partial x_j} dx_i \, dx_j = G_{ij} \, dx_i \, dx_j,$$

where G_{ij} is called Green's deformation tensor (or simply deformation tensor) or metric tensor, as defined on the deformed geometry,

$$G_{ij} = \mathbf{G}_i \cdot \mathbf{G}_j = z_{m,i} z_{m,j} = (\delta_{mi} + u_{m,i})(\delta_{mj} + u_{m,j}),$$

or

$$G_{ij} = \delta_{ij} + u_{i,j} + u_{j,i} + u_{m,i} u_{m,j}. \tag{2.4.2}$$

The difference between ds^2 and ds_0^2 is the measure of strain:

$$ds^2 - ds_0^2 = (G_{ij} - \delta_{ij}) \, dx_i \, dx_j = 2\gamma_{ij} \, dx_i \, dx_j, \tag{2.4.3}$$

where γ_{ij} is called the Green–Saint–Venant strain tensor or simply the strain tensor,

$$\gamma_{ij} = \tfrac{1}{2}(G_{ij} - \delta_{ij}). \tag{2.4.4}$$

Note that coefficient 2 in Eq. (2.4.3) is used to produce a standard definition for the strain components. However, this definition provides one-half of the so-called *engineering (total) shear strain*. Substitution of Eq. (2.4.2) into Eq. (2.4.4) yields

$$\gamma_{ij} = \tfrac{1}{2}(u_{i,j} + u_{j,i} + u_{m,i} u_{m,j}). \tag{2.4.5}$$

Let $u_1 = u$, $u_2 = v$, $u_3 = w$, $x_1 = x$, $x_2 = y$, and $x_3 = z$; thus we can write the relation in Eq. (2.4.5) as

$$\gamma_{11} = \frac{\partial u}{\partial x} + \frac{1}{2}\left[\left(\frac{\partial u}{\partial x}\right)^2 + \left(\frac{\partial v}{\partial x}\right)^2 + \left(\frac{\partial w}{\partial x}\right)^2\right], \tag{2.4.6a}$$

$$\gamma_{22} = \frac{\partial v}{\partial y} + \frac{1}{2}\left[\left(\frac{\partial u}{\partial y}\right)^2 + \left(\frac{\partial v}{\partial y}\right)^2 + \left(\frac{\partial w}{\partial y}\right)^2\right], \tag{2.4.6b}$$

$$\gamma_{33} = \frac{\partial w}{\partial z} + \frac{1}{2}\left[\left(\frac{\partial u}{\partial z}\right)^2 + \left(\frac{\partial v}{\partial z}\right)^2 + \left(\frac{\partial w}{\partial z}\right)^2\right], \tag{2.4.6c}$$

$$\gamma_{12} = \frac{1}{2}\left(\frac{\partial u}{\partial y} + \frac{\partial v}{\partial x} + \frac{\partial u}{\partial x}\frac{\partial u}{\partial y} + \frac{\partial v}{\partial x}\frac{\partial v}{\partial y} + \frac{\partial w}{\partial x}\frac{\partial w}{\partial y}\right), \tag{2.4.6d}$$

$$\gamma_{23} = \frac{1}{2}\left(\frac{\partial v}{\partial z} + \frac{\partial w}{\partial y} + \frac{\partial u}{\partial y}\frac{\partial u}{\partial z} + \frac{\partial v}{\partial y}\frac{\partial v}{\partial z} + \frac{\partial w}{\partial y}\frac{\partial w}{\partial z}\right), \tag{2.4.6e}$$

$$\gamma_{31} = \frac{1}{2}\left(\frac{\partial w}{\partial x} + \frac{\partial u}{\partial z} + \frac{\partial u}{\partial z}\frac{\partial u}{\partial x} + \frac{\partial v}{\partial z}\frac{\partial v}{\partial x} + \frac{\partial w}{\partial z}\frac{\partial w}{\partial x}\right). \tag{2.4.6f}$$

These equations represent nonlinear strain-displacement relations in Cartesian coordinates; note that $\gamma_{11} = \gamma_{xx}$, $\gamma_{22} = \gamma_{yy}$, $\gamma_{33} = \gamma_{zz}$, $2\gamma_{12} = \gamma_{xy}$, $2\gamma_{23} = \gamma_{yz}$, and $2\gamma_{31} = \gamma_{zx}$. Here, γ_{12}, γ_{23}, and γ_{31} are the *tensor shear strain* components, whereas γ_{xy}, γ_{yz}, and γ_{zx} refer to the engineering (or total) shear strains. Example 2.8 will further explain the terminology and their physical significance.

Also note that, as is obvious from Eq. (2.4.4), the strain tensor is symmetric, that is,

$$\gamma_{ij} = \gamma_{ji}.$$

Example 2.6 Calculation of Strain Components Consider a material undergoing a uniform deformation, as shown in Fig. 2.4.1. Determine the displacement field, the tangent vectors, and the strain tensor.

Solution

1. Displacement Field. The general form that describes the variation of u_i is

$$u_i = C_1 + C_2x_1 + C_3x_2 + C_4x_1x_2.$$

From the boundary conditions at A, B, and D, find $C_1 = C_2 = C_3 = 0$. Note that for $x_1 = a$ and $x_2 = b$ at C we have

$$u_1 = \alpha = C_4ab.$$

This gives $C_4 = \alpha/ab$ and thus,

$$u_1 = \frac{x_1x_2}{ab}\alpha.$$

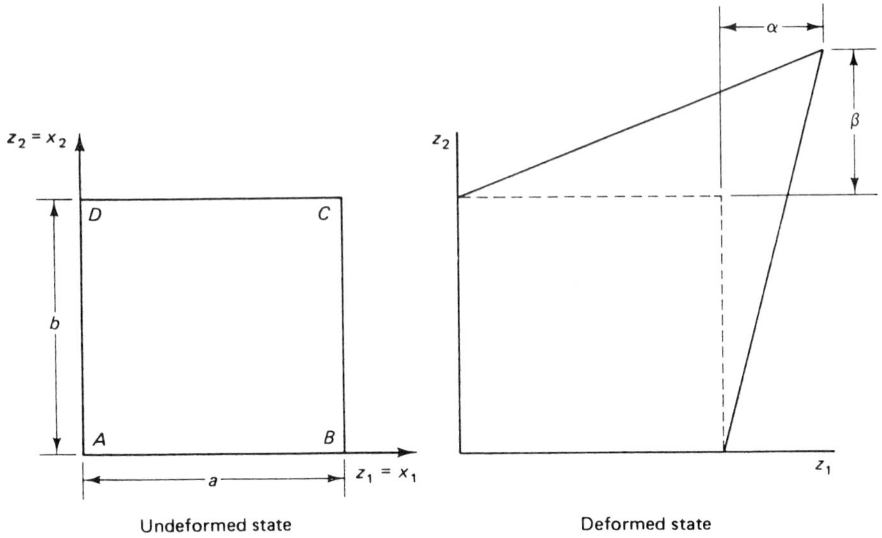

Figure 2.4.1 A rectangular block (homogeneous and isotropic material) is deformed uniformly with a large strain.

Similarly

$$u_2 = \frac{x_1 x_2}{ab}\beta.$$

Thus, the displacement vector **u** is

$$\mathbf{u} = \frac{x_1 x_2}{ab}(\alpha \mathbf{i}_1 + \beta \mathbf{i}_2).$$

2. Tangent Vectors

$$\mathbf{G}_i = \frac{\partial z_m \mathbf{i}_m}{\partial x_i} = \frac{\partial}{\partial x_i}(x_m + u_m)\mathbf{i}_m = \left(\delta_{mi} + \frac{\partial u_m}{\partial x_i}\right)\mathbf{i}_m,$$

$$\mathbf{G}_1 = \left(1 + \frac{\partial u_1}{\partial x_1}\right)\mathbf{i}_1 + \frac{\partial u_2}{\partial x_1}\mathbf{i}_2 = \left(1 + \frac{\alpha x_2}{ab}\right)\mathbf{i}_1 + \frac{\beta x_2}{ab}\mathbf{i}_2.$$

Similarly,

$$\mathbf{G}_2 = \frac{\alpha x_1}{ab}\mathbf{i}_1 + \left(1 + \frac{\beta x_1}{ab}\right)\mathbf{i}_2,$$

$$\mathbf{G}_3 = \mathbf{i}_3.$$

3. Strain Tensor

$$\gamma_{ij} = \tfrac{1}{2}(G_{ij} - \delta_{ij}),$$

$$\gamma_{11} = \frac{\alpha x_2}{ab} + \tfrac{1}{2}x_2^2\left(\frac{\alpha^2 + \beta^2}{(ab)^2}\right),$$

$$\gamma_{22} = \frac{\beta x_1}{ab} + \tfrac{1}{2}x_1^2\left(\frac{\alpha^2 + \beta^2}{(ab)^2}\right),$$

$$\gamma_{12} = \tfrac{1}{2}\left(\frac{\alpha x_1 + \beta x_2}{ab} + \frac{\alpha^2 + \beta^2}{(ab)^2}x_1x_2\right),$$

$$\gamma_{33} = \gamma_{13} = \gamma_{31} = 0.$$

Example 2.7 Small Extensional Strain Consider $\mathrm{d}\mathbf{R} = (\mathrm{d}x_1, 0, 0)$ and determine the strain γ_{11}, assuming that its magnitude is small.

Solution

$$\mathrm{d}s^2 - \mathrm{d}s_0^2 = 2\gamma_{ij}\,\mathrm{d}x_i\,\mathrm{d}x_j = 2\gamma_{11}(\mathrm{d}x_1)^2 = 2\gamma_{11}\,\mathrm{d}s_0^2.$$

Solve for the axial strain γ_{11}, thus we have

$$\gamma_{11} = \tfrac{1}{2}\left(\frac{\mathrm{d}s^2}{\mathrm{d}s_0^2} - 1\right) = \tfrac{1}{2}(\beta^2 - 1),$$

where β is defined as an *extension ratio*:

$$\beta = \frac{\mathrm{d}s}{\mathrm{d}s_0} = (1 + 2\gamma_{11})^{1/2} = 1 + \gamma_{11} - \tfrac{1}{2}\gamma_{11}^2 + \ldots$$

Neglect higher-order terms to obtain:

$$\gamma_{11} \simeq \beta - 1 = \frac{\mathrm{d}s - \mathrm{d}s_0}{\mathrm{d}s_0} = \frac{\mathrm{d}u_1}{\mathrm{d}x_1},$$

which is the same as Eq. (2.4.6a) with all the nonlinear terms neglected, and identified as the normal strain in undergraduate mechanics. It refers to a special case (small strain) which results from the general definition of the strain (large strain), as derived in the form shown in Eqs. (2.4.6a–f).

Example 2.8 Small Shear Strain The undeformed square block in Fig. 2.4.2 undergoes a shear deformation. Calculate the shear strain, assuming that its magnitude is small. Here, $\mathrm{d}\mathbf{R} = (\mathrm{d}x_1, \mathrm{d}x_2, 0)$. From Eq. (2.4.4),

$$\gamma_{12} = \tfrac{1}{2}(G_{12} - 0)$$

and

$$\cos\Psi = \frac{\mathbf{G}_1\,\mathrm{d}x_1 \cdot \mathbf{G}_2\,\mathrm{d}x_2}{|\mathbf{G}_1\,\mathrm{d}x_1 \cdot \mathbf{G}_2\,\mathrm{d}x_2|} = \frac{2\gamma_{12}\,\mathrm{d}x_1\,\mathrm{d}x_2}{\sqrt{G_{11}}\sqrt{G_{22}}\,\mathrm{d}x_1\,\mathrm{d}x_2}$$

$$= \frac{2\gamma_{12}}{\sqrt{1 + 2\gamma_{11}}\sqrt{1 + 2\gamma_{22}}}.$$

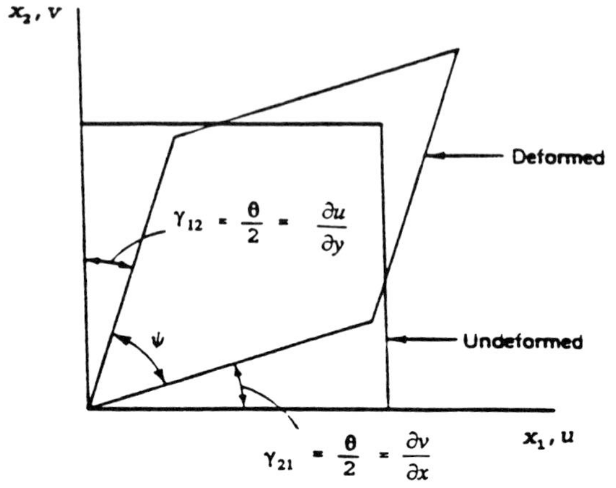

Figure 2.4.2 Shear deformation. Engineering or total shear strain $\gamma_{xy} = 2\gamma_{12} = 2\gamma_{21} = \theta$.

For small strains, $\gamma_{11} \ll 1$ and $\gamma_{22} \ll 1$, and therefore

$$\gamma_{12} \simeq \tfrac{1}{2}\cos\Psi = \tfrac{1}{2}\cos\left(\frac{\pi}{2} - \theta\right) = \tfrac{1}{2}\sin\theta \simeq \frac{\theta}{2}.$$

This implies that the shear strain γ_{12} is the tensor shear strain and is equal to one-half of θ, which is known as the engineering shear strain (the total shear or total angle change due to shears, $\gamma_{xy} = \gamma_{yx} = 2\gamma_{12} = 2\gamma_{21}$).

Once again, the small shear strain is the same as in Eq. (2.4.6a) with all nonlinear terms neglected,

$$\gamma_{21} = \gamma_{12} = \tfrac{1}{2}\left(\frac{\partial u}{\partial y} + \frac{\partial v}{\partial x}\right) \qquad \textit{tensor shear strain},$$

$$\gamma_{xy} = \gamma_{12} + \gamma_{21} = \frac{\partial u}{\partial y} + \frac{\partial v}{\partial x} \qquad \textit{total (engineering) shear strain}.$$

2.4.2 Curvilinear Coordinates

If an undeformed state is represented by curvilinear coordinates, then the position vectors are defined as

$$\mathbf{dr} = \frac{\partial \mathbf{r}}{\partial \xi_i}\,\mathrm{d}\xi_i = \mathbf{r}_{,i}\,\mathrm{d}\xi_i = \mathbf{g}_i\,\mathrm{d}\xi_i,$$

$$\mathrm{d}\mathbf{R} = \frac{\partial \mathbf{R}}{\partial \xi_i}\,\mathrm{d}\xi_i = \frac{\partial}{\partial \xi_i}(\mathbf{r} + \mathbf{u})\,\mathrm{d}\xi_i = (\mathbf{g}_i + \mathbf{u}_{,i})\,\mathrm{d}\xi_i,$$

where

$$\mathbf{u}_{,i} = (u_j \mathbf{g}^j)_{,i} = u_{j,i}\mathbf{g}^j - \Gamma^j_{ik} u_j \mathbf{g}^k,$$

or

$$\mathbf{u}_{,i} = (u^j \mathbf{g}_j)_{,i} = \mathbf{u}^j_{,i}\mathbf{g}_j + \Gamma^k_{ji} u^j \mathbf{g}_k.$$

Thus, we obtain

$$\gamma_{ij} = \tfrac{1}{2}(G_{ij} - g_{ij}) = \tfrac{1}{2}(u_{i|j} + u_{j|i} + u^m_{|i} u_{m|j}) \tag{2.4.7}$$

with

$$u_{i|j} = \frac{\partial u_i}{\partial \xi_j} - \Gamma^m_{ij} u_m, \tag{2.4.8}$$

$$u^i_{|j} = \frac{\partial u^i}{\partial \xi_j} + \Gamma^i_{jm} u^m. \tag{2.4.9}$$

We discussed the transformation of the first-order tensor components into physical components in subsection 2.3.3. Since the strain tensor is of the second order, we take a somewhat different approach. Note that the strain tensor is always a covariant form, although the coordinate lines ξ_i are expected to convect on the curvilinear coordinates and become contravariant. For this reason, it is appropriate to change ξ_i to ξ^i if confusion is likely to arise, as discussed in subsection 2.2.1. Thus, the analog of Eq. (2.4.3) for curvilinear coordinates is

$$\mathrm{d}s^2 - \mathrm{d}s_0^2 = 2\gamma_{ij}\,\mathrm{d}\xi^i\,\mathrm{d}\xi^j, \tag{2.4.10}$$

in which the repeated indices occur diagonally as they should for curvilinear coordinates, where $\mathrm{d}\xi^i$ is the contravariant component of a vector, the *physical component* $\mathrm{d}\bar{\xi}^i$ is given by

$$\mathrm{d}\bar{\xi}^i = \sqrt{g_{(ii)}}\,\mathrm{d}\xi^i. \tag{2.4.11}$$

In view of Eqs. (2.4.10) and (2.4.11), it follows that

$$\mathrm{d}s^2 - \mathrm{d}s_0^2 = 2\bar{\gamma}_{ij}\,\mathrm{d}\bar{\xi}^i\,\mathrm{d}\bar{\xi}^j,$$

where

$$\bar{\gamma}_{ij} = \frac{1}{\sqrt{g_{(ii)}g_{(jj)}}}\gamma_{ij} \tag{2.4.12a}$$

or, for curvilinear orthogonal coordinates,

$$\bar{\gamma}_{ij} = \sqrt{g^{(ii)}g^{(jj)}}\,\gamma_{ij}. \tag{2.4.12b}$$

An alternate approach is to use a second-order tensor as a product of two first-order tensors in the form

$$A_{ij} = V_i W_j$$

with

$$V_i = \frac{1}{\sqrt{g^{(ii)}}}\bar{V}_i \quad \text{and} \quad W_j = \frac{1}{\sqrt{g^{(jj)}}}\bar{W}_j.$$

Thus,

$$A_{ij} = \frac{1}{\sqrt{g^{(ii)}g^{(jj)}}}\bar{V}_i\bar{W}_j = \sqrt{g_{(ii)}g_{(jj)}}\,\bar{A}_{ij}. \tag{2.4.12c}$$

Similarly, we can show the physical component of the strain tensor in the form of Eq. (2.4.12a) by setting $A_{ij} = \gamma_{ij}$, $\bar{A}_{ij} = \bar{\gamma}_{ij}$.

Example 2.9 Explicit Forms of the Strain Tensor Evaluate Eq. (2.4.7) for cylindrical and spherical coordinates, neglecting nonlinear terms.

Cylindrical Coordinates

$$u_{i|j} = u_{i,j} - \Gamma_{ij}^m u_m,$$
$$\gamma_{ij} = \tfrac{1}{2}(u_{i,j} + u_{j,i}) - \Gamma_{ij}^m u_m.$$

The tensor components of the strain are

$$\gamma_{11} = u_{1,1} = \frac{\partial u_1}{\partial r},$$

$$\gamma_{22} = u_{2,2} - \Gamma_{22}^1 u_1 - \Gamma_{22}^2 u_2 = u_{2,2} + r u_1 = \frac{\partial u_2}{\partial \theta} + r u_1,$$

$$\gamma_{33} = u_{3,3} = \frac{\partial u_3}{\partial z},$$

$$\gamma_{12} = \tfrac{1}{2}(u_{1,2} + u_{2,1}) - \Gamma_{12}^1 u_1 - \Gamma_{12}^2 u_2$$
$$= \frac{1}{2}\left(\frac{\partial u_1}{\partial \theta} + \frac{\partial u_2}{\partial r}\right) - \frac{u_2}{r},$$

$$\gamma_{23} = \tfrac{1}{2}(u_{2,3} + u_{3,2}) = \frac{1}{2}\left(\frac{\partial u_2}{\partial z} + \frac{\partial u_3}{\partial \theta}\right),$$

$$\gamma_{31} = \tfrac{1}{2}(u_{3,1} + u_{1,3}) = \frac{1}{2}\left(\frac{\partial u_3}{\partial r} + \frac{\partial u_1}{\partial z}\right).$$

The physical components of the displacements are

$$u_1 = \bar{u}_1 = u_r, \quad u_2 = r\bar{u}_2 = r u_\theta, \quad u_3 = \bar{u}_3 = u_z.$$

Finally, the physical components of the strain tensor are as follows.

$$\bar{\gamma}_{11} = \gamma_{rr} = \sqrt{g^{(11)}g^{(11)}}\,\gamma_{11} = \frac{\partial u_r}{\partial r},$$

$$\bar{\gamma}_{22} = \gamma_{\theta\theta} = \sqrt{g^{(22)}g^{(22)}}\,\gamma_{22} = \frac{1}{r}\frac{\partial u_\theta}{\partial \theta} + \frac{u_r}{r},$$

$$\bar{\gamma}_{33} = \gamma_{zz} = \sqrt{g^{(33)}g^{(22)}}\,\gamma_{33} = \frac{\partial u_z}{\partial z},$$

$$\bar{\gamma}_{12} = \tfrac{1}{2}\gamma_{r\theta} = \sqrt{g^{(11)}g^{(22)}}\,\gamma_{12} = \tfrac{1}{2}\left(\frac{1}{r}\frac{\partial u_r}{\partial \theta} + \frac{\partial u_\theta}{\partial r}\right) - \frac{u_\theta}{r},$$

$$\bar{\gamma}_{23} = \tfrac{1}{2}\gamma_{\theta z} = \sqrt{g^{(22)}g^{(33)}}\,\gamma_{23} = \tfrac{1}{2}\left(\frac{\partial u_\theta}{\partial z} + \frac{1}{r}\frac{\partial u_z}{\partial \theta}\right),$$

$$\bar{\gamma}_{31} = \tfrac{1}{2}\gamma_{zr} = \sqrt{g^{(33)}g^{(11)}}\,\gamma_{31} = \tfrac{1}{2}\left(\frac{\partial u_z}{\partial r} + \frac{\partial u_r}{\partial z}\right).$$

Spherical Coordinates

$$\gamma_{11} = \frac{\partial u_1}{\partial R} \quad \gamma_{22} = \frac{\partial u_2}{\partial \alpha} + Ru_1,$$

$$\gamma_{33} = \frac{\partial u_3}{\partial \theta} + R(\sin^2\alpha)u_1 + (\sin\alpha\cos\alpha)u_2,$$

$$\gamma_{12} = \tfrac{1}{2}\left(\frac{\partial u_1}{\partial \alpha} + \frac{\partial u_2}{\partial R}\right) - \frac{1}{R}u_2 \quad \gamma_{31} = \tfrac{1}{2}\left(\frac{\partial u_3}{\partial R} + \frac{\partial u_1}{\partial \theta}\right) - \frac{1}{R}u_3,$$

$$\gamma_{23} = \tfrac{1}{2}\left(\frac{\partial u_2}{\partial \theta} + \frac{\partial u_3}{\partial \alpha}\right) - (\cot\alpha)u_3,$$

$$u_1 = \bar{u}_1 = u_R, \quad u_2 = R\bar{u}_2 = Ru_\alpha, \quad u_3 = R(\sin\alpha)\bar{u}_3 = R(\sin\alpha)u_\theta,$$

$$\bar{\gamma}_{11} = \gamma_{RR} = \frac{\partial u_R}{\partial R} \quad \bar{\gamma}_{22} = \gamma_{\alpha\alpha} = \frac{1}{R}\frac{\partial u_\alpha}{\partial \alpha} + \frac{u_R}{R},$$

$$\bar{\gamma}_{33} = \gamma_{\theta\theta} = \frac{1}{R\sin\alpha}\frac{\partial u_\theta}{\partial \theta} + \frac{u_R}{R} + \frac{\cot\alpha}{R}u_\theta,$$

$$\bar{\gamma}_{12} = \tfrac{1}{2}\gamma_{R\alpha} = \tfrac{1}{2}\left(\frac{1}{R}\frac{\partial u_R}{\partial \alpha} + \frac{\partial u_\alpha}{\partial R} - \frac{u_\alpha}{R}\right),$$

$$\bar{\gamma}_{23} = \tfrac{1}{2}\gamma_{\alpha\theta} = \tfrac{1}{2}\left(\frac{1}{R\sin\alpha}\frac{\partial u_\alpha}{\partial \theta} + \frac{1}{R}\frac{\partial u_\theta}{\partial \alpha} - \frac{\cot\alpha}{r}u_\theta\right),$$

$$\bar{\gamma}_{31} = \tfrac{1}{2}\gamma_{\theta R} = \tfrac{1}{2}\left(\frac{1}{R\sin\alpha}\frac{\partial u_R}{\partial \theta} + \frac{\partial u_\theta}{\partial R} - \frac{u_\theta}{R}\right).$$

For nonlinear terms $u^m_{|i}u_{m|j}$ involved in the curvilinear coordinates, similar operations are repeated, and the reader is encouraged to verify the results given in Problem 2.14.

2.5 Rate-of-Deformation Tensor

The strain tensor, which is a measure of deformation in Lagrangian coordinates in solid mechanics, is, in general, not applicable for fluid mechanics. In fluid mechanics, because the dependent variables are velocity $\mathbf{V}(\mathbf{R}, t)$ and density $\rho(\mathbf{R}, t)$ and the independent variables are $\mathbf{R}$ and t (see Fig. 2.2.3), it is logical to determine the time rate of change of $\mathrm{d}s^2 - \mathrm{d}s_0^2$ as follows.

Take the substantial derivative of $\mathrm{d}s^2 - \mathrm{d}s_0^2$, which gives

$$\frac{D}{Dt}(\mathrm{d}s^2 - \mathrm{d}s_0^2) = \frac{Dds^2}{Dt} = \frac{D}{Dt}(\mathrm{d}\mathbf{R} \cdot \mathrm{d}\mathbf{R}) = \frac{D}{Dt}(\mathrm{d}z_i\,\mathrm{d}z_j\delta_{ij})$$

$$= \delta_{ij}\,\mathrm{d}z_i\frac{Ddz_j}{Dt} + \delta_{ij}\,\mathrm{d}z_j\frac{Ddz_i}{Dt},$$

where $D\,\mathrm{d}s_0^2/Dt = 0$. This is because $\mathrm{d}s_0$ implies a line segment at $t = t_0 = 0$ prior to disturbances, and the velocity measurement is made at $t = t$, and

$$\frac{D\,\mathrm{d}z_i}{Dt} = \frac{\partial}{\partial t}\left(\frac{\partial z_i}{\partial x_j}\mathrm{d}x_j\right) + \frac{\partial}{\partial z_k}\left(\frac{\partial z_i}{\partial x_j}\mathrm{d}x_j\right)\frac{\partial z_k}{\partial t}$$

$$= \frac{\partial}{\partial x_j}\left(\frac{\partial z_i}{\partial t}\right)\mathrm{d}x_j + \frac{\partial}{\partial x_j}(\delta_{ik})\frac{\partial z_k}{\partial t}\,\mathrm{d}x_j$$

$$= \frac{\partial V_i}{\partial x_j}\mathrm{d}x_j = \frac{\partial V_i}{\partial z_k}\frac{\partial z_k}{\partial x_j}\mathrm{d}x_j = V_{i,k}\,\mathrm{d}z_k.$$

Thus, we obtain

$$\frac{D}{Dt}(\mathrm{d}s^2 - \mathrm{d}s_0^2) = 2\,\mathrm{d}_{ij}\,\mathrm{d}z_i\,\mathrm{d}z_j,$$

where d_{ij} is called the *rate-of-deformation tensor*

$$\mathrm{d}_{ij} = \tfrac{1}{2}\left(\frac{\partial V_i}{\partial z_j} + \frac{\partial V_j}{\partial z_i}\right) = \tfrac{1}{2}(V_{i,j} + V_{j,i}) \qquad (2.5.1)$$

in which partial derivatives are with respect to the current spatial coordinates z_i. This is in contrast to the case of the strain tensor (Eq. [2.4.5]), where the partial derivatives are with respect to the undeformed reference coordinates x_i. It can similarly be shown that if the domain is curvilinear (see Fig. 2.5.1), then

$$\mathrm{d}_{ij} = \tfrac{1}{2}(V_{i|j} + V_{j|i}) \qquad (2.5.2)$$

with

$$V_{i|j} = \frac{\partial V_i}{\partial \xi_j} - \Gamma_{ij}^k V_k, \qquad (2.5.3)$$

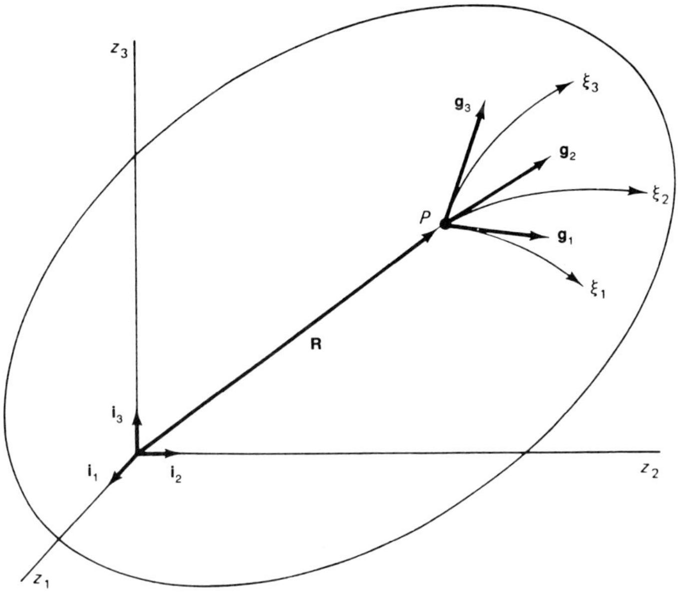

Figure 2.5.1 Eulerian curvilinear coordinates.

$$V_{j|i} = \frac{\partial V_j}{\partial \xi_i} - \Gamma^k_{ji} V_k. \tag{2.5.4}$$

Further discussion of the rate-of-deformation tensor d_{ij} appears in Chapter 5.

2.6 Coordinate Transformations for Strains

It has been and will repeatedly be demonstrated that we make use of the invariant properties of tensors in developing both constitutive and governing equations in general. In this section, we study additional properties of tensors related to coordinate transformation and strain invariants.

We begin with a simple example of coordinate transformation that involves old coordinates x_i and new coordinates x'_i, as shown in Fig. 2.6.1,

$$dx'_i = a_{ij}\, dx_j, \tag{2.6.1a}$$

where a_{ij} is the transformation matrix or proper orthogonal matrix. Conversely, we have

$$dx_j = a_{ij}\, dx'_i, \tag{2.6.1b}$$

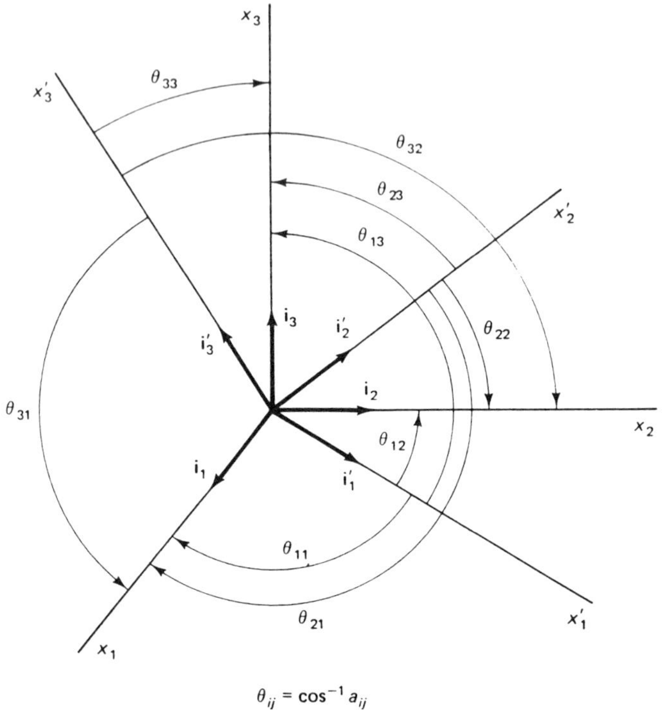

Figure 2.6.1 Coordinate transformation.

or

$$dx_i = a_{ji}\, dx'_j. \tag{2.6.1c}$$

In index notation, the transpose of a matrix arises when, as in Eqs. (2.6.1b) and (2.6.1c), the first index of a_{ij} or a_{ji} is repeated, not the second index, as pointed out in Chapter 1. Substituting Eqs. (2.6.1b) and (2.6.1c) into Eq. (2.4.3) yields

$$ds^2 - ds_0^2 = 2a_{ir}a_{js}\gamma_{rs}\, dx'_i\, dx'_j. \tag{2.6.2a}$$

In the new coordinates, we may write

$$ds^2 - ds_0^2 = 2\gamma'_{ij}\, dx'_i\, dx'_j. \tag{2.6.2b}$$

Equating Eqs. (2.6.2a) and (2.6.2b) gives

$$\gamma'_{ij} = a_{ir}a_{js}\gamma_{rs}. \tag{2.6.3}$$

This represents the transformation of the strain tensor from the old coordinates to the new coordinates.

Example 2.10 Calculation of Extensional Strain Calculate the extensional strain along the diagonal $\overline{AC}$ in Fig. 2.4.1 for $a = b$ of Example 2.6.

Solution. The direction cosines are calculated in the undeformed geometry, as implied in Eq. (2.6.3).

$$a_{ir} = \begin{bmatrix} \cos\theta & \sin\theta & 0 \\ -\sin\theta & \cos\theta & 0 \\ 0 & 0 & 1 \end{bmatrix}$$

with

$$\cos\theta = \sin\theta = \frac{1}{\sqrt{2}}.$$

The strain tensor in the new coordinates is given by

$$\gamma'_{ij} = a_{ir}a_{js}\gamma_{rs}.$$

Thus,

$$\begin{aligned}(\gamma'_{11})_{\overline{AC}} &= a_{11}a_{11}\gamma_{11} + a_{11}a_{12}\gamma_{12} + a_{12}a_{11}\gamma_{21} + a_{12}a_{12}\gamma_{22} \\ &= (\cos^2\theta)\gamma_{11} + 2(\cos\theta\sin\theta)\gamma_{12} + (\sin^2\theta)\gamma_{22} \\ &= \frac{\gamma_{11}}{2} + \gamma_{12} + \frac{\gamma_{22}}{2},\end{aligned}$$

where, for $a = b = 1$ and $x_2 = x_1$

$$\begin{aligned}\gamma_{11} &= \alpha x_1 + \tfrac{1}{2}(\alpha^2 + \beta^2)x_1^2, \\ \gamma_{12} &= \tfrac{1}{2}(\alpha + \beta)x_1 + \tfrac{1}{2}(\alpha^2 + \beta^2)x_1^2, \\ \gamma_{22} &= \beta x_1 + \tfrac{1}{2}(\alpha^2 + \beta^2)x_1^2.\end{aligned}$$

Finally,

$$(\gamma'_{11})_{\overline{AC}} = (\alpha + \beta)x_1 + (\alpha^2 + \beta^2)x_1^2.$$

This is the extensional strain distribution along the diagonal $\overline{AC}$ on the deformed geometry.

Principal Strains

At a given point in the body, it is always possible to choose a special set of axes through the point so that the shear strain components vanish. These special axes are called *principal axes* of the strain or *principal directions*. There are three planes through the point perpendicular to the three principal axes, called *principal planes*. The normal strain components on the three principal planes are called *principal strains* and will be denoted by γ_1, γ_2, and γ_3, which are equal to γ_{11}, γ_{22}, and γ_{33}, respectively, when all shear strain components are zero.

If a laboratory cube specimen is loaded perpendicular to one or more planes of this cube, the shear strains as well as the normal strains of different magnitudes develop on various planes rotated (0°–90°) from the principal axes. Consider strain components γ_{ij} on a plane arbitrarily inclined from the principal planes. Let n_j be the direction cosines that identify the inclined plane. Then, the relationship between the principal strain components γ_i and the γ_{ij} on the inclined plane is given by

$$\gamma_i = \gamma_{ij} n_j. \tag{2.6.4a}$$

If we assign a scalar $\gamma = \lambda$ for one of the principal strain components, then the principal strain γ_i may be written in an alternative form:

$$\gamma_i = \lambda n_i = \lambda \delta_{ij} n_j. \tag{2.6.4b}$$

Equate Eqs. (2.6.4a) and (2.6.4b) to yield

$$(\gamma_{ij} - \lambda \delta_{ij}) n_j = 0. \tag{2.6.5}$$

This equation can be derived directly from Eq. (2.6.3) by considering a scalar in the direction, n_i, normal to the surface of a principal plane:

$$\gamma(n) = n_i n_j \gamma_{ij}, \tag{2.6.6}$$

where n_i has the property

$$\mathbf{n} \cdot \mathbf{n} = n_i n_i = 1. \tag{2.6.7}$$

We now construct a scalar function f from the superposition of Eqs. (2.6.6) and (2.6.7):

$$f = \gamma_{ij} n_i n_j - \lambda (n_i n_i - 1), \tag{2.6.8}$$

where λ is known as the *Lagrange* multiplier, providing a compatibility with reference to Eq. (2.6.6). Physically, f represents the work required to satisfy Eq. (2.6.6) and that required to enforce the constraint condition, Eq. (2.6.7). The sign for the constraint energy is arbitrary, plus or minus, depending upon convenience, as required by the final physical consequence. The question here is: In which direction or for what normal vectors $\mathbf{n}$ will $\gamma(n)$ be a maximum? To obtain the condition of extrema, we proceed with differentiating the function f with respect to direction cosines and setting the result equal to zero, a standard procedure for finding an extremum. Differentiation of tensorial quantities with respect to another tensorial quantity requires a careful manipulation:

$$\begin{aligned} \frac{\partial}{\partial n_k}(\gamma_{ij} n_i n_j) &= \gamma_{ij}\left(\frac{\partial n_i}{\partial n_k} n_j + n_i \frac{\partial n_j}{\partial n_k}\right) = \gamma_{ij}(\delta_{ik} n_j + n_i \delta_{jk}) \\ &= \gamma_{kj} n_j + \gamma_{ik} n_i = 2\gamma_{kj} n_j. \end{aligned}$$

Likewise,

$$\frac{\partial}{\partial n_k}(\lambda n_i n_i) = \lambda\left(\frac{\partial n_i}{\partial n_k} n_i + n_i \frac{\partial n_i}{\partial n_k}\right) = \lambda(\delta_{ik} n_i + n_i \delta_{ik}) = 2\lambda n_k.$$

Note that in both of these cases the free index is k. Thus, if k is changed to i, then

$$\frac{\partial f}{\partial n_i} = 2(\gamma_{ij} - \lambda\delta_{ij})n_j = 0,$$

or

$$(\gamma_{ij} - \lambda\delta_{ij})n_j = 0,$$

which is the same as Eq. (2.6.5). This is known as the Lagrange multiplier method of determining the extremum condition for γ_{ij}.

Since n_j is arbitrary, for nontrivial solutions of Eq. (2.6.5) to exist, we must have

$$|\gamma_{ij} - \lambda\delta_{ij}| = 0. \tag{2.6.9}$$

An evaluation of the determinant of this matrix gives a cubic equation of the form, known as the Cayley–Hamilton equation,

$$-\lambda^3 + \lambda^2 I_1 - \lambda I_2 + I_3 = 0, \tag{2.6.10}$$

in which

$$I_1 = \operatorname{tr}\gamma_{ij} = \gamma_{ii}. \tag{2.6.11a}$$

Here, *tr* is the trace that implies the sum of diagonal terms in γ_{ij}.

$$I_2 = \tfrac{1}{2}(\gamma_{ii}\gamma_{jj} - \gamma_{ij}\gamma_{ij}), \tag{2.6.11b}$$

$$I_3 = |\gamma_{ij}| = \tfrac{1}{6}(\gamma_{ii}\gamma_{jj}\gamma_{kk} - 3\gamma_{ii}\gamma_{jk}\gamma_{jk} + 2\gamma_{ij}\gamma_{jk}\gamma_{ki}), \tag{2.6.11c}$$

and the functions I_1, I_2, and I_3 have the same value in any rectangular Cartesian coordinate system. For this reason, they are called invariants or, more specifically, the *principal strain invariants*, with I_1, I_2, and I_3 being the first, second, and third principal strain invariants, respectively.

It is easy to show that the principal strain invariants of the deformation tensor G_{ij} are

$$\begin{aligned}
I_1 &= G_{ii} = \delta^{ij}G_{ji} = \delta^{ij}(2\gamma_{ji} + \delta_{ji}) = 2\gamma_{ii} + 3,\\
I_2 &= \tfrac{1}{2}(\delta^{ir}\delta^{js}G_{ri}G_{sj} - \delta^{ir}\delta^{js}G_{rj}G_{si})\\
&= 2(\gamma_{ii}\gamma_{jj} - \gamma_{ij}\gamma_{ji}) + 4\gamma_{ii} + 3,\\
I_3 &= |G_{ij}|\\
&= \tfrac{4}{3}\epsilon_{ijk}\epsilon_{rst}\gamma_{ir}\gamma_{js}\gamma_{kt} + 2(\gamma_{ii}\gamma_{jj} - \gamma_{ij}\gamma_{ji}) + 2\gamma_{ii} + 1.
\end{aligned}$$

These invariants play a central role in developing constitutive equations for nonlinear materials.

The expression given by Eq. (2.6.9) is a standard eigenvalue problem. Since γ_{ij} is a 3×3 matrix, we expect to obtain three eigenvalues. They represent the three principal strains (major, intermediate, and minor). For each eigenvalue there exist three eigenvector components. Thus, there will be a total of nine eigenvector components corresponding to three eigenvalues. These eigenvectors are the direction cosines of the 3×3 matrix which determine the angles of rotation required to identify the location of principal axes or planes.

Example 2.11 Determination of Planes of Principal Strains Given the strain tensor

$$\gamma_{ij} = \begin{bmatrix} 1 & \sqrt{3} & 0 \\ \sqrt{3} & 0 & 0 \\ 0 & 0 & 1 \end{bmatrix},$$

determine (1) the principal strain invariants, (2) the principal strains, and (3) the principal directions.

1. *Principal Strain Invariants*

$$I_1 = \gamma_{ii} = 2,$$
$$I_2 = \tfrac{1}{2}(\gamma_{ii}\gamma_{jj} - \gamma_{ij}\gamma_{ij}) = -2,$$
$$I_3 = |\gamma_{ij}| = -3.$$

2. *Principal Strains*

$$\gamma_{ij} - \lambda\delta_{ij} = \begin{bmatrix} 1-\lambda & \sqrt{3} & 0 \\ \sqrt{3} & -\lambda & 0 \\ 0 & 0 & 1-\lambda \end{bmatrix},$$

$$|\gamma_{ij} - \lambda\delta_{ij}| = 0 = (1-\lambda)\left(\lambda - \frac{1+\sqrt{13}}{2}\right)\left(\lambda - \frac{1-\sqrt{13}}{2}\right),$$

$$\gamma_{(1)} = \lambda_{(1)} = \frac{1+\sqrt{13}}{2}, \quad \gamma_{(2)} = \lambda_{(2)} = 1, \quad \gamma_{(3)} = \lambda_{(3)} = \frac{1-\sqrt{13}}{2}.$$

This gives the major principal strain $= (1 + \sqrt{13})/2$, the intermediate principal strain $= 1$, and the minor principal strain $= (1 - \sqrt{13})/2$.

3. Principal Directions

Eigenvectors for $\gamma_{(1)} = (1+\sqrt{13})/2$ give the principal directions in $\mathbf{n}^{(1)}$:

$$\begin{bmatrix} 1-\dfrac{1+\sqrt{13}}{2} & \sqrt{3} & 0 \\ \sqrt{3} & -\dfrac{1+\sqrt{13}}{2} & 0 \\ 0 & 0 & 1-\dfrac{1+\sqrt{13}}{2} \end{bmatrix} \begin{bmatrix} n_1^{(1)} \\ n_2^{(1)} \\ n_3^{(1)} \end{bmatrix}$$

$$= \begin{bmatrix} \left(1-\dfrac{1+\sqrt{13}}{2}\right)n_1^{(1)} + \sqrt{3}n_2^{(1)} \\ \sqrt{3}n_1^{(1)} - \left(\dfrac{1+\sqrt{13}}{2}\right)n_2^{(1)} \\ \left(1-\dfrac{1+\sqrt{13}}{2}\right)n_3^{(1)} \end{bmatrix} = \begin{bmatrix} 0 \\ 0 \\ 0 \end{bmatrix},$$

which gives

$$n_1^{(1)} = \frac{1+\sqrt{13}}{2\sqrt{3}} n_2^{(1)}, \quad n_3^{(1)} = 0.$$

From

$$\mathbf{n}^{(1)} \cdot \mathbf{n}^{(1)} = \left(\frac{1+2\sqrt{13}+13}{12} + 1\right)(n_2^{(1)})^2 = 1$$

we obtain

$$n_2^{(1)} = n_1^{(1)} \sqrt{\frac{12}{12+(1+\sqrt{13})^2}} = \frac{1+\sqrt{13}}{2\sqrt{3}} \sqrt{\frac{12}{12+(1+\sqrt{13})^2}}.$$

Thus,

$$\mathbf{n}^{(1)} = [n_1^{(1)} \quad n_2^{(1)} \quad n_3^{(1)}] = [0.8 \quad 0.6 \quad 0].$$

Similarly, eigenvectors for $\lambda_{(2)} = 1$ can be calculated, noting that $\mathbf{n}^{(2)} \cdot \mathbf{n}^{(2)} = 1$. Thus,

$$\mathbf{n}^{(2)} = [0 \quad 0 \quad 1].$$

Eigenvectors for $\lambda_{(3)} = (1-\sqrt{13})/2$ give the principal directions in

$\mathbf{n}^{(3)}$, which may be obtained in the same manner, but perhaps more easily by taking a cross product of $\mathbf{n}^{(1)}$ and $\mathbf{n}^{(2)}$:

$$\mathbf{n}^{(3)} = \mathbf{n}^{(1)} \times \mathbf{n}^{(2)} = \begin{bmatrix} \mathbf{i}_1 & \mathbf{i}_2 & \mathbf{i}_3 \\ 0.8 & 0.6 & 0 \\ 0 & 0 & 1 \end{bmatrix} = 0.6\mathbf{i}_1 - 0.8\mathbf{i}_2.$$

Therefore,

$$a_{ij} = \begin{bmatrix} \mathbf{n}^{(1)} \\ \mathbf{n}^{(2)} \\ \mathbf{n}^{(3)} \end{bmatrix} = \begin{bmatrix} 0.8 & 0.6 & 0 \\ 0 & 0 & 1 \\ 0.6 & -0.8 & 0 \end{bmatrix}.$$

These results can be checked by:

$$\gamma'_{ij} = a_{ir}a_{js}\gamma_{rs} = \begin{bmatrix} \lambda_{(1)} & 0 & 0 \\ 0 & \lambda_{(2)} & 0 \\ 0 & 0 & \lambda_{(3)} \end{bmatrix}$$

which, as shown in Eq. (2.6.3), indicates that the direction cosines a_{ij} transform the strain tensor γ_{ij} into the principal directions, leading to the principal strains $\lambda_{(1)}$, $\lambda_{(2)}$, and $\lambda_{(3)}$. It is interesting to note that the summing implied in $a_{ir}a_{js}\gamma_{rs}$ is the same as the matrix multiplication:

$$[a][\gamma][a]^{\mathrm{T}} = \begin{bmatrix} 0.8 & 0.6 & 0 \\ 0 & 0 & 1 \\ 0.6 & -0.8 & 0 \end{bmatrix} \begin{bmatrix} 1 & \sqrt{3} & 0 \\ \sqrt{3} & 0 & 0 \\ 0 & 0 & 1 \end{bmatrix} \begin{bmatrix} 0.8 & 0 & 0.6 \\ 0.6 & 0 & -0.8 \\ 0 & 1 & 0 \end{bmatrix}$$
$$= \begin{bmatrix} 2.3 & 0 & 0 \\ 0 & 1 & 0 \\ 0 & 0 & -1.3 \end{bmatrix}.$$

These exercises indicate that if laboratory strain measurements for specimens at any orientation are provided, then it is possible to locate the planes of the principal directions on which shear strains are zero and only principal normal strains prevail.

Maximum Shear Strains

The shear strains in the directions x'_1 and x'_2 can be expressed in terms of the principal strains $\lambda_{(i)}$ by

$$\gamma'_{12} = a_{1i}a_{2i}\lambda_{(i)}. \tag{2.6.12}$$

Since x'_1 and x'_2 in Fig. 2.6.1 are orthogonal, the direction cosines satisfy the conditions

$$a_{1i}a_{2i} = 0, \tag{2.6.13a}$$

$$a_{1i}a_{1i} = 1, \tag{2.6.13b}$$

$$a_{2i}a_{2i} = 1. \tag{2.6.13c}$$

The reader may verify these requirements from a_{ij} in Eq. (1.2.12). To simultaneously satisfy the conditions given by Eqs. (2.6.12) and (2.6.13), we construct a scalar function f as a superposition of Eqs. (2.6.12) and (2.6.13),

$$f = a_{1i}a_{2i}\lambda_{(i)} - \mu a_{1i}a_{2i} - \mu_1(a_{1i}a_{1i} - 1) - \mu_2(a_{2i}a_{2i} - 1),$$

where μ, μ_1, and μ_2 represent Lagrange multipliers, similar to the case of principal strains discussed earlier. The necessary conditions for an extremum (maximum or minimum) of f with respect to the direction cosines are

$$\frac{\partial f}{\partial a_{1i}} = (\lambda_{(i)} - \mu)a_{2i} - 2\mu_1 a_{1i} = 0, \qquad (2.6.14a)$$

$$\frac{\partial f}{\partial a_{2i}} = (\lambda_{(i)} - \mu)a_{1i} - 2\mu_2 a_{2i} = 0. \qquad (2.6.14b)$$

Multiply Eq. (2.6.14a) by a_{1i} and satisfy Eqs. (2.6.13a and b) to give

$$\lambda_{(i)}a_{1i}a_{2i} - 2\mu_1 = 0,$$

or

$$\mu_1 = \tfrac{1}{2}\lambda_{(i)}a_{1i}a_{2i} = \tfrac{1}{2}\gamma'_{12}.$$

Similarly, from Eqs. (2.6.14b) and (2.6.13a and c),

$$\mu_2 = \tfrac{1}{2}\lambda_{(\mathrm{i})}a_{1i}a_{2i} = \tfrac{1}{2}\gamma'_{12}.$$

If we multiply Eq. (2.6.14a) by a_{2i} and enforce the conditions of Eqs. (2.6.13a and c), we get

$$\mu = \lambda_{(i)}a_{2i}a_{2i} = \gamma'_{22}.$$

Likewise, from Eqs. (2.6.14b) and (2.6.13a and b),

$$\mu = \lambda_{(i)}a_{1i}a_{1i} = \gamma'_{11}.$$

The substitution of μ, μ_1, and μ_2 into Eq. (2.6.14) yields

$$(\lambda_{(1)} - \gamma'_{11})a_{21} - \gamma'_{12}a_{11} = 0, \qquad (2.6.15a)$$

$$(\lambda_{(2)} - \gamma'_{11})a_{22} - \gamma'_{12}a_{12} = 0, \qquad (2.6.15b)$$

$$(\lambda_{(3)} - \gamma'_{11})a_{23} - \gamma'_{12}a_{13} = 0, \qquad (2.6.15c)$$

$$(\lambda_{(1)} - \gamma'_{11})a_{11} - \gamma'_{12}a_{21} = 0, \qquad (2.6.15d)$$

$$(\lambda_{(2)} - \gamma'_{11})a_{12} = \gamma'_{12}a_{22} = 0, \qquad (2.6.15e)$$

$$(\lambda_{(3)} - \gamma'_{11})a_{13} = \gamma'_{12}a_{23} = 0. \qquad (2.6.15f)$$

From Eqs. (2.6.15a and d), we get

$$a_{21} = \frac{\gamma'_{12}}{\lambda_{(1)} - \gamma'_{11}}a_{11}, \quad a_{21} = \frac{\lambda_{(1)} - \gamma'_{11}}{\gamma'_{12}}a_{11},$$

which leads to $a_{11}^2 = a_{21}^2$ and $a_{11} = \pm a_{21}$. Similarly, $a_{22}^2 = a_{12}^2$, $a_{21}^2 = a_{22}^2$, and $a_{13}^2 = a_{23}^2 = 0$. Thus, we note that

$$a_{11}^2 = a_{12}^2 = a_{21}^2 = a_{22}^2 = B^2.$$

From Eqs. (2.6.13b and c), we must have

$$2B^2 = 1, \quad B = \pm \frac{1}{\sqrt{2}}.$$

To satisfy the conditions in Eq. (2.6.13a), we require that the signs of all B's be the same except for one; for example, we could take

$$a_{11} = a_{21} = a_{22} = \frac{1}{\sqrt{2}}, \quad a_{12} = -\frac{1}{\sqrt{2}}.$$

Depending on the choice of signs, we obtain either positive or negative shear strains. Note also that angles corresponding to the direction cosines are odd multiples of $\cos^{-1}(1/\sqrt{2}) = 45°$ with x_i' bisecting x_i. The maximum shear strain in the $x_m x_n$ $(m \neq n)$ plane, therefore, is given by

$$\gamma'_{mn} = a_{mi} a_{ni} \lambda_{(i)} \quad (m \neq n) \tag{2.6.16}$$

$$\gamma'_{12} = \frac{1}{\sqrt{2}}\frac{1}{\sqrt{2}}\lambda_{(1)} - \frac{1}{\sqrt{2}}\frac{1}{\sqrt{2}}\lambda_{(2)} = \tfrac{1}{2}(\lambda_{(1)} - \lambda_{(2)}).$$

Similar procedures may be followed for the $x_1 x_3$ and $x_2 x_3$ planes. Thus,

$$\gamma'_{13} = \tfrac{1}{2}(\lambda_{(1)} - \lambda_{(3)}),$$
$$\gamma'_{23} = \tfrac{1}{2}(\lambda_{(2)} - \lambda_{(3)}).$$

These results indicate that the maximum shear strains represent the radii of Mohr circles, as shown in Example 2.12.

It is obvious from Eq. (2.6.16) that shear strains are zero (minimum) for all $a_{ij} = 0$ corresponding to the principal planes.

Example 2.12 Mohr Circle Representation Given the results in Example 2.11, plot the Mohr circles that show the principal strains, normal and shear strains, maximum shear strains, and orientations of planes between the principal plane and the plane on which the strain data are given.

Solution. First verify the results from elementary relations. Principal strains and orientations of planes are given by (refer to undergraduate texts on the mechanics of materials),

$$\gamma_{p1,2} = \frac{\gamma_{xx} + \gamma_{yy}}{2} \pm \sqrt{\left(\frac{\gamma_{xx} - \gamma_{yy}}{2}\right)^2 + \left(\frac{\gamma_{xy}}{2}\right)^2}$$

$$= \frac{1+0}{2} \pm \sqrt{\left(\frac{1+0}{2}\right)^2 + \left(\frac{2\sqrt{3}}{2}\right)^2} = \begin{cases} 2.3 \\ -1.3 \end{cases},$$

$$\gamma_{p1,3} = \frac{\gamma_{xx} + \gamma_{zz}}{2} \pm \sqrt{\left(\frac{\gamma_{xx} - \gamma_{zz}}{2}\right)^2 + \left(\frac{\gamma_{xz}}{2}\right)^2} = \begin{cases} 1 \\ 1 \end{cases},$$

$$\gamma_{p2,3} = \frac{\gamma_{yy} + \gamma_{zz}}{2} \pm \sqrt{\left(\frac{\gamma_{yy} - \gamma_{zz}}{2}\right)^2 + \left(\frac{\gamma_{yz}}{2}\right)^2} = \begin{cases} 1 \\ 0 \end{cases},$$

$$\tan 2\theta = \frac{\gamma_{xy}/2}{(\gamma_{xx} - \gamma_{yy})/2} = 3.46, \quad \theta = 36.94°.$$

Note that for $\theta = 36.94°$ we have $\cos\theta = 0.8$ and $\sin\theta = 0.6$. These

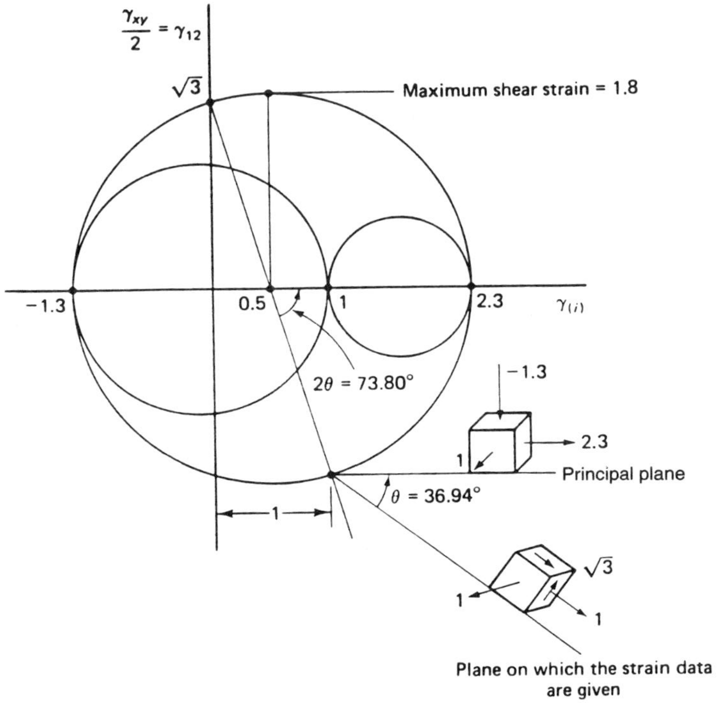

Figure 2.6.2 Mohr circles for Example 2.12. Quantities shown on the cube represent strains. Intermediate principal strain is located at the intersection of two intercircles which indicate major and minor principal strains in contact with the exterior enclosing circle. The plane on which the original strain data are given is located at an angle rotated clockwise from the principal plane. Note that the original strain data appear on the exterior circle.

results are identical to those obtained from the general approach using tensor analysis. The Mohr circle representation is shown in Fig. 2.6.2.

If the strain data are fully populated in three dimensions, rotations of planes for which the strain data are given with respect to the planes of principal planes cannot be shown in three-dimensional configurations. Thus, in this case, the usefulness of Mohr circles is limited to two dimensions.

2.7 Dilatational and Deviatoric Properties

Changes in volume are referred to as *dilatation*. For infinitesimal deformation, the dilatation is defined as

$$D = J = \frac{\mathrm{d}\Omega}{\mathrm{d}\Omega_0} = \sqrt{G} = \sqrt{|\delta_{ij} + 2\gamma_{ij}|} \simeq 1 + \gamma_{ii}. \qquad (2.7.1)$$

It follows from Eq. (2.7.1) that the volumetric strain Δ is

$$\Delta = \frac{\mathrm{d}\Omega - \mathrm{d}\Omega_0}{\mathrm{d}\Omega_0} = D - 1 = \gamma_{ii},$$

which indicates that only the extensional strains are responsible for dilatation.

Let the mean extensional strain be defined as

$$\gamma = \frac{\gamma_{ii}}{3}. \qquad (2.7.2)$$

The total strain may be given as the sum of the *deviatoric* (shear) part γ_{ij}^* and the dilatational (extensional or spherical) part $\gamma\delta_{ij}$:

$$\gamma_{ij} = \gamma_{ij}^* + \gamma\delta_{ij} = \gamma_{ij}^* + \tfrac{1}{3}\gamma_{kk}\delta_{ij},$$

and

$$\gamma_{ij}^* = \gamma_{ij} - \tfrac{1}{3}\gamma_{kk}\delta_{ij}, \qquad (2.7.3)$$

where γ_{ij}^* is known as the deviatoric strain tensor. Note that Eq. (2.7.3) holds only for a small strain assumption given by Eq. (2.7.1). Dilatational and deviatoric properties are more precisely defined in stress (see Chapter 3) and Eulerian coordinates where the geometric constraints of large strains are not involved (see Chapter 5).

For Eulerian coordinates, the deviatoric rate of deformation tensor d_{ij}^* is given by

$$\mathrm{d}_{ij}^* = \mathrm{d}_{ij} - \tfrac{1}{3}\mathrm{d}_{kk}\delta_{ij} \qquad (2.7.4)$$

in which no small strain approximations, such as in Eq. (2.7.1), are required. An application of the deviatoric rate of the deformation tensor will be discussed in subsection 5.2.2.

2.8 Compatibility Equations

At this point, we return to our six strain-displacement equations, as defined in Eq. (2.4.5) and spelled out in Eq. (2.4.6), and consider them as partial differential equations to be solved. If three displacement components u_i are given, then it is possible to determine the six strain components. The six strain components, however, may sometimes not uniquely define the three displacement components. There are more equations available than there are unknowns. Thus, the existence of unique displacements cannot be guaranteed unless certain compatibility conditions are satisfied.

Consider displacements of two points A and B. Let A_1 and B_1 denote the rectangular coordinates at A and B, respectively. We see that the displacement at B may be related to the displacement at A through the expression

$$\mathbf{u}_B = \mathbf{u}_A + \int_A^B \mathrm{d}\mathbf{u} = \mathbf{u}_A + \int_A^B \mathbf{u}_{,i}\,\mathrm{d}x_i = \mathbf{u}_A + \int_A^B \mathbf{u}_{,i}\,\mathrm{d}(x_i - B_i).$$

Integrating by parts, we get

$$\begin{aligned}\mathbf{u}_B &= \mathbf{u}_A + (B_i - A_i)\mathbf{u}_{,i} + \int_A^B (B_i - x_i)\mathbf{u}_{,ij}\,\mathrm{d}x_j \\ &= \mathbf{u}_A + (B_i - A_i)\mathbf{u}_{,i} + \int_A^B (B_i - x_i)\mathbf{G}_{i,j}\,\mathrm{d}x_j.\end{aligned} \tag{2.8.1}$$

Here, the definition given in Eq. (2.2.8) is used so that $\mathbf{u}_{,ij} = \mathbf{G}_{i,j}$. Let $\mathbf{W}_n = (B_i - x_i)\mathbf{G}_{i,n}$. Then, in a simply connected region, $\int_A^B \mathbf{W}_n\,\mathrm{d}x_n$ is independent of the path if and only if

$$\frac{\partial \mathbf{W}_n}{\partial x_m} - \frac{\partial \mathbf{W}_m}{\partial x_n} = 0.$$

Therefore, the existence of $\mathbf{u}$ is ensured only if

$$\frac{\partial}{\partial x_m}[(B_i - x_i)\mathbf{G}_{i,n}] - \frac{\partial}{\partial x_n}[(B_i - x_i)\mathbf{G}_{i,m}] = 0$$

$$(\Gamma_{nmp} - \Gamma_{mnp})\mathbf{G}^P + (B_i - x_i)[(\Gamma_{inr}\mathbf{G}^r)_{,m} - (\Gamma_{imr}\mathbf{G}^r)_{,n}] = 0$$

or

$$\begin{aligned}&\{(\Gamma_{nmp} - \Gamma_{mnp}) + \\ &\quad (B_i - x_i)[\Gamma_{inp,m} - \Gamma_{imp,n} + \mathbf{G}^{qr}(-\Gamma_{inr}\Gamma_{mpq} + \Gamma_{imr}\Gamma_{npq})]\}\mathbf{G}^P = 0,\end{aligned} \tag{2.8.2}$$

where

$$\Gamma_{nmp} = \frac{\partial^2 z_r}{\partial x_n \partial x_m}\frac{\partial z_r}{\partial x_p} = \gamma_{np,m} + \gamma_{mp,n} - \gamma_{nm,p} = \Gamma_{mnp}.$$

For $B_i \neq x_i$, Eq. (2.8.2) is satisfied if and only if the bracketed terms vanish,

$$R_{ijkq} = \gamma_{jk,qi} + \gamma_{qi,jk} - \gamma_{jq,ki} - \gamma_{ki,jq} + G^{rs}(\Gamma_{jkr}\Gamma_{qis} - \Gamma_{jqr}\Gamma_{kis}) = 0, \tag{2.8.3}$$

where R_{ijkq} is called the *Riemann—Christoffel tensor*. Because of the symmetry of γ_{ij}, we have only six independent equations corresponding to

$$R_{2112}, \quad R_{3113}, \quad R_{2332}, \quad R_{2113}, \quad R_{2331}, \quad R_{1223} = 0.$$

These represent the integrability conditions (with regard to the existence of a solution) or compatibility conditions of the strain components.

For two-dimensional space, if we neglect higher-order terms (small strains), then Eq. (2.8.3) assumes the form:

$$R_{2112} = \gamma_{22,11} + \gamma_{11,22} - \gamma_{12,12} - \gamma_{21,21},$$

or

$$R_{2112} = \frac{\partial^2 \gamma_{11}}{\partial x_2^2} + \frac{\partial^2 \gamma_{22}}{\partial x_1^2} - 2\frac{\partial^2 \gamma_{12}}{\partial x_1 \partial x_2} = 0. \tag{2.8.4}$$

This is called the *compatibility equation*. If the strain components vanish, then the integrand of Eq. (2.8.1) must vanish. This means that the displacement at B is the sum of the displacement at A and the relative displacement due to rigid-body motion.

Problems

2.1 Expand the determinant of the metric tensor (or Green's deformation tensor) and show that it is equal to the square of the determinant of the deformation gradient.

2.2 Transform the Lagrangian to Eulerian coordinates to show that the conservation of mass for fluids can be derived from the conservation of mass for solids. Include complete details of the algebra involved.

2.3 Let the motion of a fluid in steady-state flow be given in the Eulerian coordinates by

$$V_1 = kz_1, \quad V_2 = -kz_2, \quad V_3 = 0.$$

(a) Show that the motions and velocities in Lagrangian coordinates are

$$z_1 = x_1 e^{k(t-t_0)}, \qquad v_1 = kx_1 e^{k(t-t_0)},$$

$$z_2 = x_2 \mathrm{e}^{-k(t-t_0)}, \quad v_2 = -kx_2 \mathrm{e}^{-k(t-t_0)},$$

$$z_3 = x_3, \quad v_3 = 0.$$

(b) Transform the Lagrangian velocities into Eulerian coordinates to prove that the results in part a are recovered.

(c) Plot the Eulerian and Lagrangian velocity distributions.

2.4. Prove Eqs. (2.3.6) and (2.3.7) with complete details. Show that the proof of these equations may be obtained, using the determinants $|\partial z_i/\partial \xi_j|$ for Eq. (2.3.6) and $|\partial \xi_i/\partial z_j|$ for Eq. (2.3.7).

2.5 Given

$$\mathbf{v} = 3\mathbf{i}_1 + 6\mathbf{i}_2,$$
$$\mathbf{g}_1 = 4\mathbf{i}_1 - 2\mathbf{i}_2,$$
$$\mathbf{g}_2 = 3\mathbf{i}_1 + 2\mathbf{i}_2,$$
$$\mathbf{g}_3 = \mathbf{i}_3,$$

(a) calculate the metric tensors;

(b) calculate the covariant and contravariant components of $\mathbf{v}$;

(c) show that the original vector $\mathbf{v}$ as given can be recovered using the results obtained in parts a and b;

(d) draw a sketch to show all your results.

2.6 Prove:

(a) $\mathbf{g}^i = g^{ij}\mathbf{g}_j$,

(b) $g_{ik}g^{kj} = \delta_{ij}$,

(c) $\Gamma_{ijk} = \frac{1}{2}(g_{ik,j} + g_{jk,i} - g_{ij,k})$,

(d) $\Gamma^k_{ij} = \frac{1}{2}g^{km}(g_{mi,j} + g_{mj,i} - g_{ij,m})$,

(e) $\Gamma^i_{ij} = \frac{1}{2}g^{ik}g_{ik,j}$,

(f) $g^{ij} = \dfrac{1}{g}\dfrac{\partial g}{\partial g_{ij}}$,

(g) $\Gamma^i_{ij} = (\ln\sqrt{g})_{,j}$.

2.7 Prove that $g^{ij}_{,i} = 0$. Hint: Use the definition given in Eq. (2.3.4b).

2.8 Prove Eqs. (2.3.20a, b, and c).

2.9 Prove that the Laplace equation takes the following form in spherical coordinates:

$$\nabla^2\phi = \frac{\partial^2\phi}{\partial R^2} + \frac{2}{R}\frac{\partial\phi}{\partial R} + \frac{1}{R^2}\frac{\partial^2\phi}{\partial\alpha^2} + \frac{\cot\alpha}{R^2}\frac{\partial\phi}{\partial\alpha} + \frac{1}{R^2\sin^2\alpha}\frac{\partial^2\phi}{\partial\theta^2}.$$

2.10 Derive $\boldsymbol{\nabla} \cdot \mathbf{u}$, the divergence of $\mathbf{u}$, in cylindrical coordinates in terms of physical components.

2.11 Repeat Problem 2.10 for spherical coordinates.

2.12 Derive $\boldsymbol{\nabla} \times \mathbf{u}$, the curl of $\mathbf{u}$, in cylindrical coordinates in terms of physical components.

2.13 Repeat Problem 2.12 for spherical coordinates.

2.14 Show that the components of the Green–Saint–Venant strain tensor in cylindrical coordinates in terms of physical components, including the nonlinear terms, take the following forms:

$$\bar{\gamma}_{11} = \gamma_{rr} = \frac{\partial u_r}{\partial r} + \frac{1}{2}\left[\left(\frac{\partial u_r}{\partial r}\right)^2 + \left(\frac{\partial u_\theta}{\partial r}\right)^2 + \left(\frac{\partial u_z}{\partial r}\right)^2\right],$$

$$\bar{\gamma}_{22} = \gamma_{\theta\theta} = \frac{1}{r}\frac{\partial u_\theta}{\partial \theta} + \frac{1}{2}\left[\left(\frac{1}{r}\frac{\partial u_r}{\partial \theta}\right)^2 + \left(\frac{1}{r}\frac{\partial u_\theta}{\partial \theta}\right)^2 + \left(\frac{1}{r}\frac{\partial u_z}{\partial \theta}\right)^2 - \frac{2u_\theta}{r^2}\frac{\partial u_r}{\partial \theta} + \frac{2u_r}{r^2}\frac{\partial u_\theta}{\partial \theta} + \left(\frac{u_\theta}{r}\right)^2 + \frac{2u_r}{r} + \left(\frac{u_r}{r}\right)^2\right],$$

$$\bar{\gamma}_{33} = \gamma_{zz} = \frac{\partial u_z}{\partial z} + \frac{1}{2}\left[\left(\frac{\partial u_r}{\partial z}\right)^2 + \left(\frac{\partial u_\theta}{\partial z}\right)^2 + \left(\frac{\partial u_z}{\partial z}\right)^2\right],$$

$$\bar{\gamma}_{12} = \tfrac{1}{2}\gamma_{r\theta} = \frac{1}{2}\left(\frac{1}{r}\frac{\partial u_r}{\partial \theta} + \frac{\partial u_\theta}{\partial r} - \frac{u_\theta}{r} + \frac{1}{r}\frac{\partial u_r}{\partial r}\frac{\partial u_r}{\partial \theta} + \frac{1}{r}\frac{\partial u_\theta}{\partial r}\frac{\partial u_\theta}{\partial \theta} + \frac{1}{r}\frac{\partial u_z}{\partial r}\frac{\partial u_z}{\partial \theta} + \frac{u_r}{r}\frac{\partial u_\theta}{\partial r} - \frac{u_\theta}{r}\frac{\partial u_r}{\partial r}\right),$$

$$\bar{\gamma}_{23} = \tfrac{1}{2}\gamma_{\theta z} = \frac{1}{2}\left(\frac{\partial u_\theta}{\partial z} + \frac{1}{r}\frac{\partial u_z}{\partial \theta} + \frac{1}{r}\frac{\partial u_r}{\partial \theta}\frac{\partial u_r}{\partial z} + \frac{1}{r}\frac{\partial u_\theta}{\partial \theta}\frac{\partial u_\theta}{\partial z} + \frac{1}{r}\frac{\partial u_z}{\partial \theta}\frac{\partial u_z}{\partial z} - \frac{u_\theta}{r}\frac{\partial u_r}{\partial z} + \frac{u_r}{r}\frac{\partial u_\theta}{\partial z}\right),$$

$$\bar{\gamma}_{31} = \tfrac{1}{2}\gamma_{zr} = \frac{1}{2}\left(\frac{\partial u_z}{\partial r} + \frac{\partial u_r}{\partial z} + \frac{\partial u_r}{\partial z}\frac{\partial u_r}{\partial r} + \frac{\partial u_\theta}{\partial z}\frac{\partial u_\theta}{\partial r} + \frac{\partial u_z}{\partial z}\frac{\partial u_z}{\partial r}\right),$$

where $\gamma_{r\theta}$, $\gamma_{\theta z}$, and γ_{zr} are the engineering shear strains.

2.15 A two-dimensional square solid is deformed into the shape shown in Fig. P2.15. Determine:

(a) the displacement field $\mathbf{u} = \mathbf{u}(x_1, x_2, x_3)$;

(b) the tangent vector field $\mathbf{G}_i = \mathbf{G}_i(x_1, x_2, x_3)$;

(c) the components γ_{ij} of the strain tensor at the center $(x_1 = x_2 = 1/2)$;

(d) the extensional strain and shear strain along the line 30° counterclockwise from the x_1 axis with the origin $x_1 = x_2 = 0$.

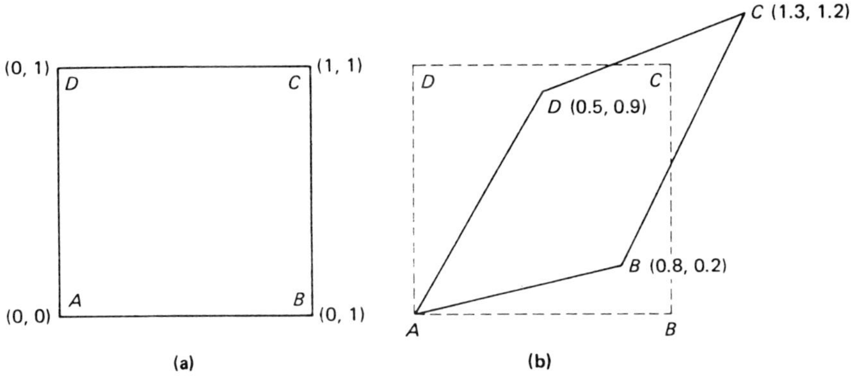

Figure P2.15 Sketch for Problem 2.13. (a) Undeformed. (b) Deformed.

2.16 Let

$$\gamma_{ij} = \begin{bmatrix} 2 & 1 & 1 \\ 1 & 2 & 0 \\ 1 & 0 & 1 \end{bmatrix}.$$

Determine:

(a) the principal strain invariants;
(b) the principal strains;
(c) the principal direction matrix a_{ij} and verify your results by recalculating the principal strains from principal direction cosines.

2.17 Given $\gamma'_{ij} = a_{ir}a_{js}\gamma_{rs}$, prove that the maximum shear strains are

$$(\gamma'_{ij})_{\max} = \tfrac{1}{2}(\lambda_{(i)} - \lambda_{(j)}) \quad i \neq j.$$

3
Equilibrium and Kinetics

3.1 General

The notion of *stress* originates from the need to quantify internal or external forces distributed, respectively, in a body or along its boundary in equilibrium. Body forces such as gravity act inside the body, whereas surface forces act on its bounding surface. Stresses are those forces distributed over an infinitesimal unit area cut out of a body in certain directions, or over an infinitesimal unit area on the bounding surface. Stresses may also arise from hydrodynamic pressure and/or velocity gradients in a fluid. Furthermore, changes in temperature in solids or fluids give rise to stresses.

Stresses may be related to strains in solids or rates of deformation in fluids through constitutive laws. Stresses are described in many different ways, depending, for example, on the coordinate systems used, the magnitudes of the strains (in solids) or the velocity gradients (in fluids), or the types of substances involved. The basic groundwork for stresses is developed in this chapter, but related subjects will be discussed throughout the remainder of this book. The study of stress-related problems, in general, is referred to as *kinetics* for the body in equilibrium.

Discussions in this chapter include definitions of forces and stresses, balance laws of linear and angular momentum, coordinate transformations of stresses, deviatoric stresses, stresses and equations of motion with large strains, and physical components of stresses. These topics, together with the concept of strain and displacement in Chapter 2, are the basis for the theories of elastic solids and fluid mechanics which are presented in Chapters 4 and 5, respectively.

3.2 Forces and Stresses

Consider a body subjected to an external load, as shown in Fig. 3.2.1a. We examine how the external load influences an interior point (Fig. 3.2.1b). The stress vector $\boldsymbol{\sigma}$ in the undeformed state in Lagrangian coordinates is defined as:

$$\boldsymbol{\sigma} = \lim_{\Delta A \to 0} \frac{\Delta \mathbf{F}}{\Delta A} = \frac{\mathrm{d}\mathbf{F}}{\mathrm{d}A}, \tag{3.2.1}$$

where $\Delta\mathbf{F}$ is the internal force activated from the external load $\mathbf{F}$ and ΔA is the area on which $\Delta\mathbf{F}$ is acting.

The internal force $\mathrm{d}\mathbf{F} = \boldsymbol{\sigma}\,\mathrm{d}A$ that acts on a point $\mathrm{d}A$ can also be said to be acting on an inclined surface $\mathrm{d}A_i$ of an infinitesimal body, as shown in Fig. 3.2.2. The infinitesimal force $\mathrm{d}\mathbf{F}$ can be written in various forms:

$$\begin{aligned} \mathrm{d}\mathbf{F} &= \boldsymbol{\sigma}\,\mathrm{d}A = \sigma_i \mathbf{i}_i \mathrm{d}A = \boldsymbol{\sigma}_i\,\mathrm{d}A_i = \sigma_{ij}\mathbf{i}_j\,\mathrm{d}A_i \\ &= (\sigma_{11}\mathbf{i}_1 + \sigma_{12}\mathbf{i}_2 + \sigma_{13}\mathbf{i}_3)\,\mathrm{d}A_1 \\ &\quad + (\sigma_{21}\mathbf{i}_1 + \sigma_{22}\mathbf{i}_2 + \sigma_{23}\mathbf{i}_3)\,\mathrm{d}A_2 \\ &\quad + (\sigma_{31}\mathbf{i}_1 + \sigma_{32}\mathbf{i}_2 + \sigma_{33}\mathbf{i}_3)\,\mathrm{d}A_3 \end{aligned} \tag{3.2.2}$$

Here the σ_i's denote components of the stress vector $\boldsymbol{\sigma}$ acting on the inclined surface $\mathrm{d}A$, the $\boldsymbol{\sigma}_i$'s are the resultant stress vectors on the

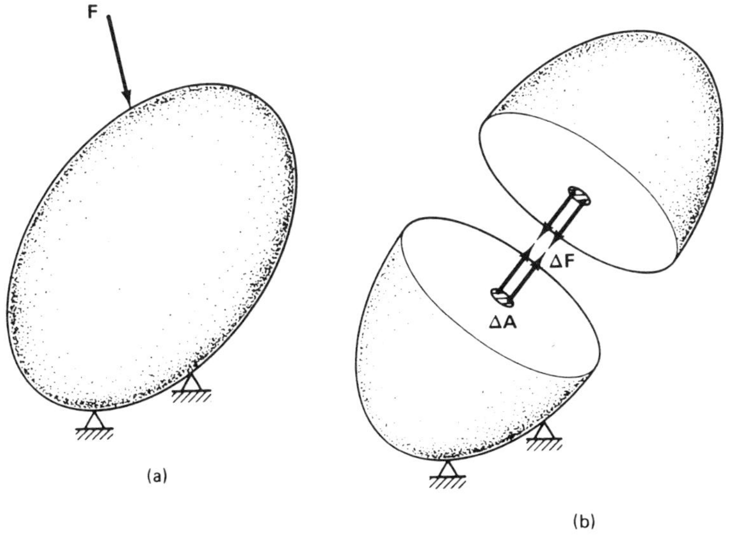

Figure 3.2.1 Body subjected to external load and interior stresses. (a) Body under load F. (b) Interior stresses.

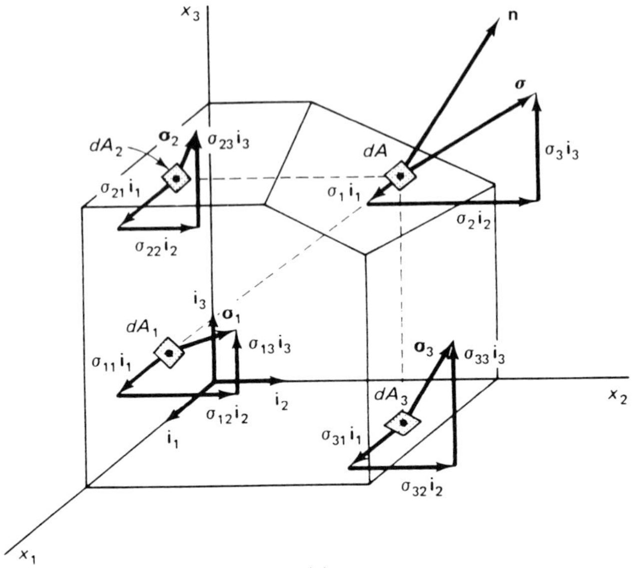

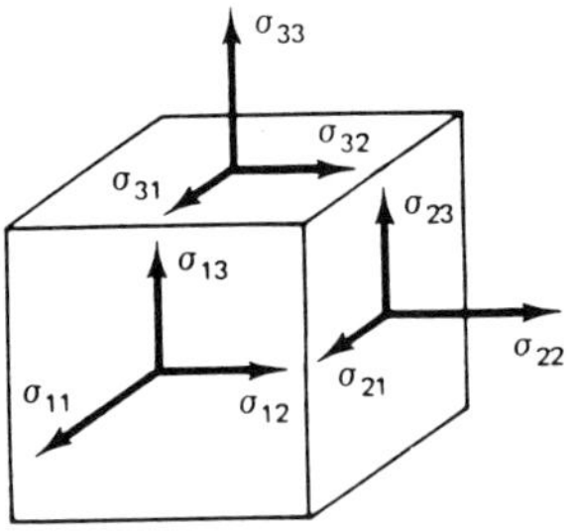

Figure 3.2.2 Undeformed boundary surface and stress components. (a) Undeformed boundary surface. (b) Components of stress in undeformed coordinates.

surface dA_i projected from dA on each of the three planes constructed on the Cartesian coordinate axes x_i, and the σ_{ij}'s are the components of the stress on these projected areas, as collected from Eq. (3.2.2):

$$\sigma_{ij} = \begin{bmatrix} \sigma_{11} & \sigma_{12} & \sigma_{13} \\ \sigma_{21} & \sigma_{22} & \sigma_{23} \\ \sigma_{31} & \sigma_{32} & \sigma_{33} \end{bmatrix},$$

where σ_{ij} is a symmetric matrix (this will be proven later). These stress components are shown schematically in Fig. 3.2.2b. The external force

$\mathbf{dF}$ results in activation of $\boldsymbol{\sigma}$ on $\mathrm{d}A$, three stress vector components $\boldsymbol{\sigma}_i$ on $\mathrm{d}A_i$ and, finally, a total of nine stress components σ_{ij} on $\mathrm{d}A_i$. Here, σ_{ij} is formally identified as the *stress tensor*.

3.3 Basic Balance Laws

3.3.1 Balance of Linear Momentum

Linear momentum $\mathbf{L}$ within a volumetric unit $\mathrm{d}\Omega$ is defined as the velocity vector summed over the incremental mass or as the product of a velocity vector with density summed over the incremental volume:

$$\mathbf{L} = \int_m \mathbf{v}\,\mathrm{d}m = \int_\Omega \mathbf{v}\rho\,\mathrm{d}\Omega.$$

The time rate of change of this linear momentum in Lagrangian coordinates is then defined as the resultant force that represents the balance of linear momentum:

$$\frac{\mathrm{d}\mathbf{L}}{\mathrm{d}t} = \frac{\mathrm{d}}{\mathrm{d}t}\int_\Omega \mathbf{v}\rho\,\mathrm{d}\Omega = \int_\Omega \rho\mathbf{F}\,\mathrm{d}\Omega + \int_\Gamma \boldsymbol{\sigma}(n)\,\mathrm{d}\Gamma. \tag{3.3.1}$$

In Eq. (3.3.1), $\mathbf{F}$ is the body force per unit mass and $\boldsymbol{\sigma}(n)$ is the stress vector normal to the surface:

$$\boldsymbol{\sigma}(n) = \boldsymbol{\sigma}_i n_i = \sigma_{ij}\mathbf{i}_j n_i, \tag{3.3.2a}$$

where each n_i is a component of a vector normal to the surface.[1] It follows from the Green–Gauss theorem that

$$\int_\Gamma \sigma_{ij}\mathbf{i}_j n_i\,\mathrm{d}\Gamma = \int_\Omega \frac{\partial \sigma_{ij}}{\partial x_i}\mathbf{i}_j\,\mathrm{d}\Omega = \int_\Omega \sigma_{ij,i}\mathbf{i}_j\,\mathrm{d}\Omega. \tag{3.3.2b}$$

Recall that the conservation of mass in solids is given by

$$\rho\,\mathrm{d}\Omega = \rho_0\,\mathrm{d}\Omega_0$$

and

$$\frac{\mathrm{d}}{\mathrm{d}t}(\rho_0\,\mathrm{d}\Omega_0) = \frac{\mathrm{d}}{\mathrm{d}t}(\rho\,\mathrm{d}\Omega) = 0.$$

With these requirements and from Eqs. (3.3.1) and (3.3.2), we obtain:

$$\int_\Omega (\sigma_{ij,i}\mathbf{i}_j + \rho\mathbf{F} - \rho\dot{\mathbf{v}})\,\mathrm{d}\Omega = 0. \tag{3.3.3a}$$

Here, some authors use direct tensors for derivatives of a stress tensor,

$$\sigma_{ij,i}\mathbf{i}_j \equiv \boldsymbol{\nabla}\cdot\mathbf{T},$$

[1] Here $\mathrm{d}\Gamma = n_i\,\mathrm{d}\Gamma_i$ requires $\boldsymbol{\sigma}(n)$ to be the normal component of $\boldsymbol{\sigma}$ in Eq. (3.2.2), as implied in Fig. 1.3.1.

where the left-hand side is clearly the vector. The right-hand side quantity $\nabla \cdot \mathbf{T}$ is called the divergence of the 'direct' stress tensor $\mathbf{T}$. Direct tensor notations without indices will be avoided in this book because the summing process is unspecified and the algebra required is unclear to the beginner.

Furthermore, the quantity $\nabla \cdot \mathbf{T}$ is supposed to be a vector (the same as the left-hand side) because $\mathbf{T}$ is a second-order tensor (not a vector), but this is not obvious from the standard vector notation (the notion that the dot product is a scalar). The reader is advised to use $\sigma_{ij,i}\mathbf{i}_j$, not $\nabla \cdot \mathbf{T}$.

For the integral equation, (3.3.3a), to be valid for all arbitrary volumes, it is necessary that the integrand vanish, so that

$$\sigma_{ij,i} + \rho F_j - \rho \dot{v}_j = 0. \tag{3.3.3b}$$

This is known as *Cauchy's first law of motion* or, simply, the equation of motion, and is associated with undeformed Cartesian coordinates.

Expand Eq. (3.3.3b) with standard notation, and write

$$\begin{aligned} \frac{\partial \sigma_{xx}}{\partial x} + \frac{\partial \sigma_{yx}}{\partial y} + \frac{\partial \sigma_{zx}}{\partial z} + \rho F_x - \rho \ddot{u}_x = 0, \\ \frac{\partial \sigma_{xy}}{\partial x} + \frac{\partial \sigma_{yy}}{\partial y} + \frac{\partial \sigma_{zy}}{\partial z} + \rho F_y - \rho \ddot{u}_y = 0, \\ \frac{\partial \sigma_{xz}}{\partial x} + \frac{\partial \sigma_{yz}}{\partial y} + \frac{\partial \sigma_{zz}}{\partial z} + \rho F_z - \rho \ddot{u}_z = 0. \end{aligned} \tag{3.3.3c}$$

These equations may also be obtained from summing all force components that act on the surfaces and the inside of an infinitesimal cube in equilibrium (see Problem 3.1).

3.3.2 Balance of Angular Momentum

The time rate of change of angular momentum is given by

$$\frac{\mathrm{d}}{\mathrm{d}t}\int_\Omega \mathbf{r} \times \rho \mathbf{v}\, \mathrm{d}\Omega = \int_\Omega \mathbf{r} \times \rho \mathbf{F}\, \mathrm{d}\Omega + \int_\Gamma \mathbf{r} \times \boldsymbol{\sigma}(n)\, \mathrm{d}\Gamma, \tag{3.3.4}$$

where

$$\begin{aligned} \int_\Gamma \mathbf{r} \times \boldsymbol{\sigma}(n)\, \mathrm{d}\Gamma &= \int_\Gamma \mathbf{r} \times \sigma_{ij}\mathbf{i}_j n_i\, \mathrm{d}\Gamma \\ &= \int_\Omega (\mathbf{r} \times \sigma_{ij}),\, {}_i\mathbf{i}_j\, \mathrm{d}\Omega \\ &= \int_\Omega \mathbf{r} \times \sigma_{ij,i}\mathbf{i}_j\, \mathrm{d}\Omega + \int_\Omega \mathbf{r}_{,i} \times \sigma_{ij}\mathbf{i}_j\, \mathrm{d}\Omega. \end{aligned} \tag{3.3.5}$$

Substituting Eq. (3.3.5) into Eq. (3.3.4) yields

$$\int_\Omega \mathbf{r} \times (\rho \dot{v}_j - \sigma_{ij,i} - \rho F_j)\mathbf{i}_j \, \mathrm{d}\Omega - \int_\Omega \mathbf{r}_{,i} \times \sigma_{ij}\mathbf{i}_j \, \mathrm{d}\Omega = 0.$$

If the balance of linear momentum is to be maintained, then we must have

$$\int_\Omega \mathbf{r}_{,i} \times \sigma_{ij}\mathbf{i}_j \, \mathrm{d}\Omega = 0.$$

Because the derivative of a position vector on the undeformed state is a unit vector (Fig. 2.2.1),

$$\mathbf{r}_{,i} = (x_k \mathbf{i}_k)_{,i} = \delta_{ki}\mathbf{i}_k = \mathbf{i}_i$$

we obtain

$$\int_\Omega \mathbf{i}_i \times \sigma_{ij}\mathbf{i}_j \, \mathrm{d}\Omega = \int_\Omega \sigma_{ij}\epsilon_{ijk}\mathbf{i}_k \, \mathrm{d}\Omega = 0, \tag{3.3.6a}$$

or

$$(\sigma_{12} - \sigma_{21})\mathbf{i}_3 + (\sigma_{31} - \sigma_{13})\mathbf{i}_2 + (\sigma_{23} - \sigma_{32})\mathbf{i}_1 = 0, \tag{3.3.6b}$$

which requires that each term in the parentheses must vanish, that is,

$$\sigma_{ij} = \sigma_{ji}. \tag{3.3.6c}$$

This is known as the balance of angular momentum, which assures the symmetry of the stress tensor. The relation given by Eq. (3.3.6) is also referred to as *Cauchy's second law of motion*.

The symmetry of the stress tensor makes it possible to write Eq. (3.3.3b) in the form

$$\sigma_{ij,j} + \rho F_i - \rho \ddot{u}_i = 0.$$

However, for conformity with Eq. (3.3.2) we maintain the form of Eq. (3.3.3) for the remainder of this book.

For curvilinear coordinates in the undeformed state, the foregoing derivations may be repeated with tangent vectors. Therefore, the surface traction normal to the boundary surface can be written in terms of the contravariant stress tensor σ^{ij} so that

$$\boldsymbol{\sigma}(n) = \sigma^{ij}\mathbf{g}_j n_i. \tag{3.3.7}$$

Here, the stress tensor must be of contravariant form to be based on the covariant curvilinear tangent vector $\mathbf{g}_j$ and the covariant normal vector component n_i. In view of Eq. (2.3.15a) or the Green–Gauss theorem, we write

$$\int_\Gamma \mathbf{V} \cdot \mathbf{n} \, \mathrm{d}\Gamma = \int_\Gamma V^i \mathbf{g}_i \cdot n_j \mathbf{g}^j \, \mathrm{d}\Gamma = \int_\Gamma V^i n_i \, \mathrm{d}\Gamma = \int_\Omega V_{|i}^{\,i} \, \mathrm{d}\Omega$$

$$= \int_\Omega \boldsymbol{\nabla} \cdot \mathbf{V} \, \mathrm{d}\Omega.$$

Similarly, by setting

$$\boldsymbol{\sigma}^i = \sigma^{ij}\mathbf{g}_j$$

we obtain

$$\int_\Gamma \sigma^{ij}\mathbf{g}_j n_i \, d\Gamma = \int_\Gamma \boldsymbol{\sigma}^i n_i \, d\Gamma = \int_\Omega \boldsymbol{\sigma}^i_{|i} \, d\Omega,$$

where

$$\begin{aligned}\boldsymbol{\sigma}^i_{|i} &= \boldsymbol{\sigma}^i_{,i} + \Gamma^i_{ij}\boldsymbol{\sigma}^j = (\sigma^{ij}\mathbf{g}_j)_{,i} + \Gamma^i_{ij}\sigma^{jm}\mathbf{g}_m \\ &= \sigma^{ij}_{,i}\mathbf{g}_j + \sigma^{ij}\Gamma^m_{ji}\mathbf{g}_m + \Gamma^i_{ij}\sigma^{jm}\mathbf{g}_m \\ &= \sigma^{ij}_{|i}\mathbf{g}_j = \frac{1}{\sqrt{g}}(\sqrt{g}\,\sigma^{ij}\mathbf{g}_j)_{,i}\end{aligned}$$

with

$$\sigma^{ij}_{|i} = \sigma^{ij}_{,i} + \Gamma^i_{im}\sigma^{mj} + \Gamma^j_{im}\sigma^{im}. \tag{3.3.8}$$

It is interesting to note that the covariant derivative of a second-order tensor expressed in Eq. (3.3.8) has a form identical to that of Eq. (2.3.20b), although it is derived in a different manner.

It follows from Eqs. (3.3.1) and (3.3.8) that

$$\int_\Omega (\sigma^{ij}_{|i}\mathbf{g}_j + \rho\mathbf{F} - \rho\dot{\mathbf{v}}) \, d\Omega = 0.$$

Since the integrand must vanish, we obtain

$$\sigma^{ij}_{|i} + \rho F^j - \rho\dot{v}^j = 0. \tag{3.3.9}$$

This is Cauchy's first law of motion, or the equation of motion, for undeformed *curvilinear* coordinates.

From operations similar to those in Eq. (3.3.4), but using the curvilinear coordinates and noting that $\mathbf{r}_{|i} = \mathbf{r}_{,i} = \mathbf{g}_i$, we obtain (see Fig. 2.3.1)

$$\int_\Omega \mathbf{r} \times (\sigma^{ij}_{|i}\mathbf{g}_j + \rho\mathbf{F} - \rho\dot{\mathbf{v}}) \, d\Omega + \int_\Omega \mathbf{g}_i \times \sigma^{ij}\mathbf{g}_j \, d\Omega = 0.$$

Thus, once again the balance of momentum requires that

$$\mathbf{g}_i \times \sigma^{ij}\mathbf{g}_j = 0$$

and it follows from Eq. (2.3.6) that the previous expression takes the form

$$\sigma^{ij}\sqrt{g}\,\epsilon_{ijk}\mathbf{g}^k = 0.$$

With algebra similar to that in Eq. (3.3.6), we obtain

$$\sigma^{ij} = \sigma^{ji}, \tag{3.3.10}$$

which is Cauchy's second law of motion in undeformed curvilinear coordinates.

The physical components $\bar{\sigma}^{ij}$ of the stress tensor may be derived in a manner similar to Eq. (2.4.12c). Thus, for cylindrical or spherical coordinates,

$$\sigma^{ij} = \sqrt{g^{(ii)}g^{(jj)}}\,\bar{\sigma}^{ij}.$$

Example 3.1. Explicit Equations of Motion for Cylindrical and Spherical Coordinates The equations of motion for small strain in cylindrical and spherical coordinates are derived in terms of physical components as:

$$\sigma^{ij}_{,i} + \Gamma^{i}_{im}\sigma^{jm} + \Gamma^{j}_{mi}\sigma^{mi} + \rho F^{j} - \rho\dot{v}^{j} = 0.$$

For Cylindrical Coordinates

$$F^1 = F_r, \qquad F^2 = \frac{F_\theta}{r}, \qquad F^3 = F_z,$$

$$\dot{v}^1 = \ddot{u}_r, \qquad \dot{v}^2 = \frac{\ddot{u}_\theta}{r}, \qquad \dot{v}^3 = \ddot{u}_z,$$

$$\sigma^{11} = \sigma_{rr}, \quad \sigma^{22} = \frac{\sigma_{\theta\theta}}{r^2}, \qquad \sigma^{33} = \sigma_{zz},$$

$$\sigma^{12} = \sigma^{21} = \frac{\sigma_{\theta r}}{r}, \quad \sigma^{13} = \sigma^{31} = \sigma_{zr}, \quad \sigma^{23} = \sigma^{32} = \frac{\sigma_{z\theta}}{r},$$

$$\frac{\partial\sigma_{rr}}{\partial r} + \frac{1}{r}\frac{\partial\sigma_{\theta r}}{\partial\theta} + \frac{\partial\sigma_{zr}}{\partial z} + \frac{\sigma_{rr} - \sigma_{\theta\theta}}{r} + \rho F_r - \rho\ddot{u}_r = 0,$$

$$\frac{\partial\sigma_{r\theta}}{\partial r} + \frac{1}{r}\frac{\partial\sigma_{\theta\theta}}{\partial\theta} + \frac{\partial\sigma_{z\theta}}{\partial z} + \frac{2}{r}\sigma_{r\theta} + \rho F_\theta - \rho\ddot{u}_\theta = 0,$$

$$\frac{\partial\sigma_{rz}}{\partial r} + \frac{1}{r}\frac{\partial\sigma_{\theta z}}{\partial\theta} + \frac{\partial\sigma_{zz}}{\partial z} + \frac{\sigma_{rz}}{r} + \rho F_z - \rho\ddot{u}_z = 0.$$

For Spherical Coordinates

$$F^1 = F_R, \qquad F^2 = \frac{F_\alpha}{R}, \qquad F^3 = \frac{F_\theta}{R\sin\alpha},$$

$$\dot{v}^1 = \ddot{u}_R, \qquad \dot{v}^2 = \frac{\ddot{u}_\alpha}{R}, \qquad \dot{v}^3 = \frac{\ddot{u}_\theta}{R\sin\alpha},$$

$$\sigma^{11} = \sigma_{RR}, \quad \sigma^{22} = \frac{\sigma_{\alpha\alpha}}{R^2}, \quad \sigma^{33} = \frac{\sigma_{\theta\theta}}{R^2\sin^2\alpha},$$

$$\sigma^{12} = \sigma^{21} = \frac{\sigma_{\alpha R}}{R}, \quad \sigma^{13} = \sigma^{31} = \frac{\sigma_{\theta R}}{R\sin\alpha}, \quad \sigma^{23} = \sigma^{32} = \frac{\sigma_{\theta\alpha}}{R^2\sin\alpha},$$

$$\frac{\partial \sigma_{RR}}{\partial R} + \frac{1}{R}\frac{\partial \sigma_{\alpha R}}{\partial \alpha} + \frac{1}{R\sin\alpha}\frac{\partial \sigma_{\theta R}}{\partial \theta} + \frac{2}{R}\sigma_{RR} + \frac{\cot\alpha}{R}\sigma_{\alpha R} - \frac{\sigma_{\alpha\alpha}}{R} - \frac{\sigma_{\theta\theta}}{R} + \rho F_R - \rho \ddot{u}_R = 0,$$

$$\frac{\partial \sigma_{R\alpha}}{\partial R} + \frac{1}{R}\frac{\partial \sigma_{\alpha\alpha}}{\partial \alpha} + \frac{1}{R\sin\alpha}\frac{\partial \sigma_{\theta\alpha}}{\partial \theta} + \frac{3}{R}\sigma_{R\alpha} + \frac{\cot\alpha}{R}\sigma_{\alpha\alpha} - \frac{(\cot\alpha)}{R}\sigma_{\theta\theta} + \rho F_\alpha - \rho \ddot{u}_\alpha = 0,$$

$$\frac{\partial \sigma_{R\theta}}{\partial R} + \frac{1}{R}\frac{\partial \sigma_{\alpha\theta}}{\partial \alpha} + \frac{1}{R\sin\alpha}\frac{\partial \sigma_{\theta\theta}}{\partial \theta} + \frac{3}{R}\sigma_{R\theta} + \frac{2}{R}(\cot\alpha)\sigma_{\alpha\theta} + \rho F_\theta - \rho \ddot{u}_\theta = 0.$$

3.4 Coordinate Transformations for Stresses

The coordinate transformations for the stress tensor can be performed in a manner similar to those for the strain tensor. Referring to Fig. 3.4.1, the old and new coordinates x_i and x_i' are related as:

$$n_i' = a_{ij}n_j \qquad \sigma_i' = \sigma_{ij}'n_j' \tag{3.4.1a}$$

and

$$\sigma_i' = a_{ir}\sigma_r = a_{ir}\sigma_{rs}n_s = a_{ir}\sigma_{rs}a_{js}n_j'. \tag{3.4.1b}$$

Equate Eqs. (3.4.1a and b) to yield

$$\sigma_{ij}' = a_{ir}a_{js}\sigma_{rs}, \tag{3.4.2}$$

in which the stress tensor σ_{rs} in the old coordinates is transformed into

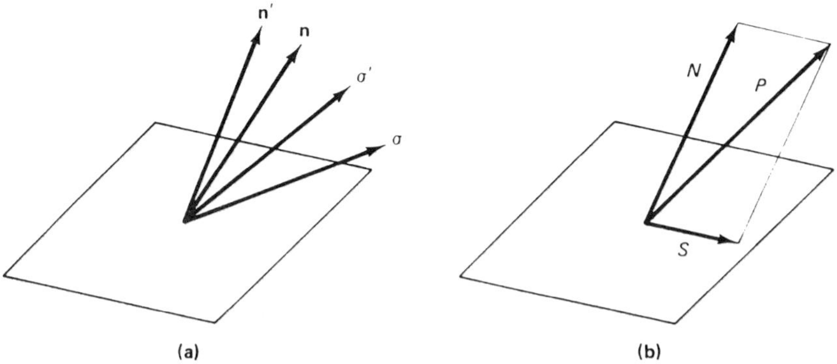

Figure 3.4.1 Coordinate transformations for normal and shear stresses. (a) Normal stresses. (b) Shear stresses.

σ'_{ij} in the new coordinates. The stress tensor σ_{ij}, rotated to the principal plane by n_j (Fig. 3.4.1a) with $\sigma = \lambda$ introduced as a scalar which corresponds to a principal stress acting on the plane n_i, may be written as

$$\sigma_{ij}n_j = \lambda n_i.$$

This relation can be written in the form

$$(\sigma_{ij} - \lambda\delta_{ij})n_j = 0. \tag{3.4.3}$$

The expression in Eq. (3.4.3) can be derived by considering a scalar in the principal direction,

$$\sigma(n) = \sigma_{ij}n_jn_i \tag{3.4.4a}$$

with the constraint

$$n_in_i = 1. \tag{3.4.4b}$$

We use the Lagrange multiplier λ similarly as in Eq. (2.6.8) to combine Eqs. (3.4.4a and b) and construct a scalar function

$$f = \sigma_{ij}n_jn_i - \lambda(n_in_i - 1)$$

to obtain the extrema of f with respect to the direction cosines n_i such that

$$\frac{\partial f}{\partial n_i} = 2(\sigma_{ij} - \lambda\delta_{ij})n_j = 0,$$

which results in Eq. (3.4.3). This is identical in form to the case of strains, Eq. (2.6.5) in Chapter 2, which leads to the eigenvalue problem,

$$|\sigma_{ij} - \lambda\delta_{ij}| = 0, \tag{3.4.5}$$

from which we obtain

$$-\lambda^3 + J_1\lambda^2 - J_2\lambda + J_3 = 0,$$

where J_1, J_2, and J_3 are called the *principal stress invariants*.

First principal stress invariant

$$J_1 = \sigma_{ii}. \tag{3.4.6a}$$

Second principal stress invariant

$$J_2 = \tfrac{1}{2}(\sigma_{ii}\sigma_{jj} - \sigma_{ij}\sigma_{ij}). \tag{3.4.6b}$$

Third principle stress invariant

$$J_3 = |\sigma_{ij}| = \tfrac{1}{6}(\sigma_{ii}\sigma_{jj}\sigma_{kk} - 3\sigma_{ii}\sigma_{jk}\sigma_{kj} + 2\sigma_{ij}\sigma_{jk}\sigma_{ki}). \tag{3.4.6c}$$

The notion that the principal stresses and their directions are identified as eigenvalues and eigenvectors of Eq. (3.4.5), respectively, is similar to our earlier study on principal strains in Chapter 2. The solution procedure is the same as given in Example 2.12.

Example 3.2 Consider the stress measurement data on a plane rotated from the plane of principal stresses given by

$$\sigma_{ij} = \begin{bmatrix} 1 & \sqrt{3} & 0 \\ \sqrt{3} & 0 & 0 \\ 0 & 0 & 1 \end{bmatrix}.$$

Determine: (1) principal stress invariants, (2) principal stresses, (3) principal directions, and (4) draw the Mohr circles for this problem.

Solution. Note that stress components are the same as in Example 2.12 for the strain components. Therefore, using the results of Example 2.12, we have the following information.

1. *Principal stress invariants*

$$I_1 = 2, \quad I_2 = -2, \quad I_3 = -3.$$

2. *Principal stresses*

$$\sigma_{(1)} = \lambda_{(1)} = \frac{1+\sqrt{13}}{2}, \quad \sigma_{(2)} = \lambda_{(2)} = 1, \quad \sigma_{(3)} = \lambda_{(3)} = \frac{1-\sqrt{13}}{2}.$$

3. *Principal direction cosine matrix*

$$a_{ij} = \begin{bmatrix} 0.8 & 0.6 & 0 \\ 0 & 0 & 1 \\ 0.6 & -0.8 & 0 \end{bmatrix}.$$

4. *Mohr circle representation* Confirm the results for step 1 from the elementary relations,

$$\sigma_{p1,2} = \frac{\sigma_{xx}+\sigma_{yy}}{2} \pm \sqrt{\left(\frac{\sigma_{xx}-\sigma_{yy}}{2}\right)^2 + \sigma_{xy}^2} = \begin{cases} 2.3 \\ -1.3 \end{cases},$$

$$\sigma_{p1,3} = \begin{cases} 1 \\ 1 \end{cases}, \quad \sigma_{p2,3} = \begin{cases} 1 \\ 0 \end{cases},$$

$$\tan 2\theta = \frac{\sigma_{xy}}{(\sigma_{xx}-\sigma_{yy})/2} = 3.46, \quad \theta = \begin{cases} 0.6 \\ 0.8 \end{cases}.$$

Note that the only difference here from Example 2.13 is the treatment of shear stresses versus shear strains:

$$\sigma_{12} = \sigma_{xy}, \qquad \sigma_{21} = \sigma_{yx},$$
$$\gamma_{12} = \tfrac{1}{2}\gamma_{xy}, \qquad \gamma_{21} = \tfrac{1}{2}\gamma_{yx}.$$

Thus, the Mohr circles will be identical to Fig. 2.6.2 except that the horizontal and vertical axes are to be labeled $\sigma_{(i)}$ and σ_{xy}, respectively.

The shear stresses acting on any plane and the maximum shear stresses are directly associated with the normal stresses on that plane and the principal stresses, respectively. Consider the normal component $N_{(i)}$ and the shear component $S_{(i)}$ of any resultant stress $P_{(i)}$ in Fig. 3.4.1b:

$$S_{(i)}^2 = P_{(i)}^2 - N_{(i)}^2. \tag{3.4.7a}$$

Since $P_i = \sigma_{ij}n_j$ and $N_i = P_{(i)}n_i$, and σ_{ij} can be transformed into the principal stress $\sigma_{(i)}$, we write

$$P_i = \sigma_{(i)}n_i, \quad N_{(i)} = \sigma_{(i)}n_in_i.$$

Substitution of these into Eq. (3.4.7a) yields

$$S_{(i)}^2 = \sigma_{(i)}^2 n_in_i - \sigma_{(i)}n_in_i\sigma_{(j)}n_jn_j. \tag{3.4.7b}$$

Note that the index with parentheses indicates the direction and the requirement that the component not be summed. We are now concerned with a direction n_i for which shear stress is a maximum or minimum, subject to the constraint given in Eq. (3.4.4b). In this respect, the Lagrange multiplier method can be used by constructing a scalar function

$$f = S_{(i)}^2 + \lambda(n_in_i - 1)$$

and we find the extremum condition,

$$\frac{\partial f}{\partial n_k} = \frac{\partial}{\partial n_k}[\sigma_{(i)}^2 n_in_i - \sigma_{(i)}n_in_i\sigma_{(j)}n_jn_j + \lambda(n_in_i - 1)] = 0$$

or

$$(\sigma_{(i)}^2 - 2\sigma_{(i)}n_jn_j\sigma_{(j)} + \lambda)n_i = 0, \tag{3.4.8}$$

which represents the requirement for the shear stresses to be either maximum or minimum.

Example 3.3. Determination of Planes of Minimum and Maximum Shear Stresses and Their Magnitudes

Solution. The solution involves two cases. Case 1 is for minimum shear stresses and case 2 is for maximum shear stresses.

Case 1. Minimum Shear Stresses

$$n_1 = \pm 1, \quad n_2 = n_3 = 0, \quad \lambda = \sigma_{(1)}^2, \quad S_{(1)} = 0,$$
$$n_2 = \pm 1, \quad n_1 = n_3 = 0, \quad \lambda = \sigma_{(2)}^2, \quad S_{(2)} = 0,$$
$$n_3 = \pm 1, \quad n_1 = n_2 = 0, \quad \lambda = \sigma_{(3)}^2, \quad S_{(3)} = 0.$$

Case 2. Maximum Shear Stresses

For $n_1 = 0$:

$$n_2 = n_3 = \pm\frac{1}{\sqrt{2}}, \quad \lambda = \sigma_{(2)}\sigma_{(3)}, \quad S_{(1)} = \tfrac{1}{2}(\sigma_{(2)} - \sigma_{(3)}).$$

For $n_2 = 0$:

$$n_1 = n_3 = \pm\frac{1}{\sqrt{2}}, \quad \lambda = \sigma_{(3)}\sigma_{(1)}, \quad S_{(2)} = \tfrac{1}{2}(\sigma_{(3)} - \sigma_{(1)}).$$

For $n_3 = 0$:

$$n_1 = n_2 = \pm\frac{1}{\sqrt{2}}, \quad \lambda = \sigma_{(1)}\sigma_{(2)}, \quad S_{(3)} = \tfrac{1}{2}(\sigma_{(1)} - \sigma_{(2)}).$$

To verify these results, we first expand Eq. (3.4.8) as follows.

Step 1:

$$[\sigma_{(1)}^2 - 2\sigma_{(1)}(\sigma_{(1)}n_1^2 + \sigma_{(2)}n_2^2 + \sigma_{(3)}n_3^2) + \lambda]n_1 = 0.$$

Step 2:

$$[\sigma_{(2)}^2 - 2\sigma_{(2)}(\sigma_{(1)}n_1^2 + \sigma_{(2)}n_2^2 + \sigma_{(3)}n_3^2) + \lambda]n_2 = 0.$$

Step 3:

$$[\sigma_{(3)}^2 - 2\sigma_{(3)}(\sigma_{(1)}n_1^2 + \sigma_{(2)}n_2^2 + \sigma_{(3)}n_3^2) + \lambda]n_3 = 0.$$

The equations must be satisfied together with the following step.

Step 4:

$$n_1^2 + n_2^2 + n_3^2 = 1.$$

For $n_1 = 0$, we have $n_2^2 = 1 - n_3^2$ and, thus, steps 2 and 3 are as shown in steps 5 and 6, respectively.

Step 5:

$$\{\sigma_{(2)}^2 - 2\sigma_{(2)}[\sigma_{(2)}(1 - n_3^2) + \sigma_{(3)}n_3^2] + \lambda\}(1 - n_3^2)^{1/2} = 0.$$

Step 6:

$$\{\sigma_{(3)}^2 - 2\sigma_{(3)}[\sigma_{(2)}(1 - n_3^2) + \sigma_{(3)} n_3^2] + \lambda\} n_3 = 0.$$

For $n_1 = 0$, $n_2 = 0$, by using step 6 with $n_3 = \pm 1$, we have

$$\sigma_{(3)}^2 - 2\sigma_{(3)}^2 + \lambda = 0 \quad \rightarrow \quad \lambda = \sigma_{(3)}^2.$$

Thus, from Eq. (3.4.7b)

$$S_{(i)}^2 = \sigma_{(3)}^2 - \sigma_{(3)}^2 = 0 \quad \rightarrow \quad S_{(i)} = S_{(3)} = 0.$$

For $n_1 = 0$, $n_3 = 0$, by using step 5 with $n_2 = \pm 1$, we get

$$\sigma_{(2)}^2 - 2\sigma_{(2)}^2 + \lambda = 0 \quad \rightarrow \quad \lambda = \sigma_{(2)}^2,$$

$$S_{(i)}^2 = \sigma_{(2)}^2 - \sigma_{(2)}^2 = 0 \quad \rightarrow \quad S_{(i)} = S_{(2)} = 0.$$

Similarly, for $n_2 = n_3 = 0$, $n_1 = \pm 1$, we have

$$\lambda = \sigma_{(1)}^2, \quad S_{(i)} = S_{(1)} = 0.$$

Physically, these results indicate that the shear stresses are zero at the principal planes with direction cosines $(\pm 1, 0, 0)$, $(0, \pm 1, 0)$, and $(0, 0, \pm 1)$. To investigate the case of $n_1 = 0$, $n_2 \neq 0$, and $n_3 \neq 0$, we return to steps 5 and 6 and write the following steps.

Step 7:

$$\sigma_{(2)}^2 - 2\sigma_{(2)}^2(1 - n_3^2) - 2\sigma_{(2)}\sigma_{(3)} n_3^2 + \lambda = 0.$$

Step 8:

$$\sigma_{(3)}^2 - 2\sigma_{(3)}\sigma_{(2)}(1 - n_3^2) - 2\sigma_{(3)}^2 n_3^2 + \lambda = 0.$$

Solve steps 7 and 8 simultaneously, to obtain:

$$n_3^2 + \tfrac{1}{2}\frac{(\sigma_{(3)} - \sigma_{(2)})^2}{(\sigma_{(3)} - \sigma_{(2)})^2} = \tfrac{1}{2} \quad n_3 = \pm\frac{1}{\sqrt{2}}.$$

It follows from step 4 that $n_2 = \pm 1/\sqrt{2}$. Substituting these results into step 8, we have

$$\sigma_{(3)}^2 - 2\sigma_{(3)}\sigma_{(2)}(\tfrac{1}{2}) - 2\sigma_{(3)}^2(\tfrac{1}{2}) + \lambda = 0,$$

which gives

$$\lambda = \sigma_{(3)}\sigma_{(2)}.$$

For these direction cosines, from (3.4.7b) we obtain

$$S_{(i)}^2 = \frac{\sigma_{(2)}^2}{2} + \frac{\sigma_{(3)}^2}{2} - \left(\frac{\sigma_{(2)}}{2} + \frac{\sigma_{(3)}}{2}\right)^2 = \frac{1}{4}(\sigma_{(2)} - \sigma_{(3)})^2.$$

Here, we define $S_{(i)}^2 = S_{(1)}^2$ and write

$$S_{(1)} = \tfrac{1}{2}(\sigma_{(2)} - \sigma_{(3)}).$$

Similarly, for $n_1 \neq 0$, $n_2 = 0$, and $n_3 \neq 0$, and $n_1 \neq 0$, $n_2 \neq 0$, and $n_3 = 0$, respectively, we have

$$n_1 = n_3 = \pm\frac{1}{\sqrt{2}}, \quad \lambda = \sigma_{(3)}\sigma_{(1)}, \quad S_{(i)} = S_{(2)} = \tfrac{1}{2}(\sigma_{(3)} - \sigma_{(1)}),$$

$$n_1 = n_2 = \pm\frac{1}{\sqrt{2}}, \quad \lambda = \sigma_{(1)}\sigma_{(2)}, \quad S_{(i)} = S_{(3)} = \tfrac{1}{2}(\sigma_{(1)} - \sigma_{(2)}).$$

Here, $S_{(1)}$, $S_{(2)}$, and $S_{(3)}$ are called the *maximum shear stresses*. They occur along the planes oriented 45° from the planes of the principal stresses because the direction cosines are $\pm 1/\sqrt{2}$ with respect to the principal planes.

3.5 The Deviatoric Stress Tensor

As was the case in Section 2.8 for the strain tensor, it is often useful to separate the stress tensor into two parts, the deviatoric part σ^*_{ij} and the spherical, or hydrostatic, part $1/3\ \sigma_{kk}\delta_{ij}$, so that

$$\sigma_{ij} = \sigma^*_{ij} + \tfrac{1}{3}\sigma_{kk}\delta_{ij},$$

or

$$\sigma^*_{ij} = \sigma_{ij} - \tfrac{1}{3}\sigma_{kk}\delta_{ij}. \tag{3.5.1}$$

This holds true, irrespective of the geometric constraints of small or large strains. This is in contrast to the deviatoric strains in which only the small strain approximation makes the relation, such as Eq. (2.7.3), valid.

If $i = j$, then we have

$$\sigma^*_{ii} = \sigma_{ii} - \tfrac{1}{3}\sigma_{kk}\delta_{ii} = \sigma_{ii} - \tfrac{1}{3}\sigma_{ii}(3) = 0,$$

which is the state of stress characterized by *pure shear*. It follows from Eq. (3.5.1) that the deviatoric stress σ^*_{ij} is equal to the total stress σ_{ij} if $i \neq j$. A state of pure shear arises if all normal stress components are the same in the expression for σ^*_{ij} in Eq. (3.5.1). The concept of deviatoric stresses is useful for viscous flows in fluid mechanics (see Chapter 5).

The principal deviatoric stress invariants are particularly important in constitutive theories of inelastic materials. To determine these quantities, we begin with the eigenvalue problems that use a process similar to Eq. (3.4.5),

$$|\sigma^*_{ij} - \sigma^*\delta_{ij}| = 0,$$

or

$$-(\sigma^*)^3 + \hat{J}_1(\sigma^*)^2 - \hat{J}_2\sigma^* + \hat{J}_3 = 0, \tag{3.5.2}$$

where the *deviatoric stress invariants* are identified as

$$\hat{J}_1 = \sigma^*_{ii} = 0, \tag{3.5.3a}$$

$$\hat{J}_2 = \sigma^2_{kk} - J_2 = \tfrac{1}{2}\sigma^*_{ij}\sigma^*_{ij}, \tag{3.5.3b}$$

$$\hat{J}_3 = J_3 - \tfrac{1}{3}\sigma_{kk}J_2 + \tfrac{2}{3}\sigma^3_{kk} = \tfrac{1}{3}\sigma^*_{ij}\sigma^*_{jk}\sigma^*_{ki}. \tag{3.5.3c}$$

Expanding $\hat{J}_2$ further, we have

$$\begin{aligned}\hat{J}_2 &= \tfrac{1}{2}[(\sigma^*_{11})^2 + (\sigma^*_{22})^2 + (\sigma^*_{33})^2] + (\sigma_{12})^2 + (\sigma_{23})^2 + \sigma^2_{31} \\ &= \tfrac{1}{6}[(\sigma_{11} - \sigma_{22})^2 + (\sigma_{22} - \sigma_{33})^2 + (\sigma_{33} - \sigma_{11})^2] + \sigma^2_{12} + \sigma^2_{23} + \sigma^2_{31} \\ &= \tfrac{3}{2}\sigma^2_{\text{oct}},\end{aligned} \tag{3.5.4}$$

where σ_{oct} denotes the octahedral shear stress acting on the octahedral plane, as shown in Fig. 3.5.1:

$$\sigma_{oct} = \tfrac{1}{3}\sqrt{[(\sigma_{11} - \sigma_{22})^2 + (\sigma_{22} - \sigma_{33})^2 + (\sigma_{33} - \sigma_{11})^2] + 6(\sigma^2_{12} + \sigma^2_{23} + \sigma^2_{31})}. \tag{3.5.5}$$

The octahedral shear stress is the resultant shear stress on a plane that makes the same angle with the three principal directions. Such a plane is called an octahedral plane; eight such planes can form an octahedron (see Fig. 3.5.1). The direction cosines of a normal-to-the-octahedral

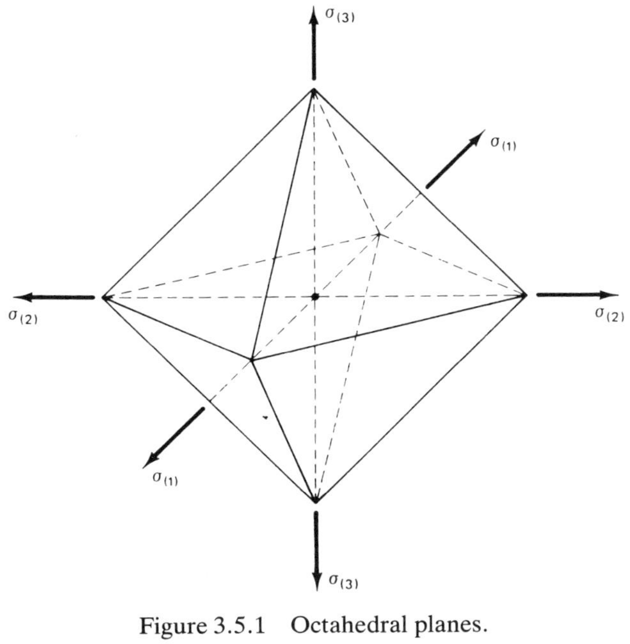

Figure 3.5.1 Octahedral planes.

plane relative to the principal axes are $1/\sqrt{3}$. Hence, using Eq. (3.4.7), we obtain the octahedral stress as the square root of $S_{(i)}^2$, with $n_1 = n_2 = n_3 = 1/\sqrt{3}$, which satisfies $n_i n_i = 1$:

$$\sigma_{oct}^2 = \tfrac{2}{9}(\sigma_{(1)}^2 + \sigma_{(2)}^2 + \sigma_{(3)}^2 - \sigma_{(1)}\sigma_{(2)} - \sigma_{(2)}\sigma_{(3)} - \sigma_{(3)}\sigma_{(1)}),$$

or

$$\begin{aligned}\sigma_{oct} &= \tfrac{1}{3}\sqrt{(\sigma_{(1)} - \sigma_{(2)})^2 + (\sigma_{(2)} - \sigma_{(3)})^2 + (\sigma_{(3)} - \sigma_{(1)})^2} \\ &= \tfrac{1}{3}\sqrt{2J_1^2 - 6J_2}\,,\end{aligned} \tag{3.5.6}$$

where the negative sign is discarded on the right-hand side because it has no physical significance. Thus, Eq. (3.5.6) reduces to Eq. (3.5.5).

In 1913, von Mises proposed that in any given material, yielding occurs at a constant value of the second deviatoric stress invariant $\hat{J}_2$ or the octahedral shear stress.

3.6 Stresses with Large Strains

Definitions of stresses with large strains on deformed surfaces are intimately associated with the deformed coordinates chosen. This has been a controversial subject in the literature. Some of the methods for defining stresses on deformed surfaces are uncompromising with computational aspects. It is the purpose of this section to review historical developments and demonstrate some practical approaches for handling stresses with large strains.

Consider a deformed surface in comparison with an undeformed surface, as shown in Fig. 3.6.1. The undeformed infinitesimal area dA_{01} constructed by $dx_2\mathbf{i}_2$ and $dx_3\mathbf{i}_3$ is

$$dA_{01} = |dx_2\mathbf{i}_2 \times dx_3\mathbf{i}_3| = |dx_2\, dx_3\mathbf{i}_1| = dx_2\, dx_3\hat{n}_1,$$

where $\hat{n}_1$ denotes a component of the unit vector normal to the surface dA_{01}. Similarly, the infinitesimal areas constructed by other coordinate lines are

$$dA_{02} = dx_3\, dx_1\hat{n}_2,$$

$$dA_{03} = dx_1\, dx_2\hat{n}_3.$$

For the deformed infinitesimal area dA_1 constructed by $dx_2\mathbf{G}_2$ and $dx_3\mathbf{G}_3$,

$$dA_1 = |dx_2\mathbf{G}_2 \times dx_3\mathbf{G}_3|$$

with

$$|\mathbf{G}_2 \times \mathbf{G}_3| = |\sqrt{G}\mathbf{G}^1| = \sqrt{G}\sqrt{\mathbf{G}^1\cdot\mathbf{G}^1} = \sqrt{GG^{11}}$$

we obtain the relationship between the deformed and undeformed

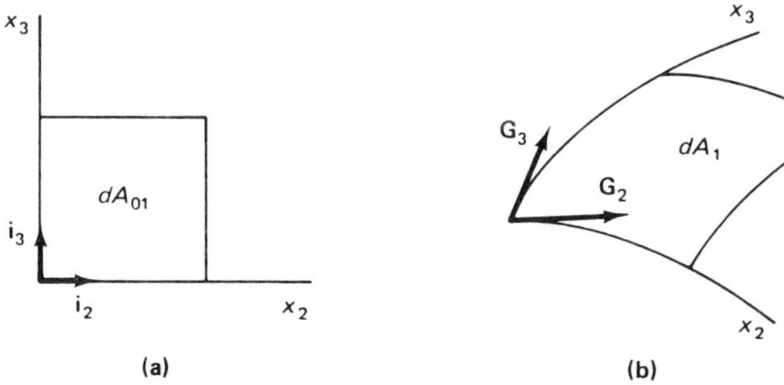

Figure 3.6.1 Undeformed and deformed areas. (a) Undeformed area. (b) Deformed area.

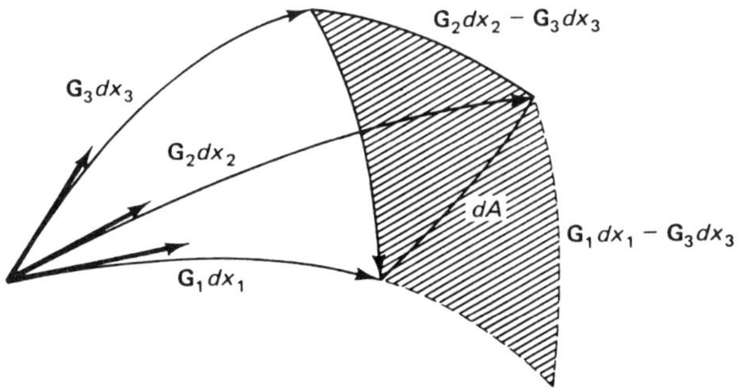

Figure 3.6.2 Deformed surface.

areas,

$$\mathrm{d}A_1 = \sqrt{GG^{11}}\,\mathrm{d}x_2\,\mathrm{d}x_3 = \sqrt{GG^{11}}\,\mathrm{d}A_{01}.$$

Thus, in general,

$$\mathrm{d}A_i = \sqrt{GG^{(ii)}}\,\mathrm{d}A_{0i} \tag{3.6.1}$$

with

$$\mathrm{d}A_{0i} = \mathrm{d}A_0\hat{n}_i, \tag{3.6.2}$$

where $\hat{n}_i$ is the component of a vector normal to the undeformed area. Note here that $\mathbf{G}^i$ is the contravariant tangent vector in the same sense as $\mathbf{g}^i$ for the curvilinear coordinates described earlier. On examining the inclined surface area $\mathrm{d}A$ of the deformed volume, as shown in Fig. 3.6.2, we find that

$$\begin{aligned}\mathrm{d}A\mathbf{n} &= (\mathbf{G}_1\,\mathrm{d}x_1 - \mathbf{G}_3\,\mathrm{d}x_3) \times (\mathbf{G}_2\,\mathrm{d}x_2 - \mathbf{G}_3\,\mathrm{d}x_3)\\ &= \sqrt{G}\mathbf{G}^1\,\mathrm{d}x_2\,\mathrm{d}x_3 + \sqrt{G}\mathbf{G}^2\,\mathrm{d}x_1\,\mathrm{d}x_3 + \sqrt{G}\mathbf{G}^3\,\mathrm{d}x_1\,\mathrm{d}x_2\\ &= \sqrt{G}\mathbf{G}^i\,\mathrm{d}A_{0i}.\end{aligned}$$

Use the relation given in Eq. (3.6.1) to obtain

$$\mathrm{d}A\mathbf{n} = \frac{\mathrm{d}A_i}{\sqrt{G^{(ii)}}}\mathbf{G}^i. \tag{3.6.3}$$

With $\mathbf{n} = n_i\mathbf{G}^i$, we have

$$\mathrm{d}A_i = \sqrt{G^{(ii)}}\,\mathrm{d}An_i. \tag{3.6.4}$$

The stress vector acting on $\mathrm{d}A$ (Fig. 3.6.3) is of the form

$$\boldsymbol{\sigma}\,\mathrm{d}A = \boldsymbol{\sigma}^i\,\mathrm{d}A_i,$$

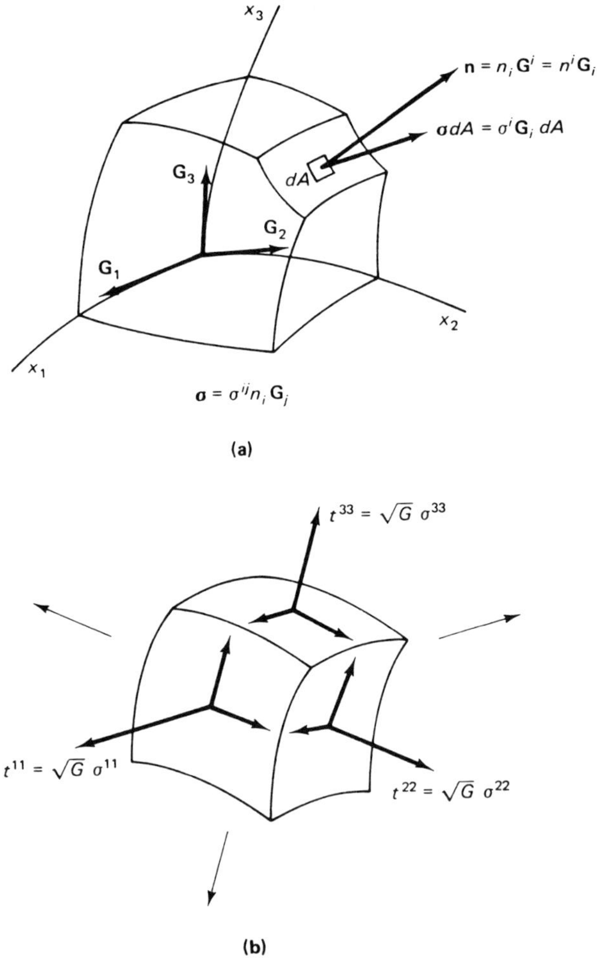

Figure 3.6.3 (a) Deformed boundary surface. (b) Components of Kirchhoff stress tensor on the deformed state, measured in terms of the undeformed area.

where $\boldsymbol{\sigma}^i$ is the contravariant stress vector and

$$\boldsymbol{\sigma} = \boldsymbol{\sigma}^i \frac{\mathrm{d}A_i}{\mathrm{d}A} = \boldsymbol{\sigma}^i \sqrt{G^{(ii)}}\, n_i. \tag{3.6.5}$$

By definition, as noted in Eq. (3.3.7), the stress vector normal to the surface is

$$\boldsymbol{\sigma} = \boldsymbol{\sigma}(n) = \sigma^{ij} n_i \mathbf{G}_j, \tag{3.6.6}$$

where σ^{ij} is the contravariant stress tensor acting on the deformed surface measured in coordinates applicable to the deformed state.

In general, the geometry of the undeformed configuration is known *a priori*. For this reason, it is convenient to use a stress tensor t^{ij} in terms of convected coordinates in the deformed state but measured per unit of undeformed area instead of deformed area (Green and Zerna 1954). Thus, we use Eqs. (3.6.1), (3.6.3), and (3.6.6) to obtain,

$$\boldsymbol{\sigma}\, \mathrm{d}A = \sigma^{ij} n_i \mathbf{G}_j\, \mathrm{d}A = \sigma^{ij} \mathbf{G}_j \sqrt{G}\, \hat{n}_i\, \mathrm{d}A_0 \tag{3.6.7a}$$

and

$$\mathbf{t}\, \mathrm{d}A_0 = t^{ij} \hat{n}_i \mathbf{G}_j\, \mathrm{d}A_0. \tag{3.6.7b}$$

Equating Eqs. (3.6.7a and b) gives

$$t^{ij} = \sqrt{G}\, \sigma^{ij}, \tag{3.6.8}$$

where σ^{ij} and t^{ij} are referred to as the Cauchy stress tensor and the Kirchhoff stress tensor, respectively. If $\sqrt{G} = 1$, then the undeformed state of stress prevails. In this case, we may place the indices in the subscript position so that $t_{ij} = \sigma_{ij}$.

An alternate way of measuring stresses in deformed states in terms of an undeformed area may be devised by defining the relation between the undeformed and deformed areas using a different approach from Eq. (3.6.1). The normal vector $\hat{\mathbf{n}}$ and the stress vector $\mathbf{T}_{(1)}$ on undeformed surface $\mathrm{d}A_0$, as related to the normal vector $\mathbf{n}$ and the stress vector $\boldsymbol{\sigma}$ on deformed surface $\mathrm{d}A_0$, are schematically shown in Fig. 3.6.4. We begin with the product of the undeformed area and the normal vector defined as:

$$\hat{\mathbf{n}}\, \mathrm{d}A_0 = \mathrm{d}x_i\, \mathrm{d}x_j \epsilon_{ijk} \mathbf{i}_k$$

or

$$\hat{n}_k \mathbf{i}_k\, \mathrm{d}A_0 = \frac{\partial x_i}{\partial z_m} \frac{\partial x_j}{\partial z_n} \mathrm{d}z_m\, \mathrm{d}z_n \epsilon_{ijk} \mathbf{i}_k.$$

Eliminate $\mathbf{i}_k$ and multiply by $\partial x_k / \partial z_p$, to get

$$\frac{\partial x_k}{\partial z_p} \hat{n}_k\, \mathrm{d}A_0 = \frac{\partial x_i}{\partial z_m} \frac{\partial x_j}{\partial z_n} \frac{\partial x_k}{\partial z_p} \mathrm{d}z_m\, \mathrm{d}z_n \epsilon_{ijk} = \frac{1}{\sqrt{G}} \epsilon_{mnp}\, \mathrm{d}z_m\, \mathrm{d}z_n. \tag{3.6.9}$$

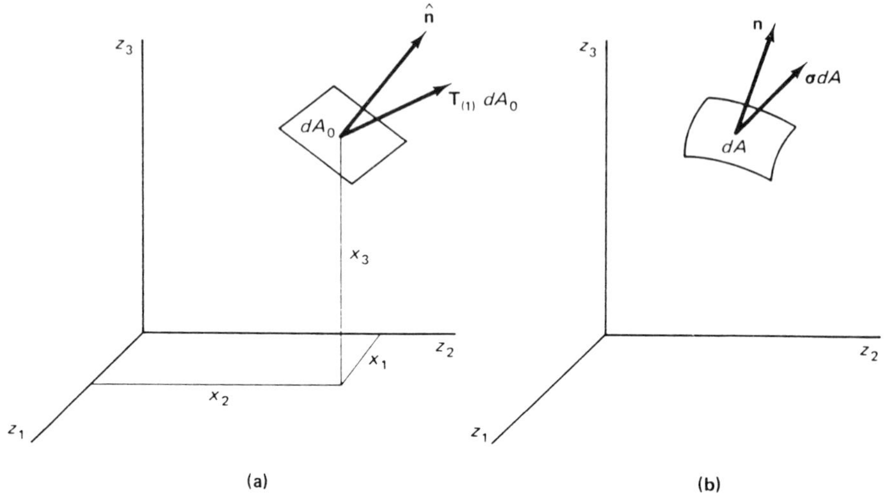

Figure 3.6.4 (a) First Piola–Kirchhoff stress vector measured in terms of the undeformed area, and (b) Cauchy stress vector measured in terms of the deformed area.

Similarly,

$$\mathbf{n}\,\mathrm{d}A = \mathrm{d}z_i\,\mathrm{d}z_j\epsilon_{ijk}\mathbf{i}_k$$

or

$$n_k\,\mathrm{d}A = \mathrm{d}z_i\,\mathrm{d}z_j\epsilon_{ijk}. \tag{3.6.10}$$

The substitution of Eq. (3.6.10) into Eq. (3.6.9) yields

$$\frac{\partial x_k}{\partial z_i}\hat{n}_k\,\mathrm{d}A_0 = \frac{1}{\sqrt{G}}n_i\,\mathrm{d}A,$$

which gives

$$n_i\,\mathrm{d}A = \sqrt{G}\frac{\partial x_k}{\partial z_i}\hat{n}_k\,\mathrm{d}A_0. \tag{3.6.11}$$

Let $\mathbf{T}_{(1)}$ and $\boldsymbol{\sigma}$ be defined as the stress vectors measured in terms of the undeformed and deformed areas, respectively. Then

$$\mathbf{T}_{(1)}\,\mathrm{d}A_0 = \boldsymbol{\sigma}\,\mathrm{d}A.$$

The foregoing relation can be written as

$$T^{ij}_{(1)}\hat{n}_i\mathbf{G}_j\,\mathrm{d}A_0 = \sigma^{ij}n_i\mathbf{G}_j\,\mathrm{d}A = \sigma^{ij}\mathbf{G}_j\sqrt{G}\frac{\partial x_k}{\partial z_i}\hat{n}_k\,\mathrm{d}A_0,$$

from which we obtain

$$T^{ij}_{(1)} = \sqrt{G}\frac{\partial x_i}{\partial z_k}\sigma^{kj}. \tag{3.6.12}$$

Here, σ^{kj} and $T^{ij}_{(1)}$ are identified as the Cauchy stress tensor and the first Piola–Kirchhoff stress tensor, respectively. Now, let Eq. (3.6.11) be multiplied by $(\partial x_k/\partial z_j)\sigma^{ji}$:

$$n_i\,\mathrm{d}A\frac{\partial x_k}{\partial z_j}\sigma^{ij} = \sqrt{G}\,\frac{\partial x_n}{\partial z_i}\hat{n}_n\,\mathrm{d}A_0\frac{\partial x_k}{\partial z_j}\sigma^{ji}.$$

We write the left-hand side of this equation for the undeformed area as:

$$\frac{\partial x_k}{\partial z_j}T^{ij}_{(1)}\hat{n}_i\,\mathrm{d}A_0 = \sqrt{G}\,\frac{\partial x_n}{\partial z_i}\frac{\partial x_k}{\partial z_j}\sigma^{ji}\hat{n}_n\,\mathrm{d}A_0.$$

Define

$$T^{jk}_{(2)} = \frac{\partial x_k}{\partial z_i}T^{ji}_{(1)},$$

then we may write

$$T^{ik}_{(2)}\hat{n}_i\,\mathrm{d}A_0 = \sqrt{G}\,\frac{\partial x_n}{\partial z_i}\frac{\partial x_k}{\partial z_j}\sigma^{ji}\hat{n}_n\,\mathrm{d}A_0,$$

which leads to

$$T^{ij}_{(2)} = \sqrt{G}\frac{\partial x_i}{\partial z_k}\frac{\partial x_j}{\partial z_n}\sigma^{kn}, \tag{3.6.13}$$

where $T^{ij}_{(2)}$ is called the second Piola–Kirchhoff stress tensor. Thus, the Cauchy stress tensor can be determined from Eq. (3.6.13):

$$\sigma^{ij} = \frac{1}{\sqrt{G}}\frac{\partial z_i}{\partial x_k}\frac{\partial z_j}{\partial x_n}T^{kn}_{(2)}. \tag{3.6.14}$$

Similarly,

$$T^{ij}_{(2)} = \sqrt{G}\left(\frac{\partial z_i}{\partial x_k}\frac{\partial z_j}{\partial x_n}\right)^{-1}\sigma^{kn}. \tag{3.6.15}$$

The substitution of Eq. (3.6.14) into Eq. (3.6.12) yields

$$T^{ij}_{(1)} = \frac{\partial z_j}{\partial x_k}T^{ik}_{(2)}. \tag{3.6.16}$$

Note that the relation given in Eq. (3.6.8) results from the definition in Eq. (3.6.2), whereas the first and second Piola–Kirchhoff stresses are based on

$$\mathrm{d}A_{0i} = \frac{\partial x_k}{\partial z_i}\hat{n}_k\,\mathrm{d}A_0. \tag{3.6.17}$$

The first Piola–Kirchhoff stress tensor is asymmetric and awkward for use in computations. The second Piola–Kirchhoff stress tensor, although symmetric, is still not practical for numerical analysis. The two are, however, useful in the development of constitutive theories. The Kirchhoff stress tensor, as defined by Eq. (3.6.8), is the most convenient in numerical analysis (Oden 1972), because calculations for deformed states can be carried out in terms of undeformed areas only by means of $\sqrt{G}$.

3.7 Equations of Motion for Large Strains

To deal with large strains, we recast the balance laws discussed in Section 3.3 in Lagrangian convective coordinates with the tangent vectors $\mathbf{G}_i$. The algebra required toward this end is identical to Eqs. (3.3.8), (3.3.9), and (3.3.10), with the substitution of $\mathbf{G}_i$ for $\mathbf{g}_i$. Thus, Cauchy's first law of motion becomes

$$\frac{1}{\sqrt{G}}(\sqrt{G}\sigma^{ij}\mathbf{G}_j)_{,i} + \rho\mathbf{F} - \rho\dot{\mathbf{v}} = 0. \tag{3.7.1}$$

Note that this equation is not practical for use in computation with the Cauchy stress tensor. By virtue of Eq. (3.6.8) and

$$\mathbf{F} = F^m\mathbf{G}_m = F^0_m\mathbf{i}_m$$
$$\dot{\mathbf{v}} = \dot{v}^m\mathbf{G}_m = \ddot{u}_m\mathbf{i}_m$$
$$\rho_0 = \sqrt{G}\rho$$

it is possible to rewrite Eq. (3.7.1) in the form

$$(t^{ij}z_{m,j})_{,i} + \rho_0 F^0_m - \rho_0\ddot{u}_m = 0, \tag{3.7.2}$$

or

$$[t^{ij}(\delta_{mj} + u_{m,j})]_{,i} + \rho_0 F^0_m - \rho_0\ddot{u}_m = 0.$$

Finally, we have an alternate Cauchy's first law of motion,

$$t^{im}_{,i} + t^{ij}_{,i}u_{m,j} + t^{ij}u_{m,ji} + \rho_0 F^0_m - \rho_0\ddot{u}_m = 0, \tag{3.7.3}$$

which contains nonlinear terms in each direction. Note also that in terms of the Kirchhoff stress tensor t^{ij} we are able to describe stresses on the deformed state in terms of the undeformed area. This is the reason why the physical components of σ^{ij} in the deformed curvilinear coordinates need not be evaluated, with σ^{ij} being replaced by t^{ij}. The nonlinear behavior due to large deformations is characterized by those terms with the first and second derivatives of displacements.

Remarks

The groundwork required for large strains as well as small strains introduced in this chapter gives rise to rigorous applications in solid mechanics. Furthermore, the use of Eulerian coordinates as well as Lagrangian coordinates facilitates the understanding of formulations in fluid mechanics in contrast to solid mechanics. With these tools on hand, we are now prepared to explore fundamental equations and the corresponding physical behavior both in solids and fluids. Specifically, we investigate how stresses and strains are related through material properties in constructing theories of constitutive equations for elastic solids and thermomechanical properties. Similarly, for fluids, we are concerned with stresses and the rate-of-deformation tensors in the derivation of the governing equations of momentum and energy related to various types of flows. These topics are presented in subsequent chapters.

Problems

3.1 Derive Cauchy's first law of motion in Cartesian coordinates and verify this law by summing the forces in all three directions on an infinitesimal cube in equilibrium. Hint: Identify all force components in terms of stresses acting on differential areas as they vary from one face to another ($\sigma_{xx}\,\mathrm{d}y\,\mathrm{d}z$ on one face and $(\sigma_{xx} + \partial\sigma_{xx}/\partial x\,\mathrm{d}x)\,\mathrm{d}y\,\mathrm{d}z$ on the other, for example) plus the inertia forces ($\rho\ddot{u}_x\,\mathrm{d}x\,\mathrm{d}y\,\mathrm{d}z$, etc.) and body forces ($\rho F_x\,\mathrm{d}x\,\mathrm{d}y\,\mathrm{d}z$, etc.) acting inside the cube.

3.2 Derive Cauchy's first law of motion in curvilinear coordinates and verify that

$$\frac{1}{\sqrt{g}}(\sqrt{g}\,\sigma^{ij}\mathbf{g}_j)_{,i} = \sigma^{ij}_{|i}\mathbf{g}_j.$$

3.3 Derive Cauchy's second law of motion in curvilinear coordinates.

3.4 Verify the results of Example 3.1 for cylindrical coordinates. Show the complete details involved in this proof.

3.5 Repeat Problem 3.4 for spherical coordinates.

3.6 Let the components of stress with reference to a set of x_i axes be given by

$$\sigma_{ij} = \begin{bmatrix} 1 & 2 & 1 \\ 2 & 4 & 2 \\ 1 & 2 & 6 \end{bmatrix}.$$

(a) Calculate the trace of the deviatoric stress tensor and show that this corresponds to the pure shear in which $\sigma^*_{ii} = 0$.

(b) Calculate the principal stress invariants and the principal stress components.

(c) Determine the principal directions and check the correctness of the results by recovering the principal stresses through coordinate transformations.

3.7 Prove that the maximum shear stresses are

$$S_{(1)} = \tfrac{1}{2}(\sigma_{(2)} - \sigma_{(3)})$$
$$S_{(2)} = \tfrac{1}{2}(\sigma_{(3)} - \sigma_{(1)})$$
$$S_{(3)} = \tfrac{1}{2}(\sigma_{(1)} - \sigma_{(2)})$$

and that they act on the planes 45° from the principal planes. Show also that shears are zero on the principal planes.

3.8 Show the detailed steps required to determine the relationship between the Cauchy and Kirchhoff stresses and derive Cauchy's first law of motion in terms of the Kirchhoff stress tensor.

3.9 Verify Eq. (3.7.3) with complete details. Rewrite Eq. (3.7.3) with standard notation (x, y, z, u, v, w, $t^{11} = t^{xx}$, $t^{12} = t^{xy}$, etc.).

4

Elastic Solids

In the previous chapters, we discussed the concepts of strain and stress without focusing on any particular material. In this chapter, we consider an elastic material. By elastic we mean that the material may be deformed but will return to its original configuration upon release of the applied loads. On the contrary, plastic materials may not return to their original positions upon the release of applied loads. Other types of materials, known as viscoelastic, exhibit time-dependent properties. Plasticity and viscoelasticity related to these inelastic materials are treated in Chung (1988) and are not covered in this book.

We begin with constitutive equations of linear elastic solids in Section 4.1, followed by Navier equations, energy principles, and the thermodynamics of solids in Sections 4.2–4. Special topics, such as finite elasticity (nonlinear elasticity), torsional stress, and fiber composites, are presented in Sections 4.5–7.

4.1 Constitutive Equations for Linear Elastic Solids

4.1.1 Three-dimensional Solids

In the theory of linear elasticity, we are concerned with an ideal material governed by Hooke's law. This law was proposed by Robert Hooke in 1678 in his essay "*Ut tensio sic vis*" as the "power of springy body is in the same proportion as the extension." Thus, the stress σ and the strain γ in one dimension are related by Young's modulus E,

$$\sigma = E\gamma,$$

which represents a linear relationship between the stress and strain. A material that obeys Hooke's law is referred to as Hookean material.

A generalized three-dimensional Hooke's law may be derived by assuming the existence of an elastic potential W which is invariant with coordinate transformations of the strain tensor γ_{ij}

$$W = W(\gamma_{ij}). \tag{4.1.1}$$

Expanding this function in Taylor series about $\gamma_{ij} = 0$, with $E_{ij} = \partial W/\partial\gamma_{ij}$, $E_{ijkm} = \partial^2 W/\partial\gamma_{ij}\partial\gamma_{km}$, $E_{ijkmnp} = \partial^3 W/\partial\gamma_{ij}\partial\gamma_{km}\partial\gamma_{np}$, etc., we obtain Clapeyron's formula,

$$W = W_0 + E_{ij}\gamma_{ij} + \tfrac{1}{2}E_{ijkm}\gamma_{ij}\gamma_{km} + \frac{1}{3!}E_{ijkmnp}\gamma_{ij}\gamma_{km}\gamma_{np} + \ldots, \tag{4.1.2}$$

where W_0 is a constant and E_{ij}, E_{ijkm}, E_{ijkmnp}, etc., denote tensorial properties required to maintain the invariant properties of W. Physically, the second term represents the energy due to residual stresses, the third term refers to the so-called *strain energy* which correspond to linear elastic deformations, and the fourth term is indicative of nonlinear behavior.

Consider δW, which is regarded as a small change in energy due to $\delta\gamma_{ij}$, as a small change in strain, related as

$$\delta W = \sigma_{ij}\delta\gamma_{ij}, \tag{4.1.3a}$$

where the symbol δ denotes the *virtual*, *incremental*, or simply small variation. The physical quantity implied in this relation may also be defined from Eq. (4.1.1) as

$$\delta W = \frac{\partial W}{\partial\gamma_{ij}}\delta\gamma_{ij}. \tag{4.1.3b}$$

Subtract Eq. (4.1.3b) from Eq. (4.1.3a) to yield

$$\left(\sigma_{ij} - \frac{\partial W}{\partial\gamma_{ij}}\right)\delta\gamma_{ij} = 0.$$

Since $\delta\gamma_{ij}$ is arbitrary, it is obvious that

$$\sigma_{ij} = \frac{\partial W}{\partial\gamma_{ij}}. \tag{4.1.4a}$$

Substitute Eq. (4.1.2) into Eq. (4.1.4a) to obtain

$$\sigma_{rs} = \frac{\partial W}{\partial\gamma_{rs}} = \frac{\partial}{\partial\gamma_{rs}}(E_{ij}\gamma_{ij} + \ldots) = E_{ij}\delta_{ir}\delta_{js} + \ldots = E_{rs} + \ldots.$$

With the indices r and s now replaced by i and j, respectively, we have

$$\sigma_{ij} = E_{ij} + E_{ijkm}\gamma_{km} + \tfrac{1}{2}E_{ijkmnp}\gamma_{km}\gamma_{np} + \ldots, \tag{4.1.4b}$$

where E_{ij} may be taken as zero at time $t = 0$ for the unstrained condition. The stress tensor results from a partial derivative of an elastic potential with respect to the strain tensor. The relation expressed in Eqs. (4.1.4a and b) is regarded as a general form of a constitutive equation for three-dimensional solids. For linear elasticity, the nonlinear terms in Eq. (4.1.4b) are neglected. Thus, the generalized Hooke's law for small strain in linear elasticity is

$$\sigma_{ij} = E_{ijkm}\gamma_{km}, \tag{4.1.5}$$

in which the linear part of the strain tensor for small strains is given by the first two terms on the right-hand side of Eq. (2.4.5),

$$\gamma_{ij} = \tfrac{1}{2}(u_{i,j} + u_{j,i}).$$

Here, E_{ijkm} is a 9×9 matrix and has a total of 81 constants ($d^n = 3^4 = 81$, with $d =$ the number of dimensions, and $n =$ orders of tensor) and must be symmetric, due to Cauchy's second law of motion. For anisotropic material, E_{ijkm} has the array of $6 \times 6 = 36$ as dictated by Eq. (4.1.5),

$$E_{ijkm} = \begin{bmatrix} E_{1111} & E_{1122} & E_{1133} & E_{1112} & E_{1123} & E_{1131} \\ & E_{2222} & E_{2233} & E_{2212} & E_{2223} & E_{2231} \\ & & E_{3333} & E_{3312} & E_{3323} & E_{3331} \\ symm. & & & E_{1212} & E_{1223} & E_{1231} \\ & & & & E_{2323} & E_{2331} \\ & & & & & E_{3131} \end{bmatrix}, \tag{4.1.6}$$

which indicates that only 21 coefficients are needed to characterize anisotropic Hookean material in general.

By means of coordinate transformation, we are able to relate the material properties in one coordinate system (old) x_i to another coordinate system (new) x'_i. Thus, by substituting Eq. (2.6.3) into the quadratic terms of Eq. (4.1.2), which correspond to the linear elastic behavior, we obtain

$$W = \tfrac{1}{2}E_{rstu}\gamma_{rs}\gamma_{tu} = \tfrac{1}{2}E_{rstu}\,a_{ir}\,a_{js}\,a_{kt}\,a_{mu}\gamma'_{ij}\gamma'_{km}. \tag{4.1.7}$$

On the other hand, the elastic potential for the new coordinates can be written as

$$W = \tfrac{1}{2}E_{ijkm}\gamma'_{ij}\gamma'_{km}. \tag{4.1.8}$$

Equations (4.1.7) and (4.1.8) together yield

$$E_{ijkm} = a_{ir}\,a_{js}\,a_{kt}\,a_{mu}E_{rstu}. \tag{4.1.9}$$

This relationship implies that the fourth-order tensor of material constants on old coordinates may be transformed into a new coordinate system through an eighth-order tensor $a_{ir}\,a_{js}\,a_{kt}\,a_{mu}$.

Monotropic Material

If the material is symmetric with respect to one plane (Fig. 4.1.1a), then the following coordinate transformation will apply:

$$x_i' = a_{ij}x_j,$$

where

$$a_{ij} = \begin{bmatrix} 1 & 0 & 0 \\ 0 & 1 & 0 \\ 0 & 0 & -1 \end{bmatrix}, \qquad (4.1.10)$$

in which the negative sign for a_{33} refers to the symmetry of the mirror

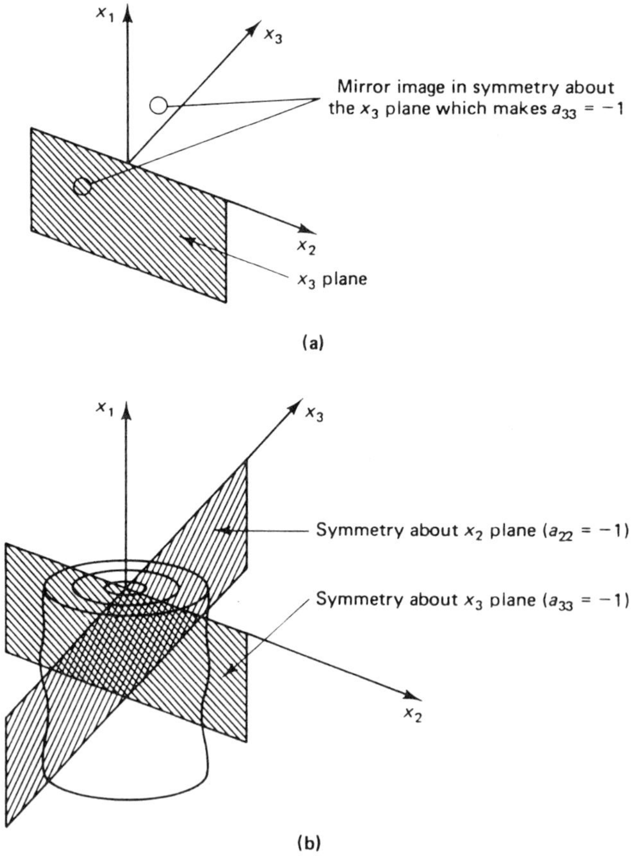

Figure 4.1.1 (a) Monotropic material: x_3 plane of symmetry, no symmetry about x_1 and x_2 planes. (b) Orthotropic material: x_2 and x_3 planes of symmetry, no symmetry about x_1 plane.

image with respect to the x_3 plane. Note that $a_{ij} = \delta_{ij}$ would have made both sides of Eq. (4.1.9) remain equal, which is the case of general anisotropy given by Eq. (4.1.6). Upon substituting Eq. (4.1.10) into Eq. (4.1.9) and expanding all terms with repeated indices, we find that E_{ijkm} is of the form

$$E_{ijkm} = \begin{bmatrix} E_{1111} & E_{1122} & E_{1133} & E_{1112} & 0 & 0 \\ & E_{2222} & E_{2233} & E_{2212} & 0 & 0 \\ & & E_{3333} & E_{3312} & 0 & 0 \\ symm. & & & E_{1212} & 0 & 0 \\ & & & & E_{2323} & E_{2331} \\ & & & & & E_{3131} \end{bmatrix}. \tag{4.1.11}$$

A material such as the one just characterized is known as monotropic material. Equation (4.1.11) shows that for monotropic materials there are 13 nonzero coefficients. Note that in Eq. (4.1.11) all terms of E_{ijkm} with the index 3 occurring an odd number of times are set equal to zero. These terms are made equal to zero to assure the validity of Eq. (4.1.9). For example, $E_{1123} = a_{1r}a_{1s}a_{2t}a_{3u}E_{rstu} = a_{11}a_{11}a_{22}a_{33}E_{1123} = (1)(1)(1)(-1)E_{1123} = -E_{1123}$. Obviously, this is not possible, and the only way the relation expressed in Eq. (4.1.9) can remain valid is if E_{1123} vanishes. All other zero entries in Eq. (4.1.11) can be verified in a similar manner.

Orthotropic Material

Material that is symmetric with respect to two planes (Fig. 4.1.1b) is called *orthotropic*. For example, if the x_2 and x_3 planes are chosen for symmetry, then

$$a_{ij} = \begin{bmatrix} 1 & 0 & 0 \\ 0 & -1 & 0 \\ 0 & 0 & -1 \end{bmatrix}. \tag{4.1.12}$$

If we proceed in a manner similar to that for the monotropic material, we obtain

$$E_{ijkm} = \begin{bmatrix} E_{1111} & E_{1122} & E_{1133} & 0 & 0 & 0 \\ & E_{2222} & E_{2233} & 0 & 0 & 0 \\ & & E_{3333} & 0 & 0 & 0 \\ symm. & & & E_{1212} & 0 & 0 \\ & & & & E_{2323} & 0 \\ & & & & & E_{3131} \end{bmatrix}. \tag{4.1.13}$$

Here, all terms of E_{ijkm} with the indices 3 and 2 occurring an odd number of times are again set equal to zero to satisfy Eq. (4.1.9). The proof is left to the reader. Note that, in this case, there are nine nonzero coefficients to be characterized. Wood is usually considered an orthotropic material.

Transversely Isotropic Material

A material is *transversely isotropic* if there is a preference for directions normal to all but one of the three axes. If this axis is x_3, then through right-handed counterclockwise rotations about the x_3 axis we must have (see Figs. 1.2.2 and 4.1.2)

$$a_{ij} = \begin{bmatrix} \cos\theta & \sin\theta & 0 \\ -\sin\theta & \cos\theta & 0 \\ 0 & 0 & 1 \end{bmatrix}. \tag{4.1.14}$$

This transformation is required in addition to two previous transformations required for monotropic and orthotropic materials. Substituting Eq. (4.1.13) into Eq. (4.1.9), we find

$$E_{1111} = (\cos^4\theta)E_{1111} + (\cos^2\theta\sin^2\theta)(2E_{1122} + 4E_{1212}) + (\sin^4\theta)E_{2222},$$

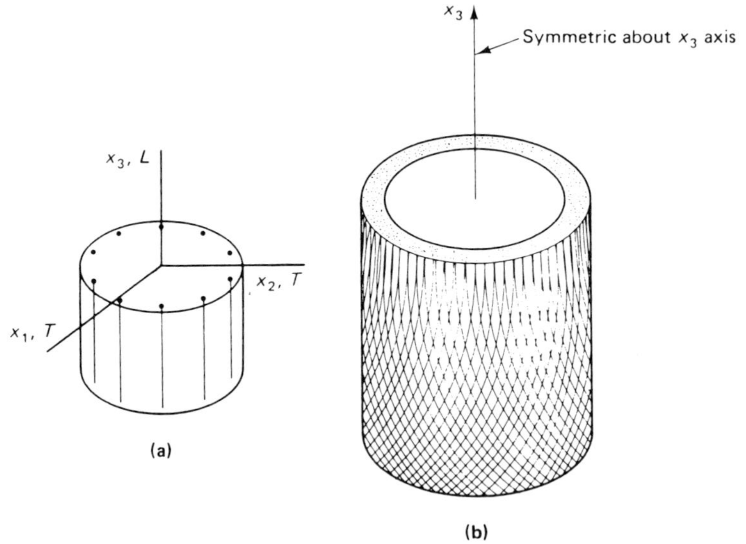

Figure 4.1.2 Transversely isotropic fiber-reinforced composites. (a) Properties in all transverse (T) directions are symmetric about the longitudinal (L) axis. (b) Fiber composites wrapped symmetrically about the x^3 axis.

$$
\begin{aligned}
E_{1122} &= (\cos^2\theta\sin^2\theta)E_{1111} + (\cos^4\theta)E_{1122} - 4(\cos^2\theta\sin^2\theta)E_{1212} \\
&\quad + (\sin^4\theta)E_{2211} + (\sin^2\theta\cos^{(2)}\theta)E_{2222}, \\
E_{1133} &= (\cos^2\theta)E_{1133} + (\sin^2\theta)E_{2233}, \\
E_{2222} &= (\sin^4\theta)E_{1111} + (\sin^2\theta\cos^2\theta)(2E_{1122} + 4E_{1212}) + (\cos^4\theta)E_{2222}, \\
E_{1212} &= (\cos^2\theta\sin^2\theta)E_{1111} - 2(\cos^2\theta\sin^2\theta)E_{1122} \\
&\quad - 2(\cos^2\theta\sin^2\theta)E_{1212} + (\cos^4\theta)E_{1212} + (\sin^2\theta\cos^2\theta)E_{2222} \\
&\quad + (\sin^4\theta)E_{1212}, \\
&\vdots
\end{aligned}
$$

etc.

These results will require that

$$E_{1111} = E_{2222}, \tag{4.1.15a}$$

$$E_{1133} = E_{2233}, \tag{4.1.15b}$$

$$E_{2323} = E_{3131}, \tag{4.1.15c}$$

$$E_{1212} = \tfrac{1}{2}(E_{1111} - E_{1122}). \tag{4.1.15d}$$

In view of these results, we obtain

$$
E_{ijkm} =
\begin{bmatrix}
E_{1111} & E_{1122} & E_{1133} & 0 & 0 & 0 \\
 & E_{2222} & E_{2233} & 0 & 0 & 0 \\
 & & E_{3333} & 0 & 0 & 0 \\
symm. & & & \tfrac{1}{2}(E_{1111} - E_{1122}) & 0 & 0 \\
 & & & & E_{2323} & 0 \\
 & & & & & E_{3131}
\end{bmatrix}.
\tag{4.1.16}
$$

Thus, there are five independent coefficients involved in transversely isotropic materials.

Isotropic Material

If the material is symmetric with respect to every plane and every axis, then the elastic properties are identical in all directions. Material that exhibits such a property is said to be *isotropic* and possesses only two nonzero coefficients.

To characterize an isotropic material, we require coordinate transformations with rotations about the x_2 and x_1 axes in addition to all previous coordinate transformations. This process will enforce symmetry about all planes and all axes. We begin with the right-handed counterclockwise rotation about the x_2 axis,

$$a_{ij} = \begin{bmatrix} \cos\theta & 0 & -\sin\theta \\ 0 & 1 & 0 \\ \sin\theta & 0 & \cos\theta \end{bmatrix}. \tag{4.1.17}$$

We follow a procedure similar to the case of transversely isotropic material, to obtain

$$\begin{aligned} E_{1111} &= E_{3333}, \\ E_{3131} &= \tfrac{1}{2}(E_{1111} - E_{1133}). \end{aligned} \tag{4.1.18}$$

Now, with the last counterclockwise rotation about the x_1 axis given by

$$a_{ij} = \begin{bmatrix} 1 & 0 & 0 \\ 0 & \cos\theta & \sin\theta \\ 0 & -\sin\theta & \cos\theta \end{bmatrix}$$

it follows that

$$E_{1122} = E_{1133}, \tag{4.1.19}$$

$$E_{3131} = \tfrac{1}{2}(E_{3333} - E_{1133}), \tag{4.1.20}$$

$$E_{2323} = \tfrac{1}{2}(E_{2222} - E_{2233}). \tag{4.1.21}$$

From these and previous results, we have

$$E_{ijkm} = \begin{bmatrix} E_{1111} & E_{1122} & E_{1133} & 0 & 0 & 0 \\ & E_{2222} & E_{2233} & 0 & 0 & 0 \\ & & E_{3333} & 0 & 0 & 0 \\ symm. & & & \mathrm{a} & 0 & 0 \\ & & & & \mathrm{b} & 0 \\ & & & & & \mathrm{c} \end{bmatrix}, \tag{4.1.22}$$

with $\mathrm{a} = \frac{1}{2}(E_{1111} - E_{1122})$, $\mathrm{b} = \frac{1}{2}(E_{2222} - E_{2233})$, $\mathrm{c} = \frac{1}{2}(E_{3333} - E_{1133})$.

If we denote $E_{1122} = E_{1133} = E_{2233} = \lambda$ and $E_{1212} = E_{2323} = E_{3131} = \mu$, then from Eqs. (4.1.19)–(4.1.21) we find that $E_{1111} = E_{2222} = E_{3333} = \lambda + 2\mu$. Substitute these definitions into Eq. (4.1.22) to yield

$$E_{ijkm} = \lambda\delta_{ij}\delta_{km} + \mu(\delta_{ik}\delta_{jm} + \delta_{im}\delta_{kj}). \tag{4.1.23}$$

These relations indicate that there are only two constants (λ and μ) involved in isotropic material.

Note that from a strictly mathematical viewpoint the most general form of an isotropic tensor of order 4 would have been

$$E_{ijkm} = \lambda\delta_{ij}\delta_{km} + \mu(\delta_{ik}\delta_{jm} + \delta_{im}\delta_{kj}) + \xi(\delta_{ik}\delta_{jm} - \delta_{im}\delta_{kj}). \tag{4.1.24}$$

However, due to symmetry in either ij or km, with the ξ terms being antisymmetric, we must set $\xi = 0$. As a result, we arrive at the relation given by Eq. (4.1.23), which represents the physical consequences of isotropy.

It is important to recognize that the physical expression for an isotropic material derived from Eqs. (4.1.1)–(4.1.22), characterized by a means of energy invariant through an elastic potential function and, finally, by the fourth-order tensor Eq. (4.1.22), coincides with the mathematical expression that depicts the fourth-order isotropic tensor, Eq. (4.1.24), but excludes the nonphysical antisymmetric terms.

The constants λ and μ in Eq. (4.1.23) are known as Lamé constants. To derive explicit forms of these constants, we proceed as follows. On substituting Eq. (4.1.23) into Eq. (4.1.5), we obtain

$$\sigma_{ij} = [\lambda\delta_{ij}\delta_{km} + \mu(\delta_{ik}\delta_{jm} + \delta_{im}\delta_{jk})]\gamma_{km}, \tag{4.1.25a}$$

or

$$\sigma_{ij} = \lambda\gamma_{kk}\delta_{ij} + 2\mu\gamma_{ij}. \tag{4.1.25b}$$

For $i = j$, expression (4.1.25) takes the form

$$\sigma_{ii} = (3\lambda + 2\mu)\gamma_{ii}. \tag{4.1.26}$$

Then substitute Eq. (4.1.26) into (4.1.25) and solve for γ_{ij}, which yields

$$\gamma_{ij} = \frac{1}{2\mu}\sigma_{ij} - \frac{\lambda\delta_{ij}}{2\mu(3\lambda + 2\mu)}\sigma_{kk}, \tag{4.1.27}$$

where λ and μ are known as Lamé constants,

$$\mu = \frac{E}{2(1 + \nu)} \tag{4.1.28a}$$

and

$$\lambda = \frac{\nu E}{(1 + \nu)(1 - 2\nu)}, \tag{4.1.28b}$$

with E and ν being the Young modulus of elasticity and the Poisson ratio, respectively. Here μ is known also as the *modulus of rigidity* or the *shear modulus*. Expand Eqs. (4.1.25b) and (4.1.27) with Eqs. (4.1.28a and b) to obtain

$$\begin{bmatrix} \sigma_{11} \\ \sigma_{22} \\ \sigma_{33} \\ \sigma_{12} \\ \sigma_{23} \\ \sigma_{31} \end{bmatrix} = \mathrm{a} \begin{bmatrix} 1 & \mathrm{b} & \mathrm{b} & 0 & 0 & 0 \\ \mathrm{b} & 1 & \mathrm{b} & 0 & 0 & 0 \\ \mathrm{b} & \mathrm{b} & 1 & 0 & 0 & 0 \\ 0 & 0 & 0 & \mathrm{c} & 0 & 0 \\ 0 & 0 & 0 & 0 & \mathrm{c} & 0 \\ 0 & 0 & 0 & 0 & 0 & \mathrm{c} \end{bmatrix} \begin{bmatrix} \gamma_{11} \\ \gamma_{22} \\ \gamma_{33} \\ 2\gamma_{12} \\ 2\gamma_{23} \\ 2\gamma_{31} \end{bmatrix} \tag{4.1.29a}$$

with

$$\mathrm{a} = \frac{E(1-\nu)}{(1+\nu)(1-2\nu)}, \ \mathrm{b} = \frac{\nu}{1-\nu}, \ \mathrm{c} = \frac{1-2\nu}{2(1-\nu)}$$

and

$$\begin{bmatrix} \gamma_{11} \\ \gamma_{22} \\ \gamma_{33} \\ \gamma_{12} \\ \gamma_{23} \\ \gamma_{31} \end{bmatrix} = \frac{1}{E}\begin{bmatrix} 1 & -\nu & -\nu & 0 & 0 & 0 \\ -\nu & 1 & -\nu & 0 & 0 & 0 \\ -\nu & -\nu & 1 & 0 & 0 & 0 \\ 0 & 0 & 0 & 1+\nu & 0 & 0 \\ 0 & 0 & 0 & 0 & 1+\nu & 0 \\ 0 & 0 & 0 & 0 & 0 & 1+\nu \end{bmatrix}\begin{bmatrix} \sigma_{11} \\ \sigma_{22} \\ \sigma_{33} \\ \sigma_{12} \\ \sigma_{23} \\ \sigma_{31} \end{bmatrix}. \tag{4.1.29b}$$

Here, the 6×6 matrices in Eqs. (4.1.29a and b) are called the elasticity matrix $[E]$ and the compliance matrix $[C]$, respectively, with $[E] = [C]^{-1}$. The reader will recall that these results are introduced in elementary mechanics courses simply as stress-strain relations, without the rigorous derivations presented here.

It is interesting to note that in terms of deviatoric strains and stresses, the expressions given by Eqs. (4.1.25) and (4.1.27) can be shown to be of the form

$$\sigma_{ij}^* = 2\mu\gamma_{ij}^*,$$

which indicates that deviatoric stresses are related to deviatoric strains only through the shear modulus μ.

Consider now the mean hydrostatic pressure defined as

$$p = \frac{-\sigma_{kk}}{3} = -\kappa\gamma_{kk} = -(\lambda + \tfrac{2}{3}\mu)\gamma_{kk}, \tag{4.1.30}$$

where the negative sign indicates the mean hydrostatic pressure in compression. The quantity κ is known as the *bulk modulus*

$$\kappa = \lambda + \tfrac{2}{3}\mu. \tag{4.1.31}$$

Substituting Eqs. (4.1.28a and b) into Eq. (4.1.30) gives the expression for mean pressure in the form

$$p = \frac{-E}{3(1-2\nu)}\gamma_{kk}.$$

At this point, we demonstrate that the Lamé constants λ and μ may be derived deductively, as guided by laboratory measurement data, namely, Young's modulus and the Poisson ratio. Toward this end, let us consider a simple case of an axial tensile loading of a bar in the x_1

direction. Then, we have

$$\sigma_{11} = \sigma, \quad \text{with all other } \sigma_{ij} = 0.$$

In view of Eq. (4.1.27), all strain components are written as

$$\gamma_{11} = \frac{\lambda + \mu}{\mu(3\lambda + 2\mu)}\sigma, \quad \gamma_{22} = \gamma_{33} = -\frac{\lambda}{2\mu(3\lambda + 2\mu)}\sigma,$$
$$\gamma_{12} = \gamma_{23} = \gamma_{13} = 0.$$

From the elementary relations

$$\gamma_{11} = \frac{\sigma}{E}, \quad \nu = -\frac{\gamma_{22}}{\gamma_{11}} = -\frac{\gamma_{33}}{\gamma_{11}}$$

it follows that

$$\frac{1}{E} = \frac{\lambda + \mu}{\mu(3\lambda + 2\mu)}, \quad \nu = \frac{\lambda}{2(\lambda + \mu)}.$$

In the case of a simple shear in the x_1x_3 and x_2x_3 planes, we have

$$\sigma_{21} = \sigma_{12} = \tau, \quad \text{all other } \sigma_{ij} = 0.$$

Thus, once again from Eq. (4.1.27),

$$2\gamma_{12} = \frac{\tau}{\mu}.$$

If a body is subjected to a uniform pressure spherically, we have

$$\sigma_{ij} = -p\delta_{ij}.$$

By virtue of Eq. (4.1.27), the dilatational strain γ_{ii} becomes

$$\gamma_{ii} = \frac{\sigma_{ii}}{3\lambda + 2\mu} = -\frac{3p}{3\lambda + 2\mu} = -\frac{p}{\kappa},$$

which gives the relation

$$\kappa = \lambda + \frac{2\mu}{3}$$

and, subsequently,

$$\lambda = \frac{\nu E}{(1 + \nu)(1 - 2\nu)}, \quad \mu = \frac{E}{2(1 + \nu)}, \quad \kappa = \frac{E}{3(1 - 2\nu)}.$$

Note that these results agree with the definitions of Lamé constants given in Eq. (4.1.28).

An incompressible material has the bulk modulus $\kappa \to \infty$, which arises when $\nu = 1/2$. For a *stable material*, we must have $\kappa > 0$, $\mu > 0$, and $E > 0$, because these constants are involved in the elastic potential function, and positive work is required to cause any deformation. Therefore, the Poisson ratio must be in the range $-1 < \nu < 1/2$. Here,

$\nu = \frac{1}{2}$ implies an incompressible material and $\nu = 0$ represents a one-dimensional deformation. Physically, however, no material is likely to have the Poisson ratio less than zero, thus leading to $0 \leq \nu \leq \frac{1}{2}$.

4.1.2 Plane Stress and Plane Strain

In many engineering applications, three-dimensional problems may be idealized as two-dimensional, or plane, problems. If one of the dimensions is small in comparison with the other dimensions, then the stress in the direction of the small dimension (assume that will be x_3) is negligible. The state of stress in this case is called *plane stress*. Thus, we may simply ignore the small dimension and perform our analysis for the two-dimensional plane of the larger dimensions only. On the other hand, if one dimension is extremely large in comparison with the other two dimensions (again, assume x_3), then it is possible that the strain in this direction will be negligible. This condition is referred to as *plane strain*. We may then be able to perform two-dimensional analysis on a sliced plane with unit thickness along the x_3 axis. Plane problems of both types are illustrated in Fig. 4.1.3.

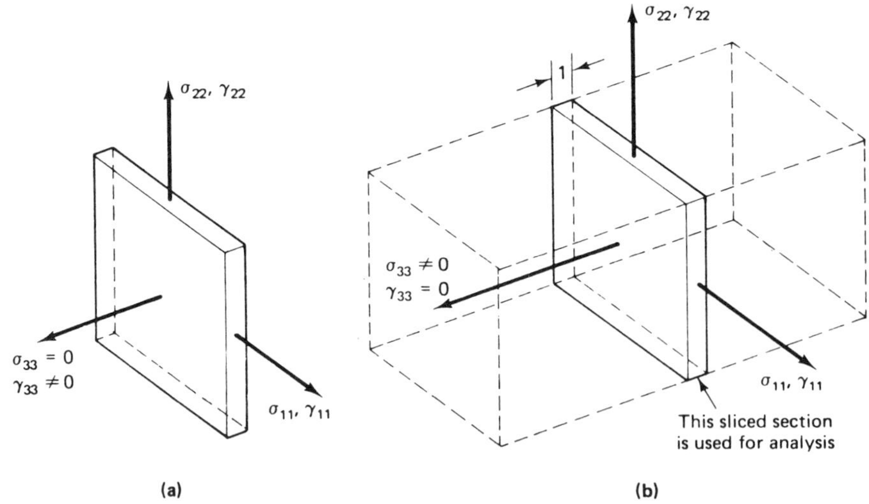

Figure 4.1.3 Plane problems with applied loads remaining in a plane: (a) Plane stress–thickness in the x_3 direction is small in comparison with other directions ($\sigma_{33} = 0$, $\gamma_{33} \neq 0$). (b) Plane strain–thickness in the x_3 direction is large in comparison with other directions so that a plane section with unit thickness in the 3–3 direction may be used for the analysis by setting $\gamma_{33} = 0$ and $\sigma_{33} \neq 0$.

For plane stress, the stress-strain relation given by Eq. (4.1.25) is modified by the conditions

$$\sigma_{33} = \sigma_{31} = \sigma_{32} = 0.$$

Thus,

$$\sigma_{33} = 0 = \lambda(\gamma_{11} + \gamma_{22} + \gamma_{33}) + 2\mu\gamma_{33},$$

from which

$$\gamma_{33} = \frac{-\lambda}{\lambda + 2\mu}(\gamma_{11} + \gamma_{22}).$$

The substitution of this last equality into Eq. (4.1.25) gives

$$\begin{bmatrix} \sigma_{11} \\ \sigma_{22} \\ \sigma_{12} \end{bmatrix} = \frac{E}{1 - \nu^2} \begin{bmatrix} 1 & \nu & 0 \\ \nu & 1 & 0 \\ 0 & 0 & \dfrac{1 - \nu}{2} \end{bmatrix} \begin{bmatrix} \gamma_{11} \\ \gamma_{22} \\ 2\gamma_{12} \end{bmatrix}. \tag{4.1.32}$$

Notice that the total engineering shear strain is seen as $\gamma_{xy} = 2\gamma_{12}$. The stress-strain relation for plane strain is obtained directly from Eq. (4.1.24) by simply retaining only the two-dimensional components:

$$\begin{bmatrix} \sigma_{11} \\ \sigma_{22} \\ \sigma_{12} \end{bmatrix} = \frac{E(1 - \nu)}{(1 + \nu)(1 - 2\nu)} \begin{bmatrix} 1 & \dfrac{\nu}{1 - \nu} & 0 \\ \dfrac{\nu}{1 - \nu} & 1 & 0 \\ 0 & 0 & \dfrac{1 - 2\nu}{2(1 - \nu)} \end{bmatrix} \begin{bmatrix} \gamma_{11} \\ \gamma_{22} \\ 2\gamma_{12} \end{bmatrix}. \tag{4.1.33}$$

In this case, the structure to be analyzed is the plane in the x_1 and x_2 directions, with a unit dimension in the x_3 direction. Thus, the results of this analysis can apply to any other plane sections in the x_3 direction.

4.2 Navier Equations

For linear elasticity, we invoke Cauchy's first law of motion for the undeformed coordinates given by Eq. (3.3.3b),

$$\sigma_{ij,i} + \rho F_j - \rho \ddot{u}_j = 0 \tag{4.2.1}$$

subject to the following boundary conditions: the Dirichlet or essential boundary condition is

$$u = \bar{u} \quad \text{on } \Gamma_1,$$

and the Neumann or natural boundary condition is

$$\sigma_{ij} n_i = \bar{S}_j \quad \text{on } \Gamma_2.$$

By virtue of the constitutive equations that result from Eq. (4.1.25) and on the exchange of indices between i and j, we arrive at the so-called Navier equation:

$$\mu u_{i,jj} + (\lambda + \mu) u_{j,ji} + \rho F_i - \rho \ddot{u}_i = 0, \tag{4.2.2a}$$

or

$$\mu \nabla^2 \mathbf{u} + (\lambda + \mu) \boldsymbol{\nabla}(\boldsymbol{\nabla} \cdot \mathbf{u}) + \rho \mathbf{F} - \rho \ddot{\mathbf{u}} = 0. \tag{4.2.2b}$$

In the absence of inertial forces, we have

$$\mu u_{i,jj} + (\lambda + \mu) u_{j,ji} + \rho F_i = 0. \tag{4.2.3}$$

For two-dimensional problems with plane strain conditions where $\gamma_{33} = \gamma_{13} = \gamma_{23} = 0$, we have

$$\sigma_{\alpha\beta} = \lambda \gamma_{\eta\eta} \delta_{\alpha\beta} + 2\mu \gamma_{\alpha\beta}$$

with α, β, $\eta = 1, 2$ and $\sigma_{33} = \lambda \gamma_{\eta\eta}$. The Navier equation assumes the form

$$\mu \left(u_{\alpha,\beta\beta} + \frac{1}{1 - 2\nu} u_{\beta,\beta\alpha} \right) + \rho F_\alpha = 0. \tag{4.2.4}$$

For plane stress conditions, however, we have

$$\sigma_{33} = \sigma_{13} = \sigma_{23} = 0$$

and

$$\gamma_{33} = -\frac{\lambda}{2\mu + \lambda}(\gamma_{11} + \gamma_{22}) = -\frac{\nu}{1 - \nu} \gamma_{\eta\eta}.$$

Therefore, the Navier equation for plane stress takes the form

$$\mu \left(u_{\alpha,\beta\beta} + \frac{1 + \nu}{1 - \nu} u_{\beta,\beta\alpha} \right) + \rho F_\alpha = 0. \tag{4.2.5}$$

Solutions of the Navier equations — Eqs. (4.2.2) and (4.2.3) for three-dimensional problems, Eq. (4.2.4) for plane strain, and Eq. (4.2.5) for plane stress — have been the subject of extensive research for more than a century. Various energy principles that facilitate solution procedures may be employed. In the presence of heat energy, the Navier equations must be modified for thermomechanical coupling, which gives rise to the concept of entropy. These and other topics will be presented in the following sections.

Curvilinear Coordinates

The Navier equations may be recast in curvilinear coordinates utilizing the vector form of Eq. (4.2.2b). First, consider $\nabla^2\mathbf{u}$ and $\boldsymbol{\nabla}(\boldsymbol{\nabla}\cdot\mathbf{u})$ in curvilinear coordinates,

$$\begin{aligned}\nabla^2\mathbf{u} &= (\boldsymbol{\nabla}\cdot\boldsymbol{\nabla})\mathbf{u} = \left(\mathbf{g}^k\frac{\partial}{\partial\xi_k}\cdot\mathbf{g}^j\frac{\partial}{\partial\xi_j}\right)\mathbf{u}\\ &= \left(g^{kj}\frac{\partial^2}{\partial\xi_k\partial\xi_j} + \mathbf{g}^k\cdot\mathbf{g}^j_{,k}\frac{\partial}{\partial\xi_j}\right)\mathbf{u}\\ &= \left(g^{kj}\frac{\partial^2}{\partial\xi_k\partial\xi_j} - \mathbf{g}^k\cdot\Gamma^j_{km}g^{mn}\mathbf{g}_n\frac{\partial}{\partial\xi_j}\right)\mathbf{u}\\ &= g^{kj}[(u^i\mathbf{g}_i)_{,kj} - \Gamma^m_{kj}(u^i\mathbf{g}_i)_{,m}] = g^{kj}\mathbf{g}_i u^i_{|kj},\end{aligned} \tag{4.2.6}$$

where

$$u^i_{|kj} = u^i_{,jk} + 2u^m_{,k}\Gamma^i_{mj} + u^m(\Gamma^i_{jm})_{,k} + u^m\Gamma^n_{mj}\Gamma^i_{nk} - u^i_{,m}\Gamma^m_{kj} - u^n\Gamma^i_{nm}\Gamma^m_{kj}$$

and

$$\begin{aligned}\boldsymbol{\nabla}(\boldsymbol{\nabla}\cdot\mathbf{u}) &= \mathbf{g}^k\frac{\partial}{\partial\xi_k}\left(\mathbf{g}^j\frac{\partial}{\partial\xi_j}\cdot u^i\mathbf{g}_i\right) = \mathbf{g}_k\frac{\partial}{\partial\xi_k}(u^i_{,i} + \mathbf{g}^j\cdot u^i\Gamma^m_{ij}\mathbf{g}_m)\\ &= \mathbf{g}^k\frac{\partial}{\partial\xi_k}(u^i_{,i} + u^i\Gamma^j_{ji}) = \mathbf{g}^k[u^i_{,ik} + u^i_{,k}\Gamma^j_{ji} + u^i(\Gamma^j_{ji})_{,k}]\\ &= g^{ki}\mathbf{g}_i u^j_{|jk},\end{aligned} \tag{4.2.7}$$

where

$$u^j_{|jk} = u^j_{,jk} + u^j_{,k}\Gamma^m_{mj} + u^j(\Gamma^m_{mj})_{,k}.$$

Using the physical components for cylindrical coordinates,

$$\Gamma^1_{22} = -r, \quad \Gamma^2_{12} = \Gamma^2_{21} = \frac{1}{r}, \quad g^{11} = g^{33} = 1, \quad g^{22} = \frac{1}{r^2},$$

$$u^1 = u_r, \quad u^2 = \frac{u_\theta}{r}, \quad u^3 = u_z$$

we obtain, for the radial direction ($i = 1$),

$$\begin{aligned}&\mu\left(\frac{\partial^2 u_r}{\partial r^2} + \frac{1}{r}\frac{\partial u_r}{\partial r} + \frac{1}{r^2}\frac{\partial^2 u_r}{\partial\theta^2} - \frac{2}{r^2}\frac{\partial u_\theta}{\partial\theta} - \frac{u_r}{r^2} + \frac{\partial^2 u_r}{\partial z^2}\right)\\ &+ (\lambda+\mu)\left(\frac{\partial^2 u_r}{\partial r^2} + \frac{1}{r}\frac{\partial^2 u_\theta}{\partial\theta\partial r} + \frac{\partial^2 u_z}{\partial r\partial z} - \frac{1}{r^2}\frac{\partial u_\theta}{\partial\theta} + \frac{1}{r}\frac{\partial u_r}{\partial r} - \frac{u_r}{r^2}\right)\\ &+ \rho F_r - \rho\ddot{u}_r = 0.\end{aligned} \tag{4.2.8a}$$

Similarly, for the tangential direction ($i = 2$):

$$\mu\left(\frac{\partial^2 u_\theta}{\partial r^2} + \frac{1}{r}\frac{\partial u_\theta}{\partial r} + \frac{1}{r^2}\frac{\partial^2 u_\theta}{\partial \theta^2} + \frac{2}{r^2}\frac{\partial u_r}{\partial \theta} + \frac{\partial^2 u_\theta}{\partial z^2} - \frac{u_\theta}{r^2}\right)$$
$$+ (\lambda + \mu)\left(\frac{1}{r}\frac{\partial^2 u_r}{\partial r \partial \theta} + \frac{1}{r^2}\frac{\partial^2 u_\theta}{\partial \theta^2} + \frac{1}{r}\frac{\partial^2 u_z}{\partial z \partial \theta} + \frac{1}{r^2}\frac{\partial u_r}{\partial \theta}\right) \tag{4.2.8b}$$
$$+ \rho F_\theta - \rho \ddot{u}_\theta = 0$$

and, finally, for the axial direction ($i = 3$):

$$\mu\left(\frac{\partial^2 u_z}{\partial r^2} + \frac{1}{r}\frac{\partial u_z}{\partial r} + \frac{1}{r^2}\frac{\partial^2 u_z}{\partial \theta^2} + \frac{\partial^2 u_z}{\partial z^2}\right)$$
$$+ (\lambda + \mu)\left(\frac{\partial^2 u_r}{\partial r \partial z} + \frac{1}{r}\frac{\partial^2 u_\theta}{\partial \theta \partial z} + \frac{\partial^2 u_z}{\partial z^2} + \frac{1}{r}\frac{\partial u_r}{\partial z}\right) + \rho F_z - \rho \ddot{u}_z = 0. \tag{4.2.8c}$$

The Navier equations for spherical coordinates may be obtained in a similar manner. The results are as follows.

R direction (radial)

$$\mu\left(\frac{\partial^2 u_R}{\partial R^2} + \frac{1}{R^2}\frac{\partial^2 u_R}{\partial \alpha^2} + \frac{1}{R^2 \sin^2\alpha}\frac{\partial^2 u_R}{\partial \theta^2} - \frac{2}{R^2}\frac{\partial u_\alpha}{\partial \alpha} - \frac{2}{R^2 \sin\alpha}\frac{\partial u_\theta}{\partial \theta}\right.$$
$$\left. - \frac{2u_R}{R^2} - \frac{2u_\alpha \cot\alpha}{R^2} + \frac{2}{R}\frac{\partial u_R}{\partial R} + \frac{\cot\alpha}{R^2}\frac{\partial u_R}{\partial \alpha}\right)$$
$$+ (\lambda + \mu)\left(\frac{\partial^2 u_R}{\partial R^2} - \frac{1}{R^2}\frac{\partial u_\alpha}{\partial \alpha} + \frac{1}{R}\frac{\partial^2 u_\alpha}{\partial \alpha \partial R} + \frac{1}{R^2 \sin\alpha}\frac{\partial u_\theta}{\partial \theta}\right.$$
$$\left. + \frac{1}{R \sin\alpha}\frac{\partial^2 u_\theta}{\partial R \partial \theta} - \frac{2u_R}{R^2} - \frac{\cot\alpha\, u_\alpha}{R^2} + \frac{2}{R}\frac{\partial u_R}{\partial R} + \frac{\cot\alpha}{R}\frac{\partial u_\alpha}{\partial R}\right)$$
$$+ \rho F_R - \rho \ddot{u}_R = 0. \tag{4.2.9a}$$

α direction (meridional)

$$\mu\left(\frac{2}{R}\frac{\partial u_\alpha}{\partial R} + \frac{\partial^2 u_\alpha}{\partial R^2} + \frac{1}{R^2}\frac{\partial^2 u_\alpha}{\partial \alpha^2} + \frac{1}{R^2 \sin^2\alpha}\frac{\partial^2 u_\alpha}{\partial \theta^2} + \frac{2}{R^2}\frac{\partial u_R}{\partial \alpha}\right.$$
$$\left. - \frac{2\cos\alpha}{R^2 \sin^2\alpha}\frac{\partial u_\theta}{\partial \theta} - \frac{u_\alpha}{R^2 \sin^2\alpha} + \frac{\cot\alpha}{R^2}\frac{\partial u_\alpha}{\partial \alpha}\right)$$
$$+ (\lambda + \mu)\left(\frac{1}{R}\frac{\partial^2 u_R}{\partial R \partial \alpha} + \frac{1}{R^2}\frac{\partial^2 u_\alpha}{\partial \alpha^2} - \frac{\cos\alpha}{R^2 \sin^2\alpha}\frac{\partial u_\theta}{\partial \theta} + \frac{1}{R^2 \sin\alpha}\frac{\partial^2 u_\theta}{\partial \alpha \partial \theta}\right.$$
$$\left. + \frac{2}{R^2}\frac{\partial u_R}{\partial \alpha} + \frac{\cot\alpha}{R^2}\frac{\partial u_\alpha}{\partial \alpha} - \frac{u_\alpha}{R^2} - \frac{u_\alpha \cot^2\alpha}{R^2}\right)$$
$$+ \rho F_\alpha - \rho \ddot{u}_\alpha = 0. \tag{4.2.9b}$$

θ direction (tangential)

$$\mu\left(\frac{2}{R}\frac{\partial u_\theta}{\partial R} + \frac{\partial^2 u_\theta}{\partial R^2} + \frac{\cot\alpha}{R^2}\frac{\partial u_\theta}{\partial \alpha} + \frac{1}{R^2}\frac{\partial^2 u_\theta}{\partial \alpha^2} + \frac{1}{R^2\sin^2\alpha}\frac{\partial^2 u_\theta}{\partial \theta^2}\right.$$
$$\left. - \frac{u_\theta}{R^2\sin^2\alpha} + \frac{2}{R^2\sin\alpha}\frac{\partial u_R}{\partial \theta} + \frac{2\cos\alpha}{R^2\sin^2\alpha}\frac{\partial u_\alpha}{\partial \theta}\right)$$
$$+ (\lambda+\mu)\left[\frac{1}{R\sin\alpha}\left(\frac{\partial^2 u_R}{\partial R\partial\theta} + \frac{1}{R}\frac{\partial^2 u_\alpha}{\partial\alpha\partial\theta} + \frac{1}{R\sin\alpha}\frac{\partial^2 u_\theta}{\partial\theta^2}\right.\right.$$
$$\left.\left. + \frac{2}{R}\frac{\partial u_R}{\partial\theta} + \frac{\cot\alpha}{R}\frac{\partial u_\alpha}{\partial\theta}\right)\right] + \rho F_\theta - \rho\ddot{u}_\theta = 0. \qquad (4.2.9c)$$

4.3 Energy Principles

Problems of elasticity are intimately related to the energy principles (or variational principles) which define equilibrium conditions. This section highlights the principles of virtual work and total potential energy. The concept of energy principles is of paramount importance in the solution of elasticity problems in engineering (Oden and Reddy, 1976).

4.3.1 Virtual Work and Total Potential Energy

The variational principle for three-dimensional linear elasticity may be derived by constructing the orthogonal projection of the residual space of governing differential equations upon the subspace spanned by δu_i, where the δu_i's are commonly known as virtual displacements. Denote the residual of Eq. (4.2.3a) as

$$R_j = \sigma_{ij,i} + \rho F_j$$

to construct an inner product, such that the virtual work summed over the domain will vanish:

$$(R_j, \delta u_j) = \int_\Omega R_j \delta u_j \, d\Omega = 0. \qquad (4.3.1a)$$

This expression is an invariant and implies the virtual work δI:

$$\delta I = \int_\Omega (\sigma_{ij,i} + \rho F_j)\delta u_j \, d\Omega = 0. \qquad (4.3.1b)$$

The integration of Eq. (4.3.1b) by parts or the use of the Green–Gauss theorem yields

$$\delta I = \int_\Gamma \sigma_{ij} n_i \delta u_j \, d\Gamma - \int_\Omega \sigma_{ij}\delta u_{j,i} \, d\Omega + \int_\Omega \rho F_j \delta u_j \, d\Omega = 0,$$

with n_i the component of a vector normal to the boundary surface where the surface traction $S_j = \sigma_{ij}n_i$ is applied. Since σ_{ij} and E_{ijkm} are symmetric, the displacement gradient $u_{k,m}$ can be written as a sum of the strain tensor γ_{km} and rotational (spin) tensor $\omega_{km} = \frac{1}{2}(u_{k,m} - u_{m,k})$

$$u_{k,m} = \gamma_{km} + \omega_{km} = \tfrac{1}{2}(u_{k,m} + u_{m,k}) + \tfrac{1}{2}(u_{k,m} - u_{m,k}), \quad (4.3.2a)$$

so that

$$E_{ijkm}u_{k,m} = E_{ijkm}(\gamma_{km} + \omega_{km}) = E_{ijkm}\gamma_{km}, \quad (4.3.2b)$$

where $E_{ijkm}\omega_{km} = 0$. Thus,

$$\delta I = \int_\Omega E_{ijkm}u_{k,m}\delta u_{j,i}\,\mathrm{d}\Omega - \int_\Omega \rho F_j \delta u_j\,\mathrm{d}\Omega - \int_\Gamma \sigma_{ij}n_i\delta u_j\,\mathrm{d}\Gamma = 0,$$

or

$$\delta I = \delta\left(\int_\Omega \tfrac{1}{2}E_{ijkm}u_{k,m}u_{j,i}\,\mathrm{d}\Omega - \int_\Omega \rho F_j u_j\,\mathrm{d}\Omega - \int_\Gamma \sigma_{ij}n_iu_j\,\mathrm{d}\Gamma\right) = 0, \quad (4.3.3)$$

where δ operates as a differential d such that the first term of Eq. (4.3.3) can be shown:

$$\begin{aligned}\delta\int_\Omega \tfrac{1}{2}E_{ijkm}u_{k,m}u_{j,i}\,\mathrm{d}\Omega &= \int_\Omega \tfrac{1}{2}E_{ijkm}\left(\frac{\partial u_{k,m}}{\partial u_{r,s}}\delta u_{r,s}u_{j,i} + u_{k,m}\frac{\partial u_{j,i}}{\partial u_{r,s}}\delta u_{r,s}\right)\mathrm{d}\Omega \\ &= \int_\Omega \tfrac{1}{2}E_{ijkm}(\delta_r^k\delta_s^m\delta u_{r,s}u_{j,i} + u_{k,m}\delta_r^j\delta_s^i\delta u_{r,s})\,\mathrm{d}\Omega \\ &= \int_\Omega \tfrac{1}{2}(E_{ijrs}u_{j,i}\delta u_{r,s} + E_{ijkm}u_{k,m}\delta u_{j,i})\,\mathrm{d}\Omega \\ &= \int_\Omega E_{ijkm}u_{k,m}\delta u_{j,i}\,\mathrm{d}\Omega.\end{aligned}$$

The last step is due to the symmetry of E_{ijrs} and the exchange of indices $r, s \to i, j$; $i, j \to k, m$. We recognize that $\partial u_{k,m}/\partial u_{r,s}$ implies a total of 81 terms of derivatives, which result in nothing but Kronecker deltas. This would have been a difficult task had the power of tensor analysis not been utilized. Here, it is important to note that ρF_j and $\sigma_{ij}n_i$ are independent of variation and only u_j is being subjected to variation. The stationary condition is a requirement of equilibrium, and it dictates that the quantity I must be equated to the integrals inside the parentheses of the right-hand side of Eq. (4.3.3), so that

$$I = \int_\Omega \tfrac{1}{2}\sigma_{ij}\gamma_{ij}\,\mathrm{d}\Omega - \int_\Omega \rho F_j u_j\,\mathrm{d}\Omega - \int_\Gamma \sigma_{ij}n_iu_j\,\mathrm{d}\Gamma. \quad (4.3.4)$$

This is known as the variational energy or the total potential energy, defined as:

$$I = U - W.$$

Here, U is referred to as the strain energy,

$$U = \int_\Omega \tfrac{1}{2}\sigma_{ij}\gamma_{ij}\,\mathrm{d}\Omega \tag{4.3.5}$$

and W is the external work,

$$W = W(b) + W(s),$$

where $W(b)$ and $W(s)$ are the external work due to body forces and surface tractions, respectively:

$$W(b) = \int_\Omega \rho F_j u_j\,\mathrm{d}\Omega \tag{4.3.6a}$$

and

$$W(s) = \int_\Gamma \sigma_{ij} n_i u_j\,\mathrm{d}\Gamma. \tag{4.3.6b}$$

The variational energy expressed in the integral form, Eq. (4.3.4), is often referred to as the *variational principle*. The integrand $(1/2)\sigma_{ij}\gamma_{ij}$ in Eq. (4.3.4), which contain derivatives (both σ_{ij} and γ_{ij} are functions of displacement gradients), is called the *variational functional*. The word 'functional' is used to distinguish this term from a 'function' which does not contain derivatives. The reader may notice that Eq. (4.3.4) is introduced in other textbooks merely as the statement for the variational principle. The derivation given here clarifies the origin.

4.3.2 Boundary Conditions

To verify the existence of boundary conditions required for a given physical problem and to derive formally the explicit forms of boundary conditions in three-dimensional elastic solids, we invoke the concept of invariant energy, as briefly discussed in Chapter 1. The representation given in Eq. (4.3.1) is manifest for the notion of invariant energy, and the results of integration by parts given by Eq. (4.3.3) or (4.3.4) reveal the existence of the Neumann (natural) boundary condition which is physically identified as the surface traction $\sigma_{ij}n_i$. The boundary integral $\int_\Gamma \sigma_{ij}n_i u_j\,\mathrm{d}\Gamma$ represents the energy required for maintaining such boundary conditions. It is important to realize that the derivation of variational energy produces only the natural boundary conditions. Thus, Dirichlet (essential) boundary conditions must be specified additionally for the solution of boundary value problems.

If one intends simply to examine the existence and explicit forms of boundary conditions in general, there is no need to seek the virtual work. Rather, it is sufficient to work with energy created due to actual deformations. Thus, we write

$$(R_j, u_j) = \int_\Omega (\sigma_{ij,i} + \rho F_j) u_j \, \mathrm{d}\Omega.$$

Integrating by parts:

$$\begin{aligned}(R_j, u_j) &= \int_\Gamma \sigma_{ij} n_i u_j \, \mathrm{d}\Gamma - \int_\Omega E_{ijkm} u_{k,m} u_{j,i} \, \mathrm{d}\Omega + \int_\Omega \rho F_j u_j \, \mathrm{d}\Omega \\ &= \int_\Gamma \sigma_{ij} n_i u_j \, \mathrm{d}\Gamma - \int_\Gamma E_{ijkm} u_k n_m u_{j,i} \, \mathrm{d}\Gamma \\ &\quad + \int_\Omega E_{ijkm} u_k u_{j,im} \, \mathrm{d}\Omega + \int_\Omega \rho F_j u_j \, \mathrm{d}\Omega \qquad (4.3.7\mathrm{a}) \\ &= \int_\Gamma (\sigma_{ij} n_i u_j - u_k n_m E_{ijkm} u_{j,i}) \, \mathrm{d}\Gamma \\ &\quad + \int_\Omega u_k E_{ijkm} u_{j,im} \, \mathrm{d}\Omega + \int_\Omega \rho F_j u_j \, \mathrm{d}\Omega.\end{aligned}$$

Here the boundary conditions include the Dirichlet (essential) boundary data u_k in addition to the Neumann boundary data $\sigma_{ij} n_i$.

For the one-dimensional case with a bar of length 1, cross-section area $A = 1$, and $\sigma_{11} = \sigma$, $u_1 = u$, and $F_1 = F$, we have

$$\begin{aligned}(R, u) &= \int_0^1 \left(\frac{\partial \sigma}{\partial x} + \rho F \right) u \, \mathrm{d}x = \sigma u \Big|_0^1 - \int_0^1 \sigma \frac{\partial u}{\partial x} \mathrm{d}x + \int_0^1 \rho F u \, \mathrm{d}x \\ &= \sigma u \Big|_0^1 - u E \frac{\partial u}{\partial x} \Big|_0^1 + \int_0^1 u E \frac{\partial^2 u}{\partial x^2} \mathrm{d}x + \int_0^1 \rho F u \, \mathrm{d}x.\end{aligned} \qquad (4.3.7\mathrm{b})$$

If we consider a bar fixed at $x = 0$ that is undergoing extensional deformation due to its own weight, we set $u = 0$ at $x = 0$, and $\sigma = 0$ at $x = 1$.

Note that each term on the right-hand side in Eq. (4.3.7b) represents energy mobilized to maintain equilibrium:

Neumann boundary condition

$(\sigma) u \Big|_0^1$ – *energy required to maintain the stress-free boundary condition $\sigma = 0$ at $x = 1$.*

Dirichlet boundary condition

$(u) E \frac{\partial u}{\partial x} \Big|_0^1 = (u) \sigma \Big|_0^1$ – *energy required to maintain the boundary condition $u = 0$ at $x = 0$,* where the quantities inside the parentheses imply boundary data. The last two terms on the right-hand side may be combined such that

$$\int_0^1 \left(E\frac{\partial^2 u}{\partial x^2} + \rho F \right) u \, \mathrm{d}x = \int_0^1 \left(\frac{\partial \sigma}{\partial x} + \rho F \right) u \, \mathrm{d}x .$$

Thus, mathematically, the right- and left-hand sides of Eq. (4.3.7b) are the same when $\sigma = 0$ and $u = 0$ for free boundaries.

Example 4.1 Derive the variational principle for a rod (with length 1 and cross-sectional area A) undergoing deformation due to its own weight.

Solution. This is a one-dimensional problem; thus, the governing equation is as follows.

Step 1.

$$\frac{\partial \sigma_{11}}{\partial x_1} + \rho F_1 = 0,$$

Step 2:

$$\sigma_{11} = \sigma_x = E\frac{\partial u}{\partial x}.$$

Substituting step 2 into step 1 yields

$$R = E\frac{\partial^2 u}{\partial x^2} + \rho F_x .$$

Thus, the virtual work becomes

$$\begin{aligned}
\delta I = (R, \delta u) &= \iint\int_0^1 \left(E\frac{\partial^2 u}{\partial x^2} + \rho F_x \right) \delta u \, \mathrm{d}x \, \mathrm{d}y \, \mathrm{d}z \\
&= A\int_0^1 \left(E\frac{\partial^2 u}{\partial x^2} + \rho F_x \right) \delta u \, \mathrm{d}x \\
&= AE\frac{\partial u}{\partial x}\delta u \Big|_0^1 - \int_0^1 A\left(E\frac{\partial u}{\partial x}\frac{\partial \delta u}{\partial x} - \rho F_x \delta u \right) \mathrm{d}x \\
&= \delta\left\{ AE\frac{\partial u}{\partial x} u \Big|_0^1 - A\int_0^1 \left[\frac{E}{2}\left(\frac{\partial u}{\partial x} \right)^2 - \rho F_x u \right] \mathrm{d}x \right\} = 0.
\end{aligned}$$

It follows that the variational principle or the total potential energy is

$$I = U - W,$$

where, with $A = 1$,

$$U = \int_0^1 \tfrac{1}{2} E \left(\frac{\partial u}{\partial x} \right)^2 \mathrm{d}x,$$

$$W = \int_0^1 \rho F_x u \, \mathrm{d}x + E \frac{\partial u}{\partial x} u \bigg|_0^1 .$$

Notice that U is the strain energy and W is the external energy consisting of the dead load due to the weight density (ρF_x) and the surface traction

$$E \frac{\partial u}{\partial x} \bigg|_0^1 ,$$

which vanishes at $x = 1$. This implies that the Neumann boundary condition

$$\left(E \frac{\partial u}{\partial x} \right)$$

at $x = 1$ is zero (stress is zero at the end of the rod). Note also that the Dirichlet boundary condition (u) at $x = 0$ (the fixed end) is zero.

Example 4.2 Reconsider the problem in Example 4.1 and determine the displacement by minimizing the total potential energy (the variational principle).

Solution
Total potential energy

Step 1:

$$I = \int_0^L \tfrac{1}{2} E \left(\frac{\partial u}{\partial x} \right)^2 \mathrm{d}x - \int_0^L \rho F_x u \, \mathrm{d}x - E \frac{\partial u}{\partial x} u \bigg|_0^L .$$

Neumann boundary condition

Step 2:

$$\frac{\partial u}{\partial x} = 0 \quad \text{at } x = L.$$

Dirichlet boundary condition

Step 3:

$$u = 0 \quad \text{at } x = 0.$$

A functional representation for u that satisfies these conditions may be given by the so-called Rayleigh–Ritz (Ritz 1909) approximations:

Step 4a:

$$u = c_i\phi_i(x) = c_1\phi_1 + c_2\phi_2 + \ldots.$$

In this case, choose a function that will satisfy both boundary conditions in steps 2 and 3.

Step 4b:

$$u = c_1(x^2 - 2xL).$$

Equilibrium is assured when the total potential energy assumes a minimum.

Step 5:

$$\delta I = \frac{\partial I}{\partial c_i}\delta c_i = 0 \quad \text{or} \quad \frac{\partial I}{\partial c_i} = 0.$$

Substitute step 4b into step 1 and, subsequently, into step 5 to yield:

Step 6:

$$\frac{\partial I}{\partial c_1} = \int_0^L [E(2x - 2L)^2 c_1 - f(x^2 - 2xL)]\,dx = 0,$$

where $f = \rho F_x$. Solving for c_1 from step 6 gives $c_1 = -f/(2E)$. Thus, substitution of this constant into step 4b leads to

$$u = -\frac{f}{2E}(x^2 - 2xL).$$

This agrees with the exact solution, which may be obtained by integrating $E(\partial^2 u/\partial x^2) + f = 0$, subject to the given boundary conditions.

Example 4.3 Derive the variational principle for the biharmonic equation

$$\frac{\partial^4\psi}{\partial x^4} + 2\frac{\partial^4\psi}{\partial x^2\partial y^2} + \frac{\partial^4\psi}{\partial y^4} = 0,$$

where ψ is the transverse displacement in plate bending.

Solution

$$\delta I = \int_\Omega \left(\frac{\partial^4\psi}{\partial x^4} + 2\frac{\partial^4\psi}{\partial x^2\partial y^2} + \frac{\partial^4\psi}{\partial y^4}\right)\delta\psi\,dx\,dy = 0.$$

Here, we caution that the cross-derivative term must be split so that integration by parts can be carried out separately with respect to x and y for each term of the right-hand side (see Eq. [1.3.11]). Note that this process is automatically taken care of if the tensorial approach is followed.

$$\begin{aligned}\delta I &= \int_\Omega \psi_{,iijj}\delta\psi \,\mathrm{d}\Omega = \int_\Gamma \psi_{,iij}n_j\delta\psi \,\mathrm{d}\Gamma - \int_\Omega \psi_{,iij}\delta\psi_{,j}\,\mathrm{d}\Omega \\ &= \int_\Gamma \psi_{,iij}n_j\delta\psi\,\mathrm{d}\Gamma - \int_\Gamma \psi_{,ii}n_j\delta\psi_{,j}\,\mathrm{d}\Gamma + \int_\Omega \psi_{,ii}\delta\psi_{,jj}\,\mathrm{d}\Omega \\ &= \delta\left[\int_\Gamma (\psi_{,iij}n_j\psi - \psi_{,ii}n_j\psi_{,j})\,\mathrm{d}\Gamma + \tfrac{1}{2}\int_\Omega \psi_{,ii}\psi_{,jj}\,\mathrm{d}\Omega\right] = 0,\end{aligned}$$

where i, $j = 1, 2$. Note that $\psi_{,iij}n_j$ and $\psi_{,ii}n_j$ in the boundary integral represent the Neumann boundary condition data. Further integrations by parts down to the zeroth-order derivative of ψ would have revealed the existence of Dirichlet boundary conditions (see Eq. [1.3.11]), but integration by parts for the derivation of the variational principle ends at the completion of the Neumann boundary data. Thus, the variational principle is obtained as

$$\begin{aligned}I &= \tfrac{1}{2}\int_\Omega \psi_{,ii}\psi_{,jj}\,\mathrm{d}\Omega + \int_\Gamma (\psi_{,iij}n_j\psi - \psi_{,ii}n_j\psi_{,j})\,\mathrm{d}\Gamma \\ &= \tfrac{1}{2}\int_\Omega \left[\left(\frac{\partial^2\psi}{\partial x^2}\right)^2 + \left(\frac{\partial^2\psi}{\partial y^2}\right)^2 + 2\frac{\partial^2\psi}{\partial x^2}\frac{\partial^2\psi}{\partial y^2}\right]\mathrm{d}\Omega \\ &\quad + \int_\Gamma \left(\frac{\partial^3\psi}{\partial x^3}n_1\psi + \frac{\partial^3\psi}{\partial x^2\partial y}n_2\psi + \frac{\partial^3\psi}{\partial y^2\partial x}n_1\psi + \frac{\partial^3\psi}{\partial y^3}n_2\psi\right. \\ &\quad \left. - \frac{\partial^2\psi}{\partial x^2}n_1\frac{\partial\psi}{\partial x} - \frac{\partial^2\psi}{\partial x^2}n_2\frac{\partial\psi}{\partial y} - \frac{\partial^2\psi}{\partial y^2}n_1\frac{\partial\psi}{\partial x} - \frac{\partial^2\psi}{\partial y^2}n_2\frac{\partial\psi}{\partial y}\right)\mathrm{d}\Gamma.\end{aligned}$$

For two-dimensional problems, the boundary integrals are line integrals that consist of third- and second-order derivatives of ψ, which represent boundary shears and moments, respectively. Once again, Dirichlet boundary conditions are the first- and zeroth-order derivatives, which indicate boundary slopes and displacements, respectively, but do not appear in the derivation of the variational principle.

For simplicity, thermoelastic considerations are not included at this state, because they require the advanced concepts that will be discussed in the next section.

4.4 Thermodynamics of Solids

4.4.1 General

If a body is heated, strains and stresses develop. Conversely, if a body is strained rapidly, then heat is generated inside the body. The system undergoing these processes can be characterized by various functions called the *state variables*. One basic state variable is temperature, which abstracts the degree of hot and cold. It is known that no system can be cooled below a certain limit. If the greatest lower bound is assigned the value zero, then the temperature is said to be absolute. The absolute temperature θ for the scale of measurement, therefore, has the property,

$$\theta > 0.$$

The absolute temperature θ is also defined as

$$\theta = T_0 + T,$$

where T_0 is the reference temperature and T is the temperature change from T_0. The heat energy supplied to a body arises from sources within the body and from the heat flux on its surfaces given by

$$Q = \int_{\Omega} \rho h \, \mathrm{d}\Omega + \int_{\Gamma} \mathbf{q} \cdot \mathbf{n} \, \mathrm{d}\Gamma, \tag{4.4.1}$$

in which h is the heat supply per unit mass, $\mathbf{q}$ is the heat flux per unit surface area, and $\mathbf{n}$ is the vector normal to the surface.

When we consider the thermal environment and the mechanical loading, our concern is to derive governing equations that will couple displacements with temperature so that their combined influence can be determined. To this end, we invoke the first and second laws of thermodynamics in the sections that follow.

4.4.2 The First Law of Thermodynamics

For the purpose of generality, we use the deformed (large strain) coordinates. This is because once the governing equations are derived in terms of large strains, then it is a simple matter to transform them to the small strain Cartesian coordinates.

We consider a motion governed by the functions z_i, σ^{ij}, F^i, ϵ, q^i, h, and η of particle $\mathbf{x}$ and time t in a *thermoelastic body*. We define the functions in the Lagrangian coordinates

$$z_i = z_i(\mathbf{x}, t) \tag{4.4.2}$$

as the components of the spatial position at the material point $\mathbf{x}$ at time t which characterize the *motion*. Here, we assume that components of the functions z_i are continuously differentiable. The contravariant stress tensor σ^{ij} and the body force per unit mass are defined in terms of the deformed convective coordinates, ϵ is the internal energy density per unit mass, q^i is the contravariant component of the heat flux, h represents the heat supply per unit mass, and η is the entropy per unit mass.

The first law of thermodynamics may be stated as follows: The time rate of change of kinetic energy and internal energy in a body is equal to the rate of work done on the body plus the changes in all other energies, such as heat, magnetic, electrical, and chemical, per unit time, or

$$\dot{K} + \dot{U} = M + \Sigma E_i. \tag{4.4.3}$$

In the absence of energies other than those due to mechanical power and heat, we write

$$\dot{K} + \dot{U} = M + Q, \tag{4.4.4}$$

where the individual variables are defined as follows.

Kinetic Energy

$$K = \tfrac{1}{2}\int_\Omega \rho \mathbf{v} \cdot \mathbf{v}\, d\Omega. \tag{4.4.5}$$

Internal Energy

$$U = \int_\Omega \rho \epsilon\, d\Omega. \tag{4.4.6}$$

Mechanical Energy

$$M = \int_\Omega \rho \mathbf{F} \cdot \mathbf{v}\, d\Omega + \int_\Gamma \boldsymbol{\sigma}(n) \cdot \mathbf{v}\, d\Gamma. \tag{4.4.7}$$

The heat energy Q is as defined in Eq. (4.4.1). Note that

$$\mathbf{v} = v^i \mathbf{G}_i, \quad \mathbf{a} = \dot{v}^i \mathbf{G}_i = a^i \mathbf{G}_i, \quad \mathbf{q} = q^i \mathbf{G}_i$$

and

$$\begin{aligned} \int_\Gamma \boldsymbol{\sigma}(n) \cdot \mathbf{v}\, d\Gamma &= \int_\Gamma \sigma^{ij} \mathbf{G}_j n_i \cdot v_k \mathbf{G}^k\, d\Gamma = \int_\Omega (\sigma^{ij} v_j)_{|i}\, d\Omega \\ &= \int_\Omega (\sigma^{ij}_{|i} v_j + \sigma^{ij} v_{j|i})\, d\Omega, \end{aligned} \tag{4.4.8}$$

where

$$\sigma^{ij} v_{j|i} = \sigma^{ij}(\delta_{mj} + u_{m,j})\dot{u}_{m,i} = \sigma^{ij} \dot{\gamma}_{ij}. \tag{4.4.9a}$$

To prove that $\sigma^{ij} v_{j|i} = \sigma^{ij}(\delta_{mj} + u_{m,j})\dot{u}_{m,i}$, we may proceed as follows. First, we must recognize that the Lagrangian velocity v_i in deformed coordinates is not equal to $\dot{u}_i$, although it is true that $\mathbf{v} = \dot{\mathbf{u}}$. To see this, we write

$$\dot{\mathbf{u}} = \mathbf{v} = v_i \mathbf{G}^i = v_i \frac{\partial x_i}{\partial z_m} \mathbf{i}_m = \dot{u}_m \mathbf{i}_m,$$

where

$$\dot{u}_m = v_i \frac{\partial x_i}{\partial z_m} \quad \text{and} \quad v_i = \frac{\partial z_m}{\partial x_i} \dot{u}_m.$$

Thus,

$$\begin{aligned} v_{j|i} &= v_{j,i} - \Gamma^r_{ji} v_r = (z_{m,j} \dot{u}_m)_{,i} - z_{s,ji} \frac{\partial x_r}{\partial z_s} \frac{\partial z_m}{\partial x_r} \dot{u}_m \\ &= z_{m,ji} \dot{u}_m + z_{m,j} \dot{u}_{m,i} - z_{m,ji} \dot{u}_m = z_{m,j} \dot{u}_{m,i} \end{aligned}$$

so that

$$\sigma^{ij} v_{j|i} = \sigma^{ij} (\delta_{mj} + u_{m,j}) \dot{u}_{m,i} = \sigma^{ij} \dot{\gamma}_{ij}.$$

This is true only if σ^{ij} is symmetric, because the time derivative of γ_{ij} in Eq. (2.4.5) is not equal to $(\delta_{mj} + u_{m,j})\dot{u}_{m,i}$. However, the product of this quantity with σ^{ij} will produce results that are identical to $\sigma^{ij}\dot{\gamma}_{ij}$. On the undeformed surface for small strain,

$$\sigma_{ij} v_{j,i} = \sigma_{ij} \dot{u}_{j,i} = \sigma_{ij} \dot{\gamma}_{ij}. \tag{4.4.9b}$$

The proof of this follows from Eq. (4.3.2b). In this expression the $u_{m,j}\dot{u}_{m,i}$ required for deformed coordinates (i.e., for large strain) is neglected in undeformed coordinates (for small strain).

Substituting Eq. (4.4.1) and Eqs. (4.4.5)–(4.4.9) into Eq. (4.4.4) yields

$$\int_\Omega (\rho \dot{v}^j - \rho F^j - \sigma^{ij}_{|i}) v_j \,\mathrm{d}\Omega + \int_\Omega (\rho \dot{\epsilon} - \sigma^{ij} \dot{\gamma}_{ij} - q^i_{|i} - \rho h) \,\mathrm{d}\Omega = 0, \tag{4.4.10}$$

where we used the relation given in Eq. (2.2.22) for the time derivative of $\rho \,\mathrm{d}\Omega$ for the conservation of mass.

Thus, for Eq. (4.4.10) to be valid for all arbitrary components of v^j and $\mathrm{d}\Omega$, it is necessary that the first integrand vanish, so that

$$\sigma^{ij}_{|i} + \rho F^j - p \dot{v}^j = 0 \tag{4.4.11}$$

and

$$\rho \dot{\epsilon} = \sigma^{ij} \dot{\gamma}_{ij} + q^i_{|i} + \rho h. \tag{4.4.12}$$

Equations (4.4.11) and (4.4.12) are Cauchy's first law of motion (or the momentum equation) and the energy equation, respectively. Returning to Eq. (4.4.10), we can see that the balance of momentum (the vanishing of the first integral) is prerequisite to the conservation of energy (the vanishing of the second integral). We conclude that the first law of thermodynamics provides conservation of momentum as well as conservation of energy.

If the undeformed area is used for the deformed state of stress, then the results in Eqs. (4.4.11) and (4.4.12) can be transformed in terms of the Kirchhoff stress t^{ij}. Thus, we use the definitions given in Section 3.7 together with

$$q_i^0 = \sqrt{G} q^i$$

to obtain

$$t^{im}_{,i} + t^{ij}_{,i} u_{m,j} + t^{ij} u_{m,ji} + \rho_0 F^0_m - \rho_0 \ddot{u}_m = 0, \qquad (4.4.13a)$$

$$\rho_0 \dot{\epsilon} = t^{ij} \dot{\gamma}_{ij} + q^0_{i,i} + \rho_0 h. \qquad (4.4.13b)$$

For Cartesian coordinates in the undeformed state of stress, the derivation of Eqs. (4.4.11) and (4.4.12) would have been simpler. However, it is obvious from Eqs. (4.4.11) and (4.4.12) that the results are in the form

$$\sigma_{ij,i} + \rho F_j - \rho \ddot{u}_j = 0 \qquad (4.4.14a)$$

and

$$\rho \dot{\epsilon} = \sigma_{ij} \dot{\gamma}_{ij} + q_{i,i} + \rho h \qquad (4.4.14b)$$

with all quantities referring to undeformed Cartesian coordinates.

4.4.3 The Second Law of Thermodynamics

The second law of thermodynamics is based on the concept of *entropy* associated with irreversible thermodynamic processes. The entropy is regarded as a measure of change of energy dissipation with respect to temperature.

We define entropy S as an additive continuous function,

$$S = \int_\Omega \rho \eta \, d\Omega,$$

where η is the entropy density per unit mass. Furthermore, the total entropy production B is defined as:

$$B = \dot{S} - \int_\Gamma \left(\frac{\mathbf{q}}{\theta}\right) \cdot \mathbf{n} \, d\Gamma - \int_\Omega \rho \left(\frac{h}{\theta}\right) d\Omega \geqslant 0. \qquad (4.4.15)$$

This expression is referred to as the second law of thermodynamics in solids, which states that the total entropy production is always greater than or equal to zero. This is also known as the *Clausius–Duhem inequality*. We may rewrite Eq. (4.4.15) as

$$\int_{\Omega}\left[\rho\dot{\eta} - \nabla\cdot\left(\frac{\mathbf{q}}{\theta}\right) - \rho\left(\frac{h}{\theta}\right)\right]\mathrm{d}\Omega \geqslant 0,$$

or

$$\rho\theta\dot{\eta} - \nabla\cdot\mathbf{q} + \frac{1}{\theta}\mathbf{q}\cdot\nabla\theta - \rho h \geqslant 0, \tag{4.4.16}$$

which is the local form of the Clausius–Duhem inequality.

The product $(-\theta\eta)$ is the irreversible heat energy due to entropy as related to temperature, with the negative sign indicating that compressive reaction results from thermal expansion (temperature rise) in a restrained body. The sum of internal energy (ϵ) and irreversible heat energy $(-\theta\eta)$ is known as Helmholtz free energy

$$\Phi = \epsilon - \theta\eta. \tag{4.4.17}$$

Substituting Eq. (4.4.17) into Eq. (4.4.12), we obtain

$$\rho\dot{\Phi} = \sigma^{ij}\dot{\gamma}_{ij} - \rho\dot{\theta}\eta - D, \tag{4.4.18}$$

where D is defined as internal dissipation,

$$D = \rho\theta\dot{\eta} - q^{i}_{|i} - \rho h. \tag{4.4.19}$$

In view of Eqs. (4.4.19) and (4.4.16), it follows that

$$D + \frac{1}{\theta}q^{i}\theta_{,i} \geqslant 0, \tag{4.4.20}$$

which is called the general dissipation inequality. For an irreversible process, we find $D > 0$, whereas $D = 0$ for a reversible process, because it is always true that

$$\frac{1}{\theta}q^{i}\theta_{,i} \geqslant 0. \tag{4.4.21}$$

For example, if the Fourier heat conduction law is used for the heat flux (the temperature gradient), then $q^{i}\theta_{,i}$ is quadratic in the temperature gradient, always rendering this product positive.

4.4.4 Constitutive Theory for Thermodynamics of Solids

To characterize the thermal deformations effectively, the constitutive theory for linear elasticity, discussed in Section 4.1, must be reinforced with the theory of thermodynamics of solids (Truesdell and Toupin 1960). Toward this end, it is necessary that the constitutive theory be developed in such a way that these governing equations are not violated:

$$\rho_0 = \rho\sqrt{G}, \tag{4.4.22a}$$

$$\sigma^{ij}_{|i} + \rho F^j - \rho \dot{v}^j = 0, \tag{4.4.22b}$$

$$\sigma^{ij} = \sigma^{ji}, \tag{4.4.22c}$$

$$\rho \dot{\epsilon} = \sigma^{ij} \dot{\gamma}_{ij} + q^i_{|i} + \rho h, \tag{4.4.22d}$$

$$D + \frac{1}{\theta} q^i \theta_{,i} \geqslant 0. \tag{4.4.22e}$$

This requirement is called *physical admissibility*. This and other rules for constitutive equations are summarized below (Gurtin 1963; Truesdell and Noll 1965).

Rule 1: Physical admissibility. All constitutive equations must be consistent with the basic physical laws of conservation of mass, balance of momentum, conservation of energy, and the Clausius–Duhem inequality.

Rule 2: Determinism. The values of the constitutive variables (σ^{ij}, q^i, η, Φ) at a material point $\mathbf{x}$ of a body at time t are determined by the histories of the motion and the temperature of all points of the body,

$$\Phi = \hat{\Phi}[z_i(x'_i, t'), \theta(x'_i, t'), \nabla\theta(x'_i, t')], \tag{4.4.23a}$$

$$\sigma^{ij} = \hat{\sigma}^{ij}[z_i(x'_i, t'), \theta(x'_i, t'), \nabla\theta(x'_i, t')], \tag{4.4.23b}$$

$$\eta = \hat{\eta}[z_i(x'_i, t'), \theta(x'_i, t'), \nabla\theta(x'_i, t')], \tag{4.4.23c}$$

$$q^i = \hat{q}^i[z_i(x'_i, t'), \theta(x'_i, t'), \nabla\theta(x'_i, t')], \tag{4.4.23d}$$

where $z_i(x'_i, t')$, $\theta(x'_i, t')$, and $\nabla\theta(x'_i, t')$ are the histories of motion and temperature with $t' \leqslant t$ and x'_i corresponding to t'. Here, we assume that all constitutive functionals depend on the same set of independent variables. This leads to the principle of equipresence.

Rule 3: Equipresence. At the outset, a quantity that appears as an independent variable in one constitutive equation should appear likewise in all constitutive equations unless ruled out by physical laws.

Rule 4: Local action (the principle of neighborhood). This rule imposes certain restrictions on the smoothness of the constitutive functionals in the neighborhood of a material point $\mathbf{x}$. The dependent constitutive variables at $\mathbf{x}$ are not appreciably affected by the values of the independent variables at material points distant from $\mathbf{x}$.

Rule 5: Material objectivity (material frame indifference). The response of a material is independent of the spatial reference frame established to describe it. This implies that the constitutive equations are invariant under observer transformations.

Rule 6: Material symmetry. The constitutive functionals for all ideal materials possess some type of isotropy group. Thus, constitutive equations must be form-invariant with respect to a group of unimodular transformations of the material's frame of reference.

Applications of these rules will be demonstrated in the following section. Additional discussion of this topic appears in Chapter 5 for Newtonian fluids.

4.4.5 Equations of Motion and Heat Conduction

We recast Eq. (4.4.23), rule 2 of the constitutive equations, in the form corresponding to a thermoelastic body:

$$\Phi = \hat{\Phi}(\gamma_{ij}, \theta, \theta_{,i}), \tag{4.4.24a}$$

$$\sigma^{ij} = \hat{\sigma}(\gamma_{ij}, \theta, \theta_{,i}), \tag{4.4.24b}$$

$$\eta = \hat{\eta}(\gamma_{ij}, \theta, \theta_{,i}), \tag{4.4.24c}$$

$$q^i = \hat{q}(\gamma_{ij}, \theta, \theta_{,i}). \tag{4.4.24d}$$

Here, the motion z_i at $t' = t$ is characterized by strains γ_{ij}. It is well known that strains and temperatures affect the free energy, but the dependency of temperature gradients on free energy remains to be verified.

It follows from Eq. (4.4.24a) that the time rate of change of free energy takes the form

$$\rho\dot{\Phi} = \rho\frac{\partial\Phi}{\partial\gamma_{ij}}\dot{\gamma}_{ij} + \rho\frac{\partial\Phi}{\partial\theta}\dot{\theta} + \rho\frac{\partial\Phi}{\partial\theta_{,i}}\dot{\theta}_{,i}.$$

Substitute the above relation into Eq. (4.4.18) to yield

$$\left(\sigma^{ij} - \rho\frac{\partial\Phi}{\partial\gamma_{ij}}\right)\dot{\gamma}_{ij} - \left(\rho\eta + \rho\frac{\partial\Phi}{\partial\theta}\right)\dot{\theta} - \rho\frac{\partial\Phi}{\partial\theta_{,i}}\dot{\theta}_{,i} - D = 0. \tag{4.4.25}$$

For a reversible process ($D = 0$) and for all arbitrary values of

$$\dot{\gamma}_{ij}, \quad \dot{\theta}, \quad \text{and} \quad \dot{\theta}_{,i},$$

it is necessary that these relations hold:

$$\sigma^{ij} = \rho\frac{\partial\Phi}{\partial\gamma_{ij}}, \tag{4.4.26}$$

$$\rho\eta = -\rho\frac{\partial\Phi}{\partial\theta}, \tag{4.4.27}$$

$$\rho\frac{\partial\Phi}{\partial\theta_{,i}} = 0. \tag{4.4.28}$$

Equation (4.4.28) implies that the free energy must be independent of the temperature gradient. Thus, we may revise Eq. (4.4.24):

$$\Phi = \hat{\Phi}(\gamma_{ij}, \theta), \tag{4.4.29a}$$

$$\sigma^{ij} = \hat{\sigma}(\gamma_{ij}, \theta), \tag{4.4.29b}$$

$$\eta = \hat{\eta}(\gamma_{ij}, \theta), \tag{4.4.29c}$$

$$q^i = \hat{q}^i(\gamma_{ij}, \theta). \tag{4.4.29d}$$

We see that the principle of equipresence (rule 3, Section 4.4.4) is to serve merely as a guide. The final form of the constitutive equation must be dictated by physical observation.

The constitutive equations (4.4.26) and (4.4.27) suggest that had a form of free energy been known, it would be possible to determine the stress tensor and entropy. To this end, we may expand Eq. (4.4.29a) into a Taylor series and retain only the second-order terms in a manner similar to Eqs. (4.1.1) and (4.1.2). In the present case, however, we have two arguments, γ_{ij} and θ. Thus, the free energy per unit volume may be written in the form:

$$\rho\Phi = \tfrac{1}{2}E^{ijkm}\gamma_{ij}\gamma_{km} - \tfrac{1}{2}\frac{c}{T_0}\theta^2 - \beta^{ij}\theta\gamma_{ij}, \tag{4.4.30}$$

where E^{ijkm} and β^{ij} are the contravariant components of the elastic modulus tensor and the thermoelastic modulus tensor, respectively, and c is the heat capacity ($c = \rho c_v$, with c_v the specific heat at constant volume). The negative signs for the last two terms on the right-hand side of Eq. (4.4.30) indicate that a temperature rise leads to a compressive reaction on the restrained body. The second term signifies the thermal energy due to temperature, and the last term is the energy due to the coupling effect between temperature and mechanical deformations. Note that Eq. (4.4.30) is equivalent to Eq. (4.1.2), the Clapeyron's formula, as extended to thermoelastic solids. Upon substitution of Eq. (4.4.30) into Eqs. (4.4.26) and (4.4.27), we obtain explicit forms of the constitutive equations for the stress tensor and entropy,

$$\sigma^{ij} = E^{ijkm}\gamma_{km} - \beta^{ij}T, \tag{4.4.31a}$$

$$\rho\eta = \frac{c}{T_0}T + \beta^{ij}\gamma_{ij}. \tag{4.4.31b}$$

Recall that the validity of constitutive relations expressed in Eqs. (4.4.26) and (4.4.27) is based on reversibility. For an irreversible process ($D > 0$), however, it is necessary to provide adequate mathematical models for the internal dissipation. For example, we may assume that $D = D^{ijkm}\dot{\gamma}_{ij}\dot{\gamma}_{km}$, with D^{ijkm} being the fourth-order tensor of damping constants as used in the theory of viscoelasticity (Chung 1988). The constitutive equation for heat flux may be given by the *Fourier heat conduction law*

$$q^i = k^{ij}T_{,j}, \tag{4.4.32}$$

where k^{ij} refers to the contravariant component of the *thermal conductivity tensor*. Note that the negative sign usually assigned to the right-hand side is absent because we are drawing the relation from the definition of Q, as given in Eq. (4.4.1).

For isotropic materials with large strains, the constitutive laws given by Eqs. (4.4.31a, b, and 4.4.32) are written in terms of contravariant metric tensors instead of Kronecker deltas as

$$\sigma^{ij} = [\lambda G^{ij}G^{km} + \mu(G^{ik}G^{jm} + G^{im}G^{jk})]\gamma_{km} - \beta G^{ij}T, \tag{4.4.33a}$$

$$q^i = kG^{ij}T_{,j}, \tag{4.4.33b}$$

$$\rho\eta = \frac{c}{T_0}T + \beta G^{ij}\gamma_{ij}. \tag{4.4.33c}$$

The foregoing relations for isotropic materials with small strains are written in the undeformed state ($\beta_{ij} = \beta\delta_{ij}$, $k_{ij} = k\delta_{ij}$),

$$\sigma_{ij} = E_{ijkm}\gamma_{km} - \beta\delta_{ij}T, \tag{4.4.34}$$

$$\rho\eta = \frac{c}{T_0}T + \beta\gamma_{ii}, \tag{4.4.35}$$

$$q_i = kT_{,i}, \tag{4.4.36}$$

where E_{ijkm} is as given by Eq. (4.1.23), k is the isotropic thermal conductivity, and the isotropic thermoelastic constant assumes the form

$$\beta = \frac{E\alpha}{1 - 2\nu} \quad \textit{for three-dimensional and plane strain cases}, \tag{4.4.37a}$$

$$\beta = \frac{E\alpha}{1 - \nu} \quad \textit{for plane stress}, \tag{4.4.37b}$$

with α being the coefficient of thermal expansion. Note that the thermoelastic constant β can be derived with $i = j$ by setting Eq.

(4.1.25) equal to the thermal stress $\sigma_{ii} = \sigma_T = \beta T$ and by defining the thermal strain as $\gamma_{ii} = \gamma_T = \alpha T$:

$$\sigma_T = \beta T = \sigma_{ii} = (3\lambda + 2\mu)\gamma_{ii} = (3\lambda + 2\mu)\alpha T,$$

which gives

$$\beta = (3\lambda + 2\mu)\alpha = \frac{E\alpha}{1 - 2\nu}.$$

For the case of plane stress, it follows from Eq. (4.1.32) that

$$\sigma_T = \beta T = \frac{E}{1 - \nu^2}(1 + \nu)\gamma_T = \frac{E\alpha T}{1 - \nu}.$$

Thus,

$$\beta = \frac{E\alpha}{1 - \nu}.$$

With these preliminaries, the equations of motion and heat conduction for an isotropic material can be derived for the reversible process as follows.

Deformed Convective Coordinates

The equation of motion in terms of the Kirchhoff stress in deformed convective coordinates using undeformed area is as given by Eq. (4.4.13a):

$$t^{im}_{,i} + t^{ij}_{,i}u_{m,j} + t^{ij}u_{m,ij} + \rho_0 F^0_m - \rho_0 \ddot{u}_m = 0 \qquad (4.4.38a)$$

with t^{ij} defined by Eqs. (3.6.8) and (4.4.33a). For heat conduction, we return to Eq. (4.4.19) and write

$$\rho\theta\dot{\eta} - q^i_{|i} - \rho h - D = 0.$$

Now set $D = 0$ for the reversible process, and substitute from Eqs. (4.4.33c) and (4.4.33b) to obtain

$$\frac{\theta}{T_0}c\dot{T} + \theta\beta(\delta_{mi} + u_{m,i})\dot{u}_{m,i} - kT_{,ii} - \rho_0 h = 0. \qquad (4.4.38b)$$

Notice that Eq. (4.4.38b) is nonlinear in both temperature and displacement. We demonstrate how this equation can be simplified in undeformed coordinates.

Undeformed Coordinates

The presentation so far has dealt with deformed coordinates that are capable of dealing with large strains. For undeformed coordinates (small-strain problems), however, similar, but simpler, algebra could have been carried out, although it is now obvious that by replacing the covariant derivatives by partial derivatives and using the notation for undeformed coordinates, we can write the equations of motion and heat conduction, respectively, as

$$\sigma_{ij,i} + \rho F_j - \rho \ddot{u}_j = 0, \tag{4.4.39a}$$

$$\frac{\theta}{T_0} c\dot{T} + \theta\beta\dot{u}_{i,i} - kT_{,ii} - \rho h = 0, \tag{4.4.39b}$$

or more explicitly, we can obtain the thermomechanically coupled Navier equation

$$\mu u_{i,jj} + (\lambda + \mu)u_{j,ij} - \alpha(3\lambda + 2\mu)T_{,i} + \rho F_i - \rho\ddot{u}_i = 0 \tag{4.4.40a}$$

and the thermomechanically coupled heat conduction equation

$$c\dot{T} + T_0\alpha(3\lambda + 2\mu)\dot{u}_{i,i} - kT_{,ii} - \rho h = 0. \tag{4.4.40b}$$

Note that, for the transient state ($T \ll T_0$), the absolute temperature is set equal to the reference temperature ($\theta \simeq T_0$), which is the common assumption accepted in the field of heat conduction. Equations (4.4.38a, b) and (4.4.40a, b) are known as thermomechanically coupled equations of motion and heat conduction for large and small strains, respectively.

Boundary Conditions

The equations of motion and heat conduction are subject to the following boundary conditions:

$$u = \hat{u} \quad \text{on } \Gamma_1, \tag{4.4.41a}$$

$$\sigma_{ij}n_i = \hat{S}_j \quad \text{on } \Gamma_2, \tag{4.4.41b}$$

$$T = \hat{T} \quad \text{on } \Gamma_3, \tag{4.4.41c}$$

$$k\frac{\partial T}{\partial n} = kT_{,i}n_i = -\hat{q} - \bar{\alpha}(T - T') \quad \text{on } \Gamma_4. \tag{4.4.41d}$$

As defined in subsection 4.3.4, the values of variable prescribed on boundaries such as u_i and T are the Dirichlet boundary conditions, whereas the gradients normal to the surface, such as $\sigma_{ij}n_i$ and $kT_{,i}n_i$, are the Neumann boundary conditions. However, the right-hand side of Eq. (4.4.41d) indicates the mixture of the value of a variable (T, T') and the heat flux (the temperature gradient, q). This is sometimes referred to as the Cauchy or Robin boundary condition, where $\hat{q}$ is the heat flux on the boundary's surface, $\bar{\alpha}$ is the heat transfer (film) coefficient, and T' denotes the ambient temperature.

The existence and explicit forms of boundary conditions equations, (4.4.41a and b), for the equations of equilibrium were shown in subsection 4.3.4. We follow the identical procedure for heat conduction. However, in boundary value problems, the first two terms of Eq. (4.4.39b) are not associated with boundary conditions. Only the second derivative terms will be involved in producing the boundary conditions. Therefore, it follows that

$$-\int_{\Omega} kT_{,ii}T \, d\Omega = -\int_{\Gamma} kT_{,i}n_i T \, d\Gamma + \int_{\Omega} kT_{,i}T_{,i} \, d\Omega. \qquad (4.4.42)$$

Further integration by parts is to verify the existence of Dirichlet boundary conditions. The Neumann boundary conditions, $kT_{,i}n_i$, as indicated by the first term on the right-hand side of Eq. (4.4.42), may consist of surface heat flux $\hat{q}$ and additional heat flux generated by the temperature difference between the surface temperature T and the ambient temperatures T' associated with heat transfer coefficient $\bar{\alpha}$, as shown in (4.4.41d).

Curvilinear Coordinates

Special care is required for the divergence of heat flux in curvilinear coordinates. Unlike $\nabla \cdot \mathbf{v}$ in Eq. (2.3.15a), it is necessary that $\nabla \cdot \mathbf{q}$ be written as:

$$\begin{aligned} \nabla \cdot \mathbf{q} &= \mathbf{g}^i \frac{\partial}{\partial \xi_i} \cdot \mathbf{g}^j q_j = g^{ij} q_{j,i} - \mathbf{g}^i \Gamma^j_{ik} \cdot \mathbf{g}^k q_j \\ &= g^{ij}(q_{j,i} - \Gamma^k_{ij} q_k) = k g^{ij}(T_{,ij} - \Gamma^k_{ji} T_{,k}). \end{aligned}$$

Thus, for cylindrical coordinates, we obtain

$$\nabla \cdot \mathbf{q} = k\nabla^2 T = k\left(\frac{\partial^2 T}{\partial r^2} + \frac{1}{r^2}\frac{\partial^2 T}{\partial \theta^2} + \frac{\partial^2 T}{\partial z^2} + \frac{1}{r}\frac{\partial T}{\partial r}\right). \qquad (4.4.43)$$

Similarly, for spherical coordinates

$$\begin{aligned} \nabla \cdot \mathbf{q} &= k\nabla^2 T \\ &= k\left(\frac{\partial^2 T}{\partial R^2} + \frac{1}{R^2}\frac{\partial^2 T}{\partial \alpha^2} + \frac{1}{R^2 \sin^2 \alpha}\frac{\partial^2 T}{\partial \theta^2} + \frac{2}{R}\frac{\partial T}{\partial R} + \frac{\cot \alpha}{R^2}\frac{\partial T}{\partial \alpha}\right). \end{aligned} \qquad (4.4.44)$$

Remarks

The relations given by Eqs. (4.4.40a and b) represent thermomechanically coupled equations of motion and heat conduction for linear thermoelasticity. Note that the equation of motion (4.4.40a) is of a hyperbolic nature, whereas the heat conduction equation (4.4.40b) is a parabolic partial differential equation. Heat conduction in some materials (helium, for example) exhibits hyperbolic behavior, a phenomenon known as *second sound* (Ackerman and Berman 1966) in which the Fourier heat conduction law, Eq. (4.4.36), must be modified. The details on second sound are beyond the scope of this book.

4.5 Finite Elasticity

If a material ceases to be linearly elastic after reaching a certain magnitude of strain and at that point exhibits nonlinear stress-strain relations, the generalized form of Hooke's law no longer applies. Here, we require a nonlinear constitutive law. Materials such as rubber fall under the category of nonlinear elasticity, commonly known as *finite elasticity*, a term that indicates the material can undergo a finite deformation and remain elastic. For an isothermal condition, the stress tensor is given by

$$\sigma^{ij} = \rho \frac{\partial \epsilon}{\partial \gamma_{ij}}.$$

Consider an elastic potential $W = \hat{W}(\gamma_{ij})$ in the form

$$W = \rho_0 \epsilon,$$

$$\rho_0 = \sqrt{G}\rho.$$

We then have the Cauchy stress tensor written in the form

$$\sigma^{ij} = \frac{1}{\sqrt{G}} \frac{\partial W}{\partial \gamma_{ij}},$$

or we can use the Kirchhoff stress tensor on the deformed surface in terms of the undeformed area to obtain

$$t^{ij} = \frac{\partial W}{\partial \gamma_{ij}}. \tag{4.5.1}$$

Since W is a function of γ_{ij} and can also be a function of the principal invariants of G_{ij}, as given by Eq. (2.7.12), we may write (Oden 1972):

$$\begin{aligned} t^{ij} &= \frac{\partial W}{\partial I_1}\frac{\partial I_1}{\partial \gamma_{ij}} + \frac{\partial W}{\partial I_2}\frac{\partial I_2}{\partial \gamma_{ij}} + \frac{\partial W}{\partial I_3}\frac{\partial I_3}{\partial \gamma_{ij}}, \\ &= 2\frac{\partial W}{\partial I_1}\delta^{ij} + 4\frac{\partial W}{\partial I_2}[\delta^{ij}(1+\gamma_{rr}) + \delta^{ir}\delta^{js}\gamma_{rs}], \\ &\quad + 2\frac{\partial W}{\partial I_3}[\delta^{ij}(1+2\gamma_{rr}) - \delta^{ir}\delta^{js}\gamma_{rs} + 2\epsilon^{imn}\epsilon^{jrs}\gamma_{mr}\gamma_{ns}]. \end{aligned} \tag{4.5.2}$$

Here W is given by:

$$W = \tfrac{1}{2}E^{ijkm}\gamma_{ij}\gamma_{km} + \tfrac{1}{6}E^{ijkmnp}\gamma_{ij}\gamma_{km}\gamma_{np} + \cdots. \tag{4.5.3}$$

If we expand the invariants in powers, then W is of the form

$$W = \sum_{r=0}^{\infty}\sum_{s=0}^{\infty}\sum_{t=0}^{\infty} C_{rst}(I_1 - 3)^r(I_2 - 3)^s(I_3 - 1)^t. \qquad (4.5.4)$$

A choice of $r = s = t = 1$ gives

$$W = C_{100}(I_1 - 3) + C_{010}(I_2 - 3) + C_{001}(I_3 - 1). \qquad (4.5.5)$$

Alternately, a polynomial approximation may be written in the form

$$\begin{aligned} W = {} & A_1K_1^2 + A_2K_2 + A_3K_1^3 + A_4K_1K_2 + A_5K_3 \\ & + A_6K_1^4 + A_7K_1^2K_2 + A_8K_1K_3 + A_9K_2^2, \end{aligned} \qquad (4.5.6)$$

in which $A_1, \ldots, A_9$ are material constants and

$$K_1 = \gamma_{ii}, \quad K_2 = \gamma_{ij}\gamma_{ij}, \quad K_3 = \gamma_{ij}\gamma_{ir}\gamma_{jr}.$$

Many materials undergo finite deformations without an appreciable change in volume. These materials are said to be *incompressible*. The most significant characteristic of incompressibility is that the stress tensor is not completely determined by the deformation. The addition of hydrostatic pressure to an incompressible elastic solid affects the stress, but the strain is not altered. The incompressibility condition is given by

$$I_3 = |G_{ij}| = 1. \qquad (4.5.7)$$

For hyperelastic materials we may write

$$W = W(\gamma_{ij}) + \lambda(I_3 - 1),$$

where λ is a Lagrange multiplier. Thus, the Kirchhoff stress tensor takes the form

$$t^{ij} = \frac{\partial W}{\partial \gamma_{ij}} + \lambda\frac{\partial I_3}{\partial \gamma_{ij}} = \frac{\partial W}{\partial \gamma_{ij}} + hG^{ij}.$$

Here, $h = 2\lambda$ is the hydrostatic pressure. In view of Eq. (4.5.7), we see that W is independent of I_3, or that

$$W = W(I_1, I_2)$$

and

$$\begin{aligned} t^{ij} &= \frac{\partial W}{\partial I_1}\frac{\partial I_1}{\partial \gamma_{ij}} + \frac{\partial W}{\partial I_2}\frac{\partial I_2}{\partial \gamma_{ij}} + \frac{\partial W}{\partial I_3}\frac{\partial I_3}{\partial \gamma_{ij}} \\ &= \frac{\partial W}{\partial I_1}\frac{\partial I_1}{\partial \gamma_{ij}} + \frac{\partial W}{\partial I_2}\frac{\partial I_2}{\partial \gamma_{ij}} + h\frac{\partial I_3}{\partial \gamma_{ij}} \\ &= 2\frac{\partial W}{\partial I_1}\delta^{ij} + 4\frac{\partial W}{\partial I_2}[\delta^{ij}(1 + \gamma_{rr}) - \delta^{ir}\delta^{js}\gamma_{rs}] \\ &\quad + 2h[\delta^{ij}(1 + 2\gamma_{rr}) - \delta^{ir}\delta^{js}\gamma_{rs} + 2\epsilon^{imn}\epsilon^{jrs}\gamma_{mr}\gamma_{ns}]. \end{aligned} \qquad (4.5.8)$$

Thus, the polynomial approximations become

$$W = \sum_{r=0}^{\infty} \sum_{s=0}^{\infty} C_{rs}(I_1 - 3)^r (I_2 - 3)^s, \quad C_{00} = 0. \tag{4.5.9}$$

For example, the well-known Mooney material (or rubbery solids) (Mooney 1940) is characterized as

$$W = C_1(I_1 - 3) + C_2(I_2 - 3), \tag{4.5.10}$$

in which $C_1 = C_{10}$ and $C_2 = C_{01}$.

In summary, the subject of finite elasticity depends on experimental data (the evaluation of constants in the expression for elastic potential) which characterize material properties. Once the appropriate forms of elastic potential are available, we then return to the Kirchhoff stress tensor of the form expressed in Eq. (4.5.8) and proceed with the analysis indicated in Section 4.4.

4.6 Torsional Stress

If a system of forces acting on a small portion of the surface of an elastic body is replaced by another statically equivalent system of force acting on the same portion of the surface, this redistribution of loading produces substantial changes in stress only in the immediate neighborhood of the loading. At the same time, stresses are essentially the same in the parts of the body which are at large distances in comparison with the linear dimensions of the surface on which the forces are changed. This principle, known as Saint-Venant, is applied to torsional problems.

Consider the cylindrical bar fixed at one end and free at the other, as shown in Fig. 4.6.1. If the free end is subjected to twisting action, the net twist angle at the free end is

$$\alpha = \theta z_1 = \theta x_1, \tag{4.6.1}$$

where θ is a twist angle per unit length. Then we have,

$$u_1 = \theta\psi(x_2, x_3), \quad u_2 = -\alpha x_3, \quad u_3 = \alpha x_2,$$

where ψ is the warping function. From $z_i = x_i + u_i$, we write:

$$z_1 = x_1 + \theta\psi(x_2, x_3), \tag{4.6.2a}$$

$$z_2 = x_2 \cos(\theta x_1) - x_3 \sin(\theta x_1), \tag{4.6.2b}$$

$$z_3 = x_2 \sin(\theta x_1) + x_3 \cos(\theta x_1). \tag{4.6.2c}$$

Since

$$\mathbf{G}_i = z_{m,i}\mathbf{i}_m$$

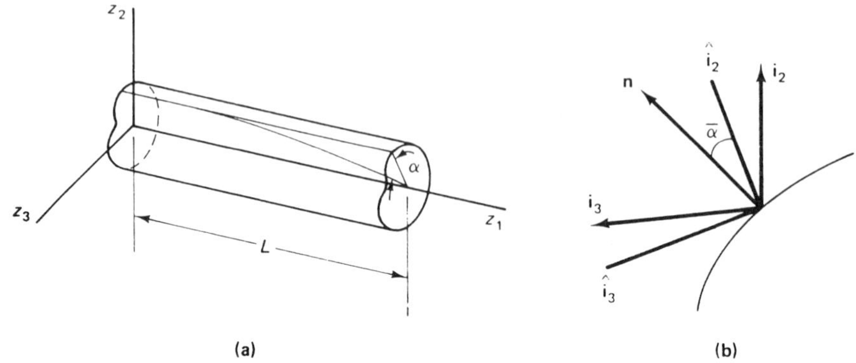

Figure 4.6.1 Cylindrical shaft in torsion. (a) Twist angle. (b) Boundaries.

we have

$$\begin{aligned} \mathbf{G}_1 &= \mathbf{i}_1 - [\theta x_2 \sin(\theta x_1) + \theta x_3 \cos(\theta x_1)]\mathbf{i}_2 \\ &\qquad + [\theta x_2 \cos(\theta x_1) - \theta x_3 \sin(\theta x_1)]\mathbf{i}_3 \\ &= \mathbf{i}_1 + \theta(-x_3 \mathbf{i}_2 + x_2 \mathbf{i}_3) \\ \mathbf{G}_2 &= \theta \psi_{,2} \hat{\mathbf{i}}_1 + \hat{\mathbf{i}}_2 \\ \mathbf{G}_3 &= \theta \psi_{,3} \hat{\mathbf{i}}_1 + \hat{\mathbf{i}}_3, \end{aligned}$$

where the $\hat{\mathbf{i}}_i$'s are the unit vectors on the deformed surface:

$$\begin{aligned} \hat{\mathbf{i}}_1 &= \mathbf{i}_1 \\ \hat{\mathbf{i}}_2 &= \mathbf{i}_2 \cos(\theta x_1) + \mathbf{i}_3 \sin(\theta x_1) \\ \hat{\mathbf{i}}_3 &= -\mathbf{i}_2 \sin(\theta x_1) + \mathbf{i}_3 \cos(\theta x_1). \end{aligned}$$

The components of the strain tensor are

$$\gamma_{ij} = \tfrac{1}{2}(G_{ij} - \delta_{ij}),$$

$$\gamma_{11} = \frac{\theta^2}{2}(x_2^2 + x_3^2), \quad \gamma_{22} = \frac{\theta^2}{2}\psi_{,2}^2, \quad \gamma_{33} = \frac{\theta^2}{2}\psi_{,3}^2,$$

$$\gamma_{12} = \frac{\theta}{2}(\psi_{,2} - x_3), \quad \gamma_{13} = \frac{\theta}{2}(\psi_{,3} + x_2), \quad \gamma_{23} = \frac{\theta^2}{2}\psi_{,2}\psi_{,3}.$$

For a small unit twist angle, we may neglect θ^2. Thus,

$$\gamma_{11} = \gamma_{22} = \gamma_{33} = \gamma_{23} = 0, \tag{4.6.3a}$$

$$\sigma_{11} = \sigma_{22} = \sigma_{33} = \sigma_{23} = 0, \tag{4.6.3b}$$

$$\gamma_{12} = \frac{\theta}{2}(\psi_{,2} - x_3), \quad \gamma_{13} = \frac{\theta}{2}(\psi_{,3} + x_2), \tag{4.6.3c}$$

$$\sigma_{12} = G\theta(\psi_{,2} - x_3), \quad \sigma_{13} = G\theta(\psi_{,3} + x_2), \tag{4.6.3d}$$

where $G = \mu$. With these approximations, we may conclude that

$$\mathbf{G}_i = \hat{\mathbf{i}}_i, \quad \boldsymbol{\sigma}_i = \sigma_{ij}\hat{\mathbf{i}}_j$$

and

$$\frac{\partial \sigma_{ij}}{\partial x_i}\hat{\mathbf{i}}_j + \rho_0 F_i \hat{\mathbf{i}}_i = 0. \tag{4.6.4}$$

Let

$$\mathbf{n} = n_i \hat{\mathbf{i}}_i = (\cos\bar{\alpha})\hat{\mathbf{i}}_2 + (\sin\bar{\alpha})\hat{\mathbf{i}}_3; \tag{4.6.5}$$

then the surface traction becomes

$$\boldsymbol{\sigma} = \sigma_{ij} n_i \hat{\mathbf{i}}_j = (\sigma_{21}\cos\bar{\alpha} + \sigma_{31}\sin\bar{\alpha})\hat{\mathbf{i}}_1. \tag{4.6.6}$$

Since the bar is subjected to twisting only at the ends ($x_1 = 0,\ L$), we have

$$\frac{\partial \sigma_{i1}}{\partial x_i} = 0 \tag{4.6.7}$$

and the traction vanishes on the lateral surface:

$$\sigma_{21}\cos\bar{\alpha} + \sigma_{31}\sin\bar{\alpha} = 0. \tag{4.6.8}$$

In view of Eqs. (4.6.3d) and (4.6.7), it follows that

$$\frac{\partial \sigma_{21}}{\partial x_2} + \frac{\partial \sigma_{31}}{\partial x_3} = \psi_{,22} + \psi_{,33} = 0. \tag{4.6.9}$$

From Eqs. (4.6.8) and (4.6.6) on the boundary (see Fig. 4.6.1), we have

$$(\psi_{,2} - x_3)\cos\bar{\alpha} + (\psi_{,3} + x_2)\sin\bar{\alpha} = 0. \tag{4.6.10}$$

It is obvious that the warping function ψ must satisfy the Laplace equation (4.6.9) and the boundary condition (4.6.10).

If $\bar{n}$ denotes distance along a normal to the boundary element $\mathrm{d}\bar{n}$, we have, from Eq. (4.6.10),

$$\frac{\partial \psi}{\partial \bar{n}} = \frac{\partial \psi}{\partial x_2}\frac{\mathrm{d}x_2}{\mathrm{d}\bar{n}} + \frac{\partial \psi}{\partial x_3}\frac{\mathrm{d}x_3}{\mathrm{d}\bar{n}} = x_3\cos\bar{\alpha} - x_2\sin\bar{\alpha}.$$

In view of Eqs. (4.6.3), (4.6.9), and (4.6.10), the resultant force on a cross section vanishes:

$$\int_\Omega \sigma_{21}\,\mathrm{d}\Omega = \int_\Omega \sigma_{31}\,\mathrm{d}\Omega = 0. \tag{4.6.11}$$

The resultant torsional moment T on the ends of the bar due to the assumed stress distribution is

$$\begin{aligned} T &= \int_\Omega (x_2\sigma_{13} - x_3\sigma_{12})\,\mathrm{d}\Omega \\ &= G\theta \int_\Omega (x_2^2 + x_3^2 + x_2\psi_{,3} - x_3\psi_{,2})\,\mathrm{d}\Omega = G\theta J, \end{aligned} \tag{4.6.12}$$

in which $G = \mu$, and

$$J = \int_\Omega (x_2^2 + x_3^2 + x_2\psi_{,3} - x_3\psi_{,2})\,\mathrm{d}\Omega \tag{4.6.13}$$

is called a torsional constant. Let us assume that ψ may be given by a constant:

$$\psi = C.$$

The boundary condition, Eq. (4.6.10), becomes

$$-x_3 \cos\bar{\alpha} + x_2 \sin\bar{\alpha} = 0.$$

Integrating with respect to $\bar{\alpha}$, we have

$$x_3 \sin\bar{\alpha} + x_2 \cos\bar{\alpha} = 0,$$

$$x_3 \frac{\mathrm{d}x_3}{\mathrm{d}\bar{n}} + x_2 \frac{\mathrm{d}x_2}{\mathrm{d}\bar{n}} = 0,$$

$$\frac{\mathrm{d}}{\mathrm{d}\bar{n}}\left(\frac{x_2^2 + x_3^2}{2}\right) = 0,$$

or

$$x_2^2 + x_3^2 = \mathit{constant},$$

which is an expression for a circle. Thus, the function of $\psi = C$ gives the solution of a circular bar. Substituting this function into Eq. (4.6.13) and assuming that $r = r_0$ gives

$$J = \int_\Omega (x_2^2 + x_3^2)\,\mathrm{d}\Omega = \tfrac{1}{2}\pi r_0^4,$$

which corresponds to the polar moment of inertia. The shear stresses are

$$\sigma_{13} = G\theta x_2 = \frac{Tx_2}{J}, \quad \sigma_{12} = -G\theta x_3 = -\frac{Tx_3}{J}. \tag{4.6.14}$$

Prandtl (1904) proposed that Eq. (4.6.9) would be satisfied if we chose a stress function ϕ such that

$$\sigma_{21} = G\theta\phi_{,3}, \quad \sigma_{31} = -G\theta\phi_{,2}, \tag{4.6.15a}$$

$$\gamma_{21} = \frac{\theta}{2}\phi_{,3}, \quad \gamma_{31} = -\frac{\theta}{2}\phi_{,2}. \tag{4.6.15b}$$

Let us consider a section of a bar which is simply connected. If the compatibility conditions of Eq. (2.9.3) are satisfied, then

$$R_{2331} = \gamma_{33,12} + \gamma_{12,33} - \gamma_{31,32} - \gamma_{32,31},$$
$$= (\phi_{,22} + \phi_{,33})_{,3} = 0,$$

$$R_{1223} = -(\phi_{,22} + \phi_{,33})_{,2} = 0.$$

Upon integration, we have

$$\phi_{,22} + \phi_{,33} = C. \tag{4.6.16}$$

The constant C can be determined by equating the right-hand sides of Eqs. (4.6.3c) and (4.6.15b):

$$\phi_{,3} = \psi_{,2} - x_3, \quad \phi_{,2} = -\psi_{,3} - x_2, \tag{4.6.17}$$

and by substituting these into Eq. (4.6.16), which gives $C = -2$. As a result, we have

$$\phi_{,22} + \phi_{,33} = -2 \tag{4.6.18}$$

subject to the boundary condition $\phi = 0$ on the surface. This is known as the strain compatibility equation for a solid undergoing torsional deformations. Solution of Eq. (4.6.18), together with Eqs. (4.6.15) and (4.6.12), will provide the state of stress and associated torsional properties.

It has been shown that, viewed from a position of the z_1 axis, a straight line of the cross section appears to remain straight, but the line is warped because particles of the line displace axially, which is an application of the Saint–Venant principle.

Example 4.4 Consider a bar with an elliptical cross section, as in Fig. 4.6.2. Determine the tangential stress and warping function.

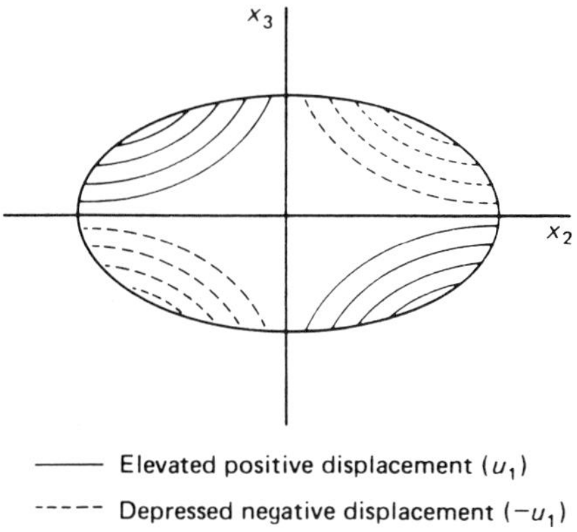

Figure 4.6.2 Lines of constant warping, application of Saint–Venant principle.

Solution. Let the boundary of the cross section be given by:

Step 1:

$$\left(\frac{x_2}{a}\right)^2 + \left(\frac{x_3}{b}\right)^2 - 1 = 0.$$

The solution of Eq. (4.6.17) yields

Step 2:

$$\phi = k\left[\left(\frac{x_2}{a}\right)^2 + \left(\frac{x_3}{b}\right)^2 - 1\right],$$

which satisfies the boundary condition $\phi = 0$ on C such that

Step 3:

$$k = -\frac{a^2 b^2}{a^2 + b^2}.$$

It follows from Eq. (4.6.15a) that the tangential stress is of the form:

Step 4:

$$\sigma_1 = \sqrt{(\sigma_{21})^2 + (\sigma_{31})^2} = \frac{2G\theta}{1 + \left(\frac{a}{b}\right)^2} \sqrt{(x_2)^2 + \left(\frac{a}{b}\right)^4 (x_3)^2}.$$

The warping function ψ is determined from Eq. (4.6.17) and step 2.

Step 5:

$$\psi_{,2} = \frac{b^2 - a^2}{a^2 + b^2} x_3.$$

Step 6:

$$\psi_{,3} = \frac{b^2 - a^2}{a^2 + b^2} x_2.$$

The integration of steps 5 and 6 yields

$$\psi = \frac{b^2 - a^2}{a^2 + b^2} x_2 x_3.$$

Contour lines of constant displacement along the x_3 axis are hyperbolas, as shown in Fig. 4.6.2. The solid lines in the figure indicate where u_1 is positive, and the dotted lines where u_1 is negative.

4.7 Fiber Composites

4.7.1 General

In Section 4.1 we presented a general approach that led to constitutive equations for anisotropic material. Subsequently, an isotropic solid was chosen in Section 4.2 to derive the governing equations for linear elasticity. Modern technology, in aerospace industries in particular, has prompted the use of composite materials with metal or glass fibers. Composite materials consist of two or more materials bonded together to exhibit desirable strength-weight properties which cannot be attained by the individual materials alone. The resulting behavior of such materials is extremely complicated on a microscopic scale. In the design of fiber-composite materials, the concept of micromechanics may be used although, in general, the macromechanics approach will be adequate for most engineering problems.

4.7.2 Coordinate Transformations for Transversely Isotropic Materials

The fiber composites may be orthotropic, as characterized by Eq. (4.1.12), but transversely isotropic (Eq. 4.1.16), as shown in Fig. 4.1.2. Suppose that the fiber is oriented in the direction x_1' (see Fig. 4.7.1),

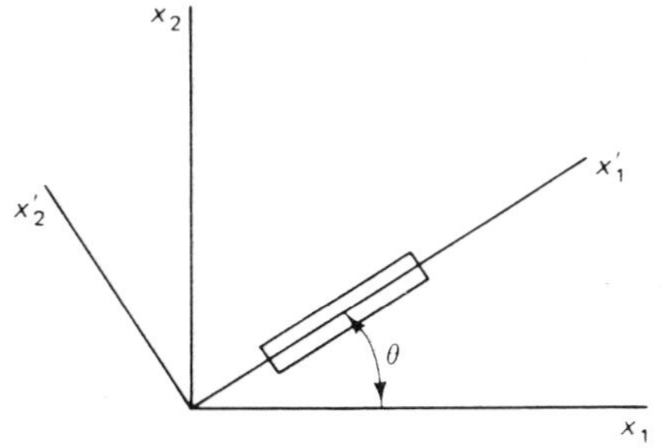

Figure 4.7.1 Coordinate transformations for a fiber composite solid.

then the coordinate transformation for a simple fiber composite solid is governed by Eqs. (4.1.13) and (2.7.3), so that the strain components with the x_1'-axis coinciding with the longitudinal direction of the fiber are written as:

$$\begin{bmatrix} \gamma_{11}' \\ \gamma_{22}' \\ \gamma_{33}' \\ \gamma_{12}' \\ \gamma_{23}' \\ \gamma_{31}' \end{bmatrix} = \begin{bmatrix} \overset{*}{c}^2 & \overset{*}{s}^2 & 0 & 2\overset{*}{c}\overset{*}{s} & 0 & 0 \\ \overset{*}{s}^2 & \overset{*}{c}^2 & 0 & -2\overset{*}{c}\overset{*}{s} & 0 & 0 \\ 0 & 0 & 1 & 0 & 0 & 0 \\ -\overset{*}{c}\overset{*}{s} & \overset{*}{s}\overset{*}{c} & 0 & \overset{*}{c}^2 - \overset{*}{c}^2 & 0 & 0 \\ 0 & 0 & 0 & 0 & \overset{*}{c} & -\overset{*}{s} \\ 0 & 0 & 0 & 0 & \overset{*}{s} & \overset{*}{c} \end{bmatrix} \begin{bmatrix} \gamma_{11} \\ \gamma_{22} \\ \gamma_{33} \\ \gamma_{12} \\ \gamma_{23} \\ \gamma_{31} \end{bmatrix}, \qquad (4.7.1)$$

with $\overset{*}{c} = \cos\theta$, $\overset{*}{s} = \sin\theta$.

Suppose that the fiber-reinforced cylinder shown in Fig. 4.7.2a has the local coordinates $x_1' = r$, $x_2' = T$, and $x_3' = L$, and the global coordinates $x_1 = r$, $x_2 = z$, and $x_3 = \theta$. The rotation about the axis $x_1' = x_1 = r$ is represented by

$$\begin{bmatrix} x_1' \\ x_2' \\ x_3' \end{bmatrix} = \begin{bmatrix} 1 & 0 & 0 \\ 0 & c & s \\ 0 & -s & c \end{bmatrix} \begin{bmatrix} x_1 \\ x_2 \\ x_3 \end{bmatrix}, \qquad (4.7.2)$$

where $c = \cos\alpha$ and $s = \sin\alpha$. The relation between global strains (γ_{rr}, γ_{zz}, $\gamma_{\theta\theta}$, and γ_{rz}) and local strains (γ_{rr}, γ_{TT}, γ_{LL}, γ_{rT}, γ_{rL}, and γ_{TL}) is given by Eqs. (2.7.3) and (4.7.2) as (see Fig. 4.7.2c):

$$\begin{bmatrix} \gamma_{rr} \\ \gamma_{TT} \\ \gamma_{LL} \\ \gamma_{rT} \\ \gamma_{rL} \\ \gamma_{TL} \end{bmatrix} = \begin{bmatrix} 1 & 0 & 0 & 0 \\ 0 & c^2 & s^2 & 0 \\ 0 & s^2 & c^2 & 0 \\ 0 & 0 & 0 & 0 \\ 0 & 0 & 0 & -s \\ 0 & -2cs & 2cs & 0 \end{bmatrix} \begin{bmatrix} \gamma_{rr} \\ \gamma_{zz} \\ \gamma_{\theta\theta} \\ \gamma_{rz} \end{bmatrix}. \qquad (4.7.3)$$

Note that the tensorial shear components have been transformed into total engineering shear components:

$$(\gamma_{ij})_{Eng.} = 2(\gamma_{ij})_{Tensor} \quad if\ i \neq j.$$

Let us now consider an axisymmetric body with the generator inclined at an angle ϕ from the vertical axis, as shown in Fig. 4.7.2b. The local and global coordinates are labeled x_i' and x_i. Consider the intermediate coordinates x_i related by the global coordinates $x_1 = r$, $x_2 = z$, and $x_3 = \theta$,

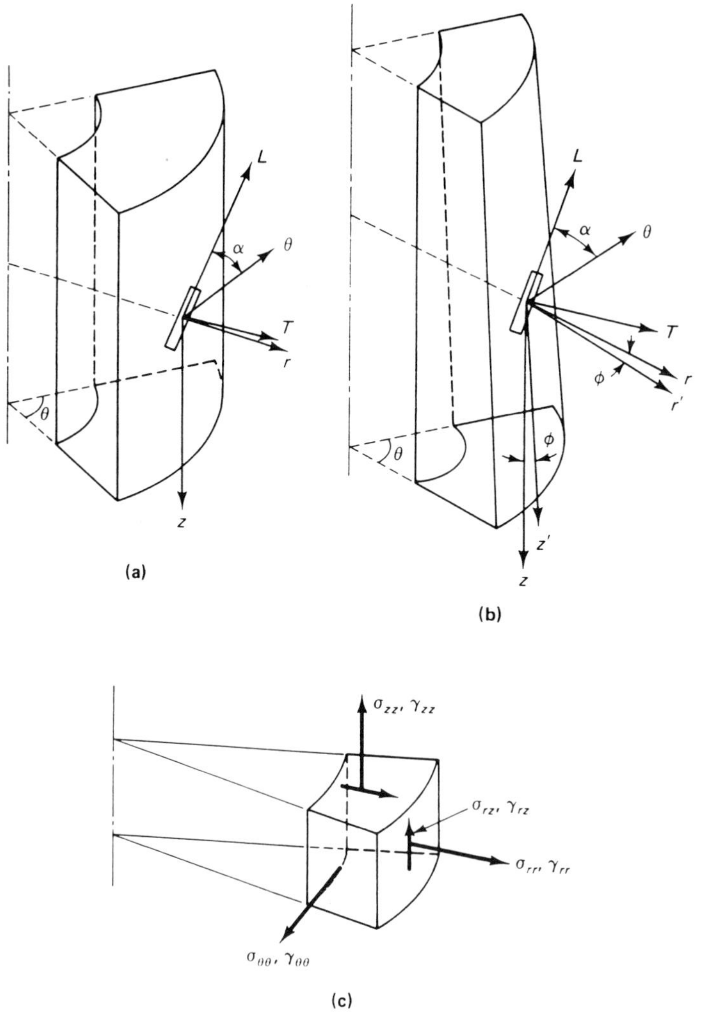

Figure 4.7.2 (a) Coordinate transformations for axisymmetric cylindrical fiber composites. (b) General axisymmetric body for transversely isotropic materials. (c) Stress and strain components.

$$\begin{bmatrix} x_1^* \\ x_2^* \\ x_3^* \end{bmatrix} = \begin{bmatrix} \bar{c} & \bar{s} & 0 \\ -\bar{s} & \bar{c} & 0 \\ 0 & 0 & 1 \end{bmatrix} \begin{bmatrix} x_1 \\ x_2 \\ x_3 \end{bmatrix}, \tag{4.7.4}$$

with $\bar{c} = \cos\phi$ and $\bar{s} = \sin\phi$. The intermediate coordinates are also related by the local coordinates $x_1' = r$, $x_2' = T$, $x_3' = L$:

$$\begin{bmatrix} x_1' \\ x_2' \\ x_3' \end{bmatrix} = \begin{bmatrix} 1 & 0 & 0 \\ 0 & c & s \\ 0 & -s & c \end{bmatrix} \begin{bmatrix} x_1^* \\ x_2^* \\ x_3^* \end{bmatrix}. \tag{4.7.5}$$

With rotations about the axis $x_3 = \theta = x_3'$ and, subsequently, about the axis $x_1' = r = x_1^*$, we obtain the transformation for the x_i and x_i' coordinates,

$$\begin{aligned} \begin{bmatrix} x_1' \\ x_2' \\ x_3' \end{bmatrix} &= \begin{bmatrix} 1 & 0 & 0 \\ 0 & c & s \\ 0 & -s & c \end{bmatrix} \begin{bmatrix} \bar{c} & \bar{s} & 0 \\ -\bar{s} & \bar{c} & 0 \\ 0 & 0 & 1 \end{bmatrix} \begin{bmatrix} x_1 \\ x_2 \\ x_3 \end{bmatrix} \\ &= \begin{bmatrix} \bar{c} & \bar{s} & 0 \\ -cs & c\bar{s} & s \\ s\bar{s} & -s\bar{c} & c \end{bmatrix} \begin{bmatrix} x_1 \\ x_2 \\ x_3 \end{bmatrix}. \end{aligned} \tag{4.7.6}$$

It follows from Eqs. (2.7.3) and (4.7.6) that the relation between the local and global strains for the general axisymmetric body is given by

$$\begin{bmatrix} \gamma_{rr} \\ \gamma_{TT} \\ \gamma_{LL} \\ \gamma_{rT} \\ \gamma_{rL} \\ \gamma_{TL} \end{bmatrix} = \begin{bmatrix} \bar{c}^2 & \bar{s}^2 & 0 & \overline{cs} \\ c^2\bar{s}^2 & c^2\bar{s}^2 & s^2 & -c^2\overline{cs} \\ s^2\bar{s}^2 & s^2\bar{c}^2 & c^2 & -s^2\overline{sc} \\ -2c\overline{cs} & 2s\overline{cs} & 0 & c\bar{c}^2 - c\bar{s}^2 \\ 2s\overline{cs} & -2s\overline{cs} & 0 & s\bar{s}^2 - s\bar{c}^2 \\ -2cs\bar{s}^2 & -2ss\bar{c}^2 & 2sc & 2cs\overline{cs} \end{bmatrix} \begin{bmatrix} \gamma_{rr} \\ \gamma_{zz} \\ \gamma_{\theta\theta} \\ \gamma_{rz} \end{bmatrix}, \tag{4.7.7}$$

in which the tensor shear strains have been converted to the engineering (total) shear strain components.

4.7.3 *Characterization of Material Constants for Orthotropic and Transversely Isotropic Composites*

To determine explicit forms of elastic constants for an orthotropic material, we define the Poisson ratio as

$$\nu_{(ij)} = -\frac{\gamma_{(ij)}}{\gamma_{(ii)}}. \tag{4.7.8}$$

Here, the index i represents the direction in which the stress is applied, and j denotes the direction transverse to i. The strain-stress relations are of the form,

$$\gamma_{ij} = C_{ijkm}\sigma_{km}, \tag{4.7.9}$$

where the compliance C_{ijkm} is given by

$$C_{ijkm} = \begin{bmatrix} \frac{1}{E_1} & \frac{-\nu_{21}}{E_2} & \frac{-\nu_{31}}{E_3} & 0 & 0 & 0 \\ & \frac{1}{E_2} & \frac{-\nu_{32}}{E_3} & 0 & 0 & 0 \\ & & \frac{1}{E_3} & 0 & 0 & 0 \\ symm. & & & \frac{1}{\mu_{12}} & 0 & 0 \\ & & & & \frac{1}{\mu_{23}} & 0 \\ & & & & & \frac{1}{\mu_{31}} \end{bmatrix}. \tag{4.7.10}$$

Note that $E_{(i)}$ is Young's modulus in the direction $x_{(i)}$, and $\mu_{(ij)}$ is the shear modulus in the plane (ij). Symmetry of compliance requires that

$$\frac{\nu_{(ij)}}{E_{(i)}} = \frac{\nu_{(ji)}}{E_{(j)}}.$$

Furthermore, an inverse of the compliance matrix equation (4.7.10) is equal to the elastic matrix:

$$E_{ijkm} = (C_{ijkm})^{-1}. \tag{4.7.11}$$

Thus, we obtain

$$\begin{aligned} E_{1111} &= \frac{1 - \nu_{23}\nu_{32}}{E_2 E_3 J}, \quad E_{1122} = \frac{\nu_{21} + \nu_{31}\nu_{23}}{E_2 E_3 J} = \frac{\nu_{12} + \nu_{32}\nu_{13}}{E_1 E_3 J}, \\ E_{1133} &= \frac{\nu_{31} + \nu_{21}\nu_{32}}{E_2 E_3 J} = \frac{\nu_{13} + \nu_{12}\nu_{23}}{E_1 E_2 J}, \\ E_{2222} &= \frac{1 - \nu_{13}\nu_{31}}{E_1 E_3 J}, \quad E_{2233} = \frac{\nu_{32} + \nu_{12}\nu_{31}}{E_1 E_3 J} = \frac{\nu_{23} + \nu_{21}\nu_{13}}{E_1 E_2 J}, \\ E_{3333} &= \frac{1 - \nu_{12}\nu_{21}}{E_1 E_2 J}, \quad E_{1212} = \mu_{12}, \quad E_{2323} = \mu_{23}, \\ E_{3131} &= \mu_{31} \end{aligned} \tag{4.7.12}$$

where

$$J = \frac{1 - \nu_{12}\nu_{21} - \nu_{23}\nu_{32} - \nu_{31}\nu_{13} - 2\nu_{21}\nu_{32}\nu_{13}}{E_1 E_2 E_3}.$$

For transversely isotropic solids (see Fig. 4.1.2), the results are simplified by Eqs. (4.1.14) and (4.1.15). Let

$$E_1 = E_2 = E_T, \quad E_3 = E_L, \quad \mu_{12} = G_T, \quad \mu_{23} = \mu_{31} = G_{TL},$$
$$\nu_{21} = \nu_{12} = \nu_{IT}, \quad \nu_{31} = \nu_{13} = \nu_{32} = \nu_{23} = \nu_{TL},$$

then we have

$$\begin{aligned} E_{1111} &= E_{2222} = E_T(E_L - E_T\nu_{TL}^2)k, \\ E_{1122} &= E_T(E_L\nu_{TT} + E_T\nu_{TL}^2)k, \\ E_{1133} &= E_{2233} = E_TE_L\nu_{TL}(1 + \nu_{TT})k, \\ E_{3333} &= E_L^2(1 - \nu_{TT}^2)k, \\ E_{1212} &= G_T, \quad E_{2323} = E_{3131} = G_{TL}, \end{aligned} \tag{4.7.13}$$

where

$$k = [E_L(1 - \nu_{TT}^2) - 2E_T\nu_{TL}^2(1 + \nu_{TT})]^{-1}.$$

For the case of a plane stress, components of the fourth-order tensor of elastic constants are simplified to

$$\begin{aligned} E_{1111} &= \frac{E_1}{1 - \nu_{12}\nu_{21}}, \quad E_{1122} = \frac{\nu_{12}E_2}{1 - \nu_{12}\nu_{21}} = \frac{\nu_{21}E_1}{1 - \nu_{12}\nu_{21}}, \\ E_{2222} &= \frac{E_2}{1 - \nu_{12}\nu_{21}}, \quad E_{1212} = G_{12}. \end{aligned} \tag{4.7.14}$$

It was shown how coordinate transformations between the strains of local and global coordinates for a simple composite solid, axisymmetric cylinders, and general axisymmetric bodies can be performed. The transformed strain components will then be substituted into the generalized Hooke's law for composites,

$$\sigma_{ij} = E_{ijkm}\gamma'_{km}. \tag{4.7.15}$$

Here, γ'_{km} is computed from Eqs. (4.7.1), (4.7.3), and (4.7.7) for a simple composite solid, axisymmetric cylinders, and general axisymmetric bodies, respectively. Furthermore, the fourth-order tensor E_{ijkm} takes the forms given in Eqs. (4.7.12), (4.7.13), and (4.7.14) for orthotropic materials, general transversely isotropic fiber-reinforced solids, and the plane stress case of transversely isotropic materials, respectively.

Once the anisotropic materials of the type given by Eq. (4.7.15) are characterized, then it is a simple matter to substitute Eq. (4.7.15) into Cauchy's first law of motion or the Navier equation for solutions of partial differential equations.

Problems

4.1 Show all components of E_{ijkm} in a 9×9 matrix and demonstrate that only 36 components in a 6×6 matrix survive, as shown in Eq. (4.1.6).

4.2 Derive the constitutive equation or stress-strain relation for a linear isotropic material, Eq. (4.1.25). Start from the concept of elastic potential, the generalized Hooke's law, and perform all coordinate transformations. Demonstrate all intermediate results for various anisotropic materials.

4.3 From laboratory test data for the Young's modulus E and the Poisson ratio ν, show that the Lamé constants λ and μ may be derived in the form

$$\lambda = \frac{\nu E}{(1+\nu)(1-2\nu)}, \quad \mu = \frac{E}{2(1+\nu)}.$$

4.4 From the stress tensor derived in Problem 4.2, write all components of the stress and strain tensors in matrix form in terms of Young's modulus and Poisson's ratio.

4.5 Show that the deviatoric stresses are related to deviatoric strains only through the shear modulus.

4.6 Derive the stress-strain relation for plane stress and plane strain.

4.7 If the plane strain data consist of $\gamma_{11} = 0.02$, $\gamma_{12} = \gamma_{21} = 0.01$, and $\gamma_{22} = 0.03$, what are the normal and shear stresses along the x_1' axis rotated $\theta = 30°$ counterclockwise about the x_3 axis? The material constants are $E = 3 \times 10^6$ psi and $\nu = 0.3$.

4.8 Derive the Navier equations, (4.2.2), (4.2.4), and (4.2.5), and expand the results into three-dimensional, plane strain, and plane stress problems in terms of standard variables, x, y, z, u, v, and w.

4.9 Rewrite Eq. (4.2.2b) using physical components in cylindrical coordinates and verify the results given in Eqs. (4.2.8a, b, c).

4.10 Derive the variational principle (total potential energy) from the Cauchy's first law of motion in three dimensions.

4.11 Repeat Problem 4.10 for one dimension and show that the strain energy per unit volume is the area of the triangle under the stress-strain curve.

4.12 Use the first and second laws of thermodynamics to derive the equations of motion and heat conduction for large and small strains. Expand the results for small strains into three-dimensional, plane strain and plane stress problems in terms of standard variables, x, y, z, u, v, and w.

4.13 Repeat Problem 4.12 for large strains.

4.14 Prove Equations (4.4.37a and b).

4.15 Derive the governing equation for the torsion of a bar of the form

$$\frac{\partial^2 \phi}{\partial y^2} + \frac{\partial^2 \phi}{\partial z^2} = -2.$$

4.16 Derive all elements of the fourth-order tensor of material constants for transversely isotropic materials.

4.17 Verify the coordinate transformations required between the local and global coordinates for: (a) simple composite solids, (b) axisymmetric composite cylinders, and (c) general axisymmetric composite bodies.

5
Newtonian Fluids

5.1 Introduction

In contrast with the study of solid mechanics, in which we are preoccupied with deformed geometries, our main purpose in fluid mechanics is to view fluid particles in motion as continua and to determine the velocities at any given point in space. For this purpose we invoke the Eulerian coordinates whose properties are detailed in subsection 2.2.2. Most fluids, whether gases or liquids, are called *Newtonian fluids*, in which the stress tensor is linearly proportional to the velocity gradients. On the other hand, certain chemical fluids, polymers, rheologically complex fluids, and suspensions, among others, are referred to as *non-Newtonian fluids*, in which the stress tensor is nonlinearly proportional to the velocity gradients.

The subject of Newtonian fluids is generally referred to as fluid mechanics, which encompasses diverse topics, such as the motion of airplanes and missiles through the atmosphere, satellites through the outer atmosphere, submarines and ships through water, the flow of liquids and gases through ducts, the transfer of heat and mass by fluid motion, propagation of sound through gases and liquids, the study of ocean waves and tides, the study of air masses in the atmosphere, astrophysical, geophysical, and meteorological problems, and reacting fluids.

Fluids, like all matter, are made up of molecules. Thus, the properties of fluid motion, such as those observed, may be studied on the basis of the mechanics of the molecules that compose the fluid. Although such a procedure may be feasible in principle, it will be a formidable task to achieve the solution of practical problems. As mentioned in Chapter 1, we are generally not interested in the details of the mechanics of the molecules. We wish to establish relations between various macroscopically observable quantities that pertain to a fluid at rest or in motion. Such observable properties are mean values in space and time obtained by taking the average over a sufficiently large volume that contains a large number of molecules and over a sufficiently long time compared to a certain time related to the mechanics of the molecules. For instance, at normal temperature and pressure, a 10^{-2} cc volume of air will contain about 2.7×10^7 molecules, which is a large number. This being so, it is a reasonable approximation to regard a fluid, whether at rest or in motion, as a continuous distribution of matter. We then speak of the fluid as a continuous medium or as a continuum.

The properties of *density* (ρ), *pressure* (p), and *temperature* (T) at a point in a liquid or a gas in static equilibrium are well known. At any point, it is the magnitude of the normal stress acting on an elemental plane area passing through that point. In a fluid at rest, only normal stresses occur and, in general, they are 'compressive' in nature. When only normal stresses occur, it is easy to show that, at any point, they should be equal in all directions.

The fact that only normal stresses occur for a fluid in static equilibrium contrasts with what happens to a solid in static equilibrium in which both normal and shear stresses prevail. The absence of shear stresses in a state of static equilibrium distinguishes a fluid from a solid and may be considered the property that defines a fluid.

All fluids undergo changes in volume under changes of pressure and temperature. The ability to change the volume of a mass of fluid is known as *compressibility*. It is well known that gases are more easily compressed than liquids. When a fixed mass of a fluid undergoes changes in volume, its density also changes, a phenomenon also known as compressibility. Under normal circumstances, the density changes in liquids due to pressure changes are small, whereas density changes due to temperature differences can be substantial. If the temperature differences are sufficiently small, the density changes in a liquid are almost nil and the liquid may then be regarded as an incompressible fluid. An incompressible fluid is one whose elements undergo no changes in volume or density. Compressibility is easy to notice in gases. In certain circumstances, however, the changes in volume or density of an element of a gas may be negligibly small. In such a case, as a reasonable approximation, the gas may be regarded as an incompressible fluid.

It is a matter of experience that even smoothly shaped bodies moving with a constant velocity through an otherwise undisturbed fluid encounter a resistance to their motion. Similarly, a fluid flowing through a pipe offers resistance. Thus, for a fluid in motion, shear and normal stresses occur on any elemental plane passing through a point in the fluid. The phenomenon of internal friction or *viscosity* is associated with these shear stresses. Such stresses give rise to a resistance to the nonuniform motion of a fluid, which results in *velocity gradients*. Shear stresses are, in general, linearly proportional to the velocity gradients through *viscosity constants*. A fluid of this type is Newtonian, as defined earlier.

When a fluid in static equilibrium is heated nonuniformly, heat may be transferred (without causing motion of the fluid) from points at which the temperature is high to those at which it is low by what is called *thermal conduction*. Observations show that under usual circumstances the *heat flux*, which is the amount of heat transferred across the surface element per unit time per unit area in the direction of the normal to the element, is proportional to the spatial rate of change of the temperature at that point in the direction of the normal. The heat flow occurs in the direction of decreasing temperature. Also, the static equilibrium of a fluid in which the temperature is not constant is unstable unless certain conditions are fulfilled. Instability of this kind leads to the appearance of convective currents in the fluid, which tend to mix the fluid in such a way that the temperature is equalized, a phenomenon known as *thermal convection*. Heat transfer also occurs in the form of *radiation* through participating or nonparticipating media. A process during which no heat is transferred to or from a system is known as an *adiabatic process*.

At this stage, it is convenient to introduce the concept of a perfect gas, which is the simplest working fluid in thermodynamics and, hence, is useful in the detailed study of thermodynamic processes. Measurements of the thermal properties of gases show that for low densities the thermal equation of state approaches the same form for all gases, namely,

$$pV = RT \tag{5.1.1}$$

or

$$p = \rho RT, \tag{5.1.2}$$

where p is the pressure, V is the specific volume, and R is the specific gas constant. Equation (5.1.1) or (5.1.2) represents the *equation of state* for a perfect (ideal) gas, known as a *calorically perfect* gas in which specific heats are constant. In practice most fluids have been observed to follow the perfect gas law given by Eq. (5.1.1) or (5.1.2). In general, pressure is a function of both density and temperature. If pressure is a function of only density, independent of temperature, then the fluid is *baratropic*.

In what follows, we present various subjects in fluid mechanics and heat transfer. We begin in Section 5.2 with the theory of constitutive equations for the stress tensor. Some historical developments are briefly reviewed. Section 5.3 gives the derivations of the most general forms of the governing equations in fluid mechanics using the first and second laws of thermodynamics. On the basis of these general forms, explicit equations for compressible viscous flow are derived for Cartesian and curvilinear coordinates in Section 5.4. We discuss an ideal flow in terms of velocity potential and stream functions, and the Bernoulli equation in Section 5.5.

Applied subjects in fluid mechanics include rotational flows and vorticity transport equations, turbulence, boundary layer, convective heat transfer, high-speed aerodynamics, acoustics, and reacting flows. These topics are presented in Sections 5.6–5.12.

5.2 Constitutive Equations

The basic principles for the development of constitutive theories for fluids are the same as those for solids. We proceed in accordance with the rules of physical admissibility, determinism, equipresence, local action, material objectivity, and material symmetry, as discussed in Chapter 4. First we review historical developments and examine transformation laws of stress tensors and the relations between the stresses and velocity gradients, thus arriving at the results analogous to those obtained for solids, and providing the basis for the most rigorous governing equations in compressible viscous flows.

5.2.1 Historical Review

To begin, a brief historical review of fluid mechanics is in order. Archimedes studied fluid behavior as early as 250 B.C., but it was not until the eighteenth century that Daniel Bernoulli described a simple constitutive law of inviscid flow,

$$\sigma_{ij} = -p\delta_{ij}. \qquad (5.2.1)$$

However, this law was inadequate for some fluid flows, as noted by Newton,[1] Cauchy, Navier, Saint-Venant, Stokes, and others. In 1821, for example, Navier proposed that additional terms proportional to velocity gradients ($V_{i,j}$) must be added to Eq. (5.2.1). In 1831 Poisson specified such terms to be of the form

$$\lambda d_{kk}\delta_{ij} + 2\mu d_{ij}, \tag{5.2.2}$$

where λ and μ are viscosity constants and d_{ij} is the rate-of-deformation tensor given by Eq. (2.5.1):

$$d_{ij} = \tfrac{1}{2}(V_{i,j} + V_{j,i}). \tag{5.2.3}$$

The derivatives in d_{ij}, as derived in Eq. (2.5.1), are with respect to the current spatial coordinates z_i instead of the undeformed reference coordinates x_i. Recall that this result arises as a consequence of the transformation between x_i and z_i. However, since no such mathematical operations are encountered in fluid mechanics problems other than in the derivation of d_{ij}, we henceforth replace z_i by x_i as the independent variable in the derivatives of all dependent variables, following the traditional notation in fluid mechanics ($\partial V_i/\partial x_j$ instead of $\partial V_i/\partial z_i$).

It is interesting to note that Eq. (5.2.2) is similar in form to the stress tensor for isotropic solids in the undeformed state. The idea in Eq. (5.2.2) was suggested by Newton in 1687. Subsequently, in 1845, Stokes proposed that, in general, the following relation should hold:

$$3\lambda + 2\mu = 0. \tag{5.2.4}$$

However, this assertion was disputed by others during the early part of the twentieth century on the basis of experimental data on some special types of fluids which are dependent on higher orders of velocity gradients (see Lamb [1879] for the classical fluid mechanics of the seventeenth to nineteenth centuries). Reiner (1945) studied a general theory of fluids using the principle of objectivity in which the fluid flow is assumed to be dependent on an isotropic function of the rate-of-deformation tensor d_{ij}. Reiner's theory was later criticized by Rivlin (1948), and subsequently, Rivlin and Eriksen (1955) offered a more general theory in which the time rate of change of d_{ij} of higher orders is added to Reiner's theory. This work was later modified by Green and Rivlin (1957), who included a dependence on the total history of deformation. Coleman and Noll (1959) and Noll (1966) proposed a

[1] Newton was the first to suggest that stress arises because of hydrostatic pressure and velocity gradients. Unfortunately, he left no records of an explicit mathematical form for the stress. Nevertheless, the term *Newtonian fluid* has been used to honor his original idea.

theory of simple fluids in which they considered the stress tensors to be dependent only on the history of deformation. A rather complete and systematic development of this theory is given by Truesdell and Noll (1965). These topics belong to non-Newtonian fluid mechanics, which is treated in Chung (1988).

5.2.2 The Stress Tensor

The stress tensor for fluids incapable of sustaining shear stresses is given by Eq. (5.2.1). When thermodynamic considerations are taken into account, we have

$$\sigma_{ij} = -\pi(\rho)\delta_{ij}, \tag{5.2.5}$$

where $\pi(\rho)$ is the thermodynamic pressure. To generalize Eq. (5.2.5) and to allow shears, in 1845 Stokes proposed that

$$\sigma_{ij} = -\pi\delta_{ij} + f_{ij}(\mathbf{d}), \tag{5.2.6}$$

where $f_{ij}(\mathbf{d})$ indicates a function of the rate-of-deformation tensor d_{ij}. If f_{ij} is linear, as in Eq. (5.2.3), then the fluid is said to be Newtonian.

One of the most widely used constitutive equations for viscous fluids is known as the Reiner–Rivlin equation,

$$\sigma_{ij} = -\pi\delta_{ij} + f_{ij} \tag{5.2.7}$$

with

$$f_{ij} = \alpha_0\delta_{ij} + \alpha_1 d_{ij} + \alpha_2 d_{ik}d_{kj}, \tag{5.2.8}$$

where α_0, α_1, and α_2 are functions of the invariants of d_{ij}

$$I_1 = d_{ii} = d(1) + d(2) + d(3), \tag{5.2.9a}$$

$$\begin{aligned} I_2 &= \tfrac{1}{2}(d_{ii}d_{jj} - d_{ij}d_{ij}) \\ &= d(1)d(2) + d(2)d(3) + d(3)d(1), \end{aligned} \tag{5.2.9b}$$

$$I_3 = |d_{ij}| = d(1)d(2)d(3) \tag{5.2.9c}$$

which are in the form similar to the invariants of the strain tensor γ_{ij} discussed in Section 2.7.

The function $f_{ij}(\mathbf{d})$ in Eq. (5.2.6) may be written as

$$f_{ij} = f_{ij}(d_{ij}, d_{ik}d_{kj}, d_{ik}d_{km}d_{mj}, \ldots). \tag{5.2.10}$$

It is possible for f_{ij} to be dependent on the past motion, as in certain polymer solutions. Fluids for which this is true are able to ‘remember’ their preceding configurations. On the other hand, if we neglect all higher orders of d_{ij}, then we get

$$f_{ij} = f_{ij}(d_{ij}) \tag{5.2.11a}$$

or

$$f_{ij} = f_{ij}\left(\frac{\partial V_i}{\partial x_j}\right). \tag{5.2.11b}$$

A fluid that obeys this relation, known as a Newtonian fluid, is unable to remember its preceding motion, although it may be able to remember its immediately preceding state of deformation history. With this in mind, we write $\partial V_i/\partial x_j$ as a sum of symmetric and antisymmetric parts (similarly as in Eq. [4.3.2a]):

$$\frac{\partial V_i}{\partial x_j} = \frac{1}{2}\left(\frac{\partial V_i}{\partial x_j} + \frac{\partial V_j}{\partial x_i}\right) + \frac{1}{2}\left(\frac{\partial V_i}{\partial x_j} - \frac{\partial V_j}{\partial x_i}\right)$$

or

$$\frac{\partial V_i}{\partial x_j} = d_{ij} + w_{ij}, \tag{5.2.12}$$

where w_{ij}, known as the rotational (spin) tensor (per unit time), is given by

$$w_{ij} = \frac{1}{2}\left(\frac{\partial V_i}{\partial x_j} - \frac{\partial V_j}{\partial x_i}\right) \tag{5.2.13}$$

and is antisymmetric, representing an angular motion. Thus, Eq. (5.2.11) may be revised for a new coordinate configuration as:

$$\bar{f}_{ij} = f_{ij}(d_{ij}, w_{ij}). \tag{5.2.14}$$

By virtue of the principle of material objectivity (frame indifference), as discussed in subsection 4.4.4, the orthogonal transformation is

$$\bar{f}_{ij} = a_{ir}a_{js}f_{rs} \tag{5.2.15}$$

and similarly,

$$\bar{d}_{ij} = a_{ir}a_{js}d_{rs}. \tag{5.2.16}$$

However, from

$$a_{ir}a_{jr} = \delta_{ij} \tag{5.2.17}$$

and its time derivative

$$\frac{\mathrm{d}}{\mathrm{d}t}(a_{ir}a_{jr}) = \dot{a}_{ir}a_{jr} + a_{ir}\dot{a}_{jr} = 0$$

we have

$$\begin{aligned} \bar{w}_{ij} &= \tfrac{1}{2}(\dot{a}_{ir}a_{jr} - a_{ir}\dot{a}_{jr}) + a_{ir}a_{js}w_{rs} \\ &= \dot{a}_{ir}a_{jr} + a_{ir}a_{js}w_{rs} \\ &= a_{ir}a_{js}(w_{rs} + \dot{a}_{rm}a_{sm}). \end{aligned} \tag{5.2.18}$$

This indicates that w_{rs} is not frame-indifferent, because orthogonal transformation is not available, as seen by the antisymmetry of $\dot{a}_{rm}a_{sm}$. In view of Eqs. (5.2.14) and (5.2.18), we have

$$a_{ir}a_{js}f_{rs} = f_{ij}(a_{ir}a_{js}d_{rs},\ a_{ir}a_{js}(w_{rs} + \dot{a}_{rm}a_{sm})). \tag{5.2.19}$$

Let us now consider the case of $a_{ir} = \delta_{ir}$, $\dot{a}_{ij} \neq 0$ (nonzero angular velocity). Then, Eq. (5.2.19) becomes

$$f_{ij} = f_{ij}(d_{ij},\ w_{ij} + \dot{a}_{ij}). \tag{5.2.20}$$

Consequently, this requires that angular motion vanish if the frame indifference is to be recovered[2] and if the stress is to be independent of rigid-body motion, as dictated by Eq. (5.2.11) or by the relation

$$a_{ir}a_{js}f_{rs} = f_{ij}(a_{ir}a_{js}d_{rs}). \tag{5.2.21}$$

Equation (5.2.21) requires that the fluid be isotropic, as deduced from Eqs. (5.2.15) and (5.2.16).

The explicit form of f_{ij} is thus written in terms of an isotropic tensor, E_{ijkm}, derived in Section 4.1,

$$f_{ij} = E_{ijkm}d_{km} = (\lambda\delta_{ij}\delta_{km} + \mu(\delta_{ik}\delta_{jm} + \delta_{im}\delta_{jk}))d_{km}$$

or

$$f_{ij} = \lambda d_{kk}\delta_{ij} + 2\mu d_{ij}. \tag{5.2.22}$$

Here, f_{ij} may be referred to as *excess stress*, which is contributed by velocity gradients, now denoted as τ_{ij}, the shear stress tensor:

$$\tau_{ij} = \lambda d_{kk}\delta_{ij} + 2\mu d_{ij}. \tag{5.2.23}$$

Note that this remarkable expression for the 'isotropic' stress tensor was derived from the most general concept of invariant energy (elastic potential) for originally 'anisotropic' solids in subsection 4.1.4 in terms of the strain tensor rather than the deformation rate tensor. The expression given by Eq. (5.2.23) can be obtained using a similar approach and taking the substantial time derivative. Such derivation is seldom mentioned in the fluid mechanics literature. Instead, Eq. (5.2.23) is accepted merely as a definition.

In fluid mechanics, λ and μ are called the dilatational viscosity constant and the shear viscosity constant, respectively, which we referred to as Lamé constants in elastic solids in Chapter 4. To establish the relation between these two constants, we invoke Stokes'

[2] Although the frame indifference is one criterion for constitutive equations, it can be asserted that the present results are the consequence of satisfying all other criteria discussed in subsection 4.4.4, directly and indirectly.

hypothesis. Toward this end, we set $i = j$ in Eq. (5.2.23):

$$\sigma_{ii} = -3\pi(\rho) + 3\lambda d_{ii} + 2\mu d_{ii}. \tag{5.2.24}$$

We denote the mean pressure as

$$-p = \tfrac{1}{3}\sigma_{ii} \tag{5.2.25}$$

with the negative sign indicating the mean hydrodynamic pressure in compression. Now, combining Eqs. (5.2.24) and (5.2.25), we obtain

$$\pi(\rho) - p = (\lambda + \tfrac{2}{3}\mu)d_{ii}. \tag{5.2.26}$$

According to Stokes, writing in 1845, the thermodynamic pressure $\pi(\rho)$ is approximately equal to the mean pressure p, which is true for most types of liquids and gases. For all arbitrary values of d_{ii}, this leads to

$$\lambda + \tfrac{2}{3}\mu = 0, \tag{5.2.27}$$

which agrees with Eq. (5.2.4). Thus, the dilatational viscosity constant can be eliminated in Eq. (5.2.23) by using the relation

$$\lambda = -\tfrac{2}{3}\mu. \tag{5.2.28}$$

Substituting Eq. (5.2.28) and the relation $\pi(\rho) = p$ into Eq. (5.2.23) gives the total stress tensor,

$$\sigma_{ij} = -p\delta_{ij} + \tau_{ij}, \tag{5.2.29}$$

where τ_{ij}, the shear stress tensor, takes the form

$$\tau_{ij} = 2\mu(d_{ij} - \tfrac{1}{3}d_{kk}\delta_{ij}). \tag{5.2.30}$$

Thus,

$$\sigma_{ij} = -p\delta_{ij} + 2\mu d_{ij}^{*}, \tag{5.2.31}$$

where

$$d_{ij}^{*} = d_{ij} - \tfrac{1}{3}d_{kk}\delta_{ij}, \tag{5.2.32}$$

which is the deviatoric part of the rate-of-deformation tensor, similar to deviatoric strains presented in Section 2.7. This implies that τ_{ij} is due only to a deviatoric part of the rate-of-deformation tensor.

We will return to the general form of f_{ij} in Eq. (5.2.10) or Eq. (5.2.8) to consider three scalar invariants of **d** in the form

$$f(d(1), d(2), d(3)) = \alpha_0 + \alpha_1 d(1) + \alpha_2 d^2(1), \tag{5.2.33a}$$

$$f(d(2), d(3), d(1)) = \alpha_0 + \alpha_1 d(2) + \alpha_2 d^2(2), \tag{5.2.33b}$$

$$f(d(3), d(1), d(2)) = \alpha_0 + \alpha_1 d(3) + \alpha_2 d^2(3), \tag{5.2.33c}$$

where α_0, α_1, and α_2 are functions of three scalar invariants of **d** and are symmetric functions of $d(1)$, $d(2)$, and $d(3)$. Thus, we may choose

$$\alpha_0 = \alpha_0(I_1, I_2, I_3), \tag{5.2.34a}$$

$$\alpha_1 = \alpha_1(I_1, I_2, I_3), \tag{5.2.34b}$$

$$\alpha_2 = \alpha_2(I_1, I_2, I_3), \tag{5.2.34c}$$

where the invariants I_1, I_2, and I_3 are given by Eq. (5.2.9). Therefore, Eq. (5.2.33) may be written

$$\text{diag}\,[\mathbf{f}(1), \mathbf{f}(2), \mathbf{f}(3)] = \alpha_0\mathbf{I} + \alpha_1\,\text{diag}\,[d(1), d(2), d(3)] + \alpha_2\,\text{diag}\,[d^2(1), d^2(2), d^2(3)] \tag{5.2.35}$$

or

$$\mathbf{f} = \alpha_0\mathbf{I} + \alpha_1\mathbf{d} + \alpha_2\mathbf{d}^2, \tag{5.2.36}$$

which is identical to Eq. (5.2.8). In general, for compressible flows, the stress tensor is a function of density ρ and temperature T in addition to the rate-of-deformation tensor:

$$\mathbf{f} = \hat{\mathbf{f}}(\mathbf{d}, \rho, T). \tag{5.2.37}$$

On the other hand, for incompressible flows, the stress is independent of density:

$$\mathbf{f} = \hat{\mathbf{f}}(\mathbf{d}, T). \tag{5.2.38}$$

Furthermore, α_0, α_1, and α_2 are independent of I_1 and the density for incompressible flows. Note that the absence of ρ and T in Eq. (5.2.11) should not have affected the derivation of Eq. (5.2.37), because ρ and T are frame-independent.

The classical Newtonian fluid results from Eq. (5.2.36) by setting $\alpha_2 = 0$ and

$$\alpha_0 = \hat{\alpha}_0 I_1 = \hat{\alpha}_0 d_{ii}, \quad \alpha_1 = 2\mu.$$

These modifications lead to

$$f_{ij} = \hat{\alpha}_0 d_{kk}\delta_{ij} + 2\mu d_{ij} \tag{5.2.39}$$

or

$$f_{ij} = \hat{\alpha} d_{kk}\delta_{ij} + 2\mu(d_{ij} - \tfrac{1}{3}d_{kk}\delta_{ij}), \tag{5.2.40}$$

where

$$\hat{\alpha} = \hat{\alpha}_0 + \tfrac{2}{3}\mu.$$

This is known as the *coefficient of bulk viscosity* and it is assumed to be negligible in most fluids. Thus, it follows that the stress tensor contributed by shears assumes the form

$$f_{ij} = \tau_{ij} = 2\mu(d_{ij} - \tfrac{1}{3}d_{kk}\delta_{ij}). \tag{5.2.41}$$

This corresponds to the shear stress for the Newtonian fluid given by Eq. (5.2.30).

It is seen that the stress tensor in Newtonian fluids consists of hydrodynamic pressure and excess stress due to velocity gradients or the deviatoric part of the rate-of-deformation tensor associated with the shear viscosity constant $\mu(\text{N-sec/m}^2)$. This constant is known as dynamic viscosity. Sometimes the kinematic viscosity, defined as $\nu = \mu/\rho$ (m^2/sec), is used for convenience in fluid mechanics and heat transfer problems.

5.3 The Navier–Stokes System of Equations

5.3.1 The First Law of Thermodynamics

It was shown in Section 3.3 that the balance (or conservation) of linear momentum leads to the equations of motion. It was also shown in Section 4.4 that the first law of thermodynamics yields the conservation of mass, momentum, and energy. Exactly the same approaches may be pursued for fluid mechanics, except that here we must use Eulerian coordinates, as indicated in subsection 2.2.2.

Just as in Eq. (4.4.4) for the Lagrangian coordinate system, the first law of thermodynamics for the Eulerian coordinate system is written with the exception that substantial derivatives now replace the ordinary time derivatives,

$$\frac{DK}{Dt} + \frac{DU}{Dt} = M + Q. \tag{5.3.1}$$

We use non-Cartesian (or curvilinear) coordinates here for the purpose of generality and because transformation into Cartesian coordinates is trivial once the derivations are performed in curvilinear coordinates. We define the following:

Kinetic Energy

$$K = \tfrac{1}{2}\int_\Omega \rho \mathbf{V} \cdot \mathbf{V}\, d\Omega = \tfrac{1}{2}\int_\Omega \rho V^i V_i\, d\Omega. \tag{5.3.2}$$

Internal Energy

$$U = \int_\Omega \rho\epsilon\, d\Omega, \tag{5.3.3}$$

where ϵ is the internal energy density. For an ideal gas, we have

$$\epsilon = c_v T, \tag{5.3.4a}$$

or

$$\epsilon = H - \frac{p}{\rho}, \tag{5.3.4b}$$

where $H = c_pT$ is the enthalpy, T the temperature, p the pressure, ρ the local density, and c_v and c_p the specific heat at constant volume and constant pressure, respectively.

Mechanical Power

$$\begin{aligned} M &= \int_\Omega \rho \mathbf{F} \cdot \mathbf{V}\, \mathrm{d}\Omega + \int_\Gamma \boldsymbol{\sigma}(n) \cdot \mathbf{V}\, \mathrm{d}\Gamma \\ &= \int_\Omega \rho F^i V_i\, \mathrm{d}\Omega + \int_\Gamma \sigma^{ij} n_i V_j\, \mathrm{d}\Gamma, \end{aligned} \tag{5.3.5}$$

where $\mathbf{F}$ is the body force and $\boldsymbol{\sigma}(n)$ is the surface traction normal to the boundary surface given by

$$\boldsymbol{\sigma}(n) = \sigma^{ij} n_i \mathbf{g}_j. \tag{5.3.6}$$

Heat Energy

$$\begin{aligned} Q &= \int_\Omega \rho h\, \mathrm{d}\Omega + \int_\Gamma \mathbf{q} \cdot \mathbf{n}\, \mathrm{d}\Gamma \\ &= \int_\Omega \rho h\, \mathrm{d}\Omega + \int_\Gamma q^i n_i\, \mathrm{d}\Gamma, \end{aligned} \tag{5.3.7}$$

where h is the heat supply and $\mathbf{q}$ is the heat flux, as defined in subsection 4.4.2. Note (see Eq. [2.2.29]) that

$$\frac{D\, \mathrm{d}\Omega}{Dt} = \frac{DJ}{Dt} \mathrm{d}\Omega_0 = J \boldsymbol{\nabla} \cdot \mathbf{V}\, \mathrm{d}\Omega_0 = \boldsymbol{\nabla} \cdot \mathbf{V}\, \mathrm{d}\Omega \tag{5.3.8}$$

and use the tensorial manipulations presented in earlier chapters, to obtain

$$\begin{aligned} &\frac{D}{Dt} \int_\Omega \frac{\rho}{2} \mathbf{V} \cdot \mathbf{V}\, \mathrm{d}\Omega \\ &\quad = \int_\Omega \frac{D}{Dt}\left(\frac{\rho}{2} \mathbf{V} \cdot \mathbf{V}\right) \mathrm{d}\Omega + \int_\Omega \frac{\rho}{2} \mathbf{V} \cdot \mathbf{V} \frac{D\, \mathrm{d}\Omega}{Dt} \\ &\quad = \int_\Omega \frac{\partial}{\partial t}\left(\frac{\rho}{2} \mathbf{V} \cdot \mathbf{V}\right) \mathrm{d}\Omega + \int_\Omega (\mathbf{V} \cdot \boldsymbol{\nabla})\left(\frac{\rho}{2} \mathbf{V} \cdot \mathbf{V}\right) \mathrm{d}\Omega \\ &\qquad + \int_\Omega \frac{\rho}{2} (\mathbf{V} \cdot \mathbf{V})(\boldsymbol{\nabla} \cdot \mathbf{V})\, \mathrm{d}\Omega \\ &\quad = \tfrac{1}{2} \int_\Omega \left[\frac{\partial}{\partial t}(\rho V^i V_i) + (\rho V^i V_i)_{|j} V^j + (\rho V^i V_i) V^j_{|j}\right] \mathrm{d}\Omega \end{aligned} \tag{5.3.9}$$

$$= \tfrac{1}{2}\int_\Omega \Big[V^i V_i \frac{\partial \rho}{\partial t} + \rho \frac{\partial}{\partial t}(V^i V_i) + \rho_{|j} V^i V_i V^j$$

$$+ \rho (V^i V_i)_{|j} V^j + \rho V^i V_i V^j_{|j} \Big] \, d\Omega$$

$$= \int_\Omega \left\{ \tfrac{1}{2} V^i V_i \left[\frac{\partial \rho}{\partial t} + (\rho V^j)_{|j} \right] + \left(\rho \frac{\partial V^i}{\partial t} + \rho V^i_{|j} V^j \right) V_i \right\} d\Omega.$$

Similarly, the substantial derivative of internal energy becomes

$$\frac{D}{Dt}\int_\Omega \rho\epsilon \, d\Omega = \int_\Omega \frac{D(\rho\epsilon)}{Dt} \, d\Omega + \int_\Omega \rho\epsilon \frac{D \, d\Omega}{Dt}$$

$$= \int_\Omega \epsilon \left[\frac{\partial \rho}{\partial t} + (\rho V^i)_{|i} \right] d\Omega + \int_\Omega \left(\rho \frac{\partial \epsilon}{\partial t} + \rho \epsilon_{|i} V^i \right) d\Omega. \tag{5.3.10}$$

The surface integral takes the form

$$\int_\Gamma \boldsymbol{\sigma}(n) \cdot \mathbf{V} \, d\Gamma = \int_\Gamma \sigma^{ij} n_i \mathbf{g}_j \cdot V_k \mathbf{g}^k \, d\Gamma = \int_\Gamma \sigma^{ij} n_i V_j \, d\Gamma$$

$$= \int_\Omega (\sigma^{ij} V_j)_{|i} \, d\Omega \tag{5.3.11}$$

$$= \int_\Omega (\sigma^{ij}_{|i} V_j + \sigma^{ij} V_{j|i}) \, d\Omega.$$

Substitute Eqs. (5.3.2)–(5.3.11) into Eq. (5.3.1) and adjust the indices to obtain

$$\int_\Omega E \left(\frac{\partial \rho}{\partial t} + (\rho V^i)_{|i} \right) d\Omega$$

$$+ \int_\Omega V^j \left(\rho \frac{\partial V^j}{\partial t} + \rho V^j_{|i} V^i - \rho F^j - \sigma^{ij}_{|i} \right) d\Omega \tag{5.3.12}$$

$$+ \int_\Omega \left(\rho \frac{\partial \epsilon}{\partial t} + \rho \epsilon_{|i} V^i - \sigma^{ij} V_{j|i} - q^i_{|i} - \rho h \right) d\Omega = 0,$$

where E is the total (stagnation) energy,

$$E = \epsilon + \tfrac{1}{2} V^i V_i.$$

At this point, for Eq. (5.3.12) to vanish, all integrands must vanish[3] such that, for all arbitrary values of E, V^j, and $d\Omega$,

$$\frac{\partial \rho}{\partial t} + (\rho V^i)_{|i} = 0, \tag{5.3.13}$$

$$\rho \frac{\partial V^j}{\partial t} + \rho V^j_{|i} V^i - \sigma^{ij}_{|i} - \rho F^j = 0, \tag{5.3.14}$$

[3] This argument will be verified independently by Eqs. (5.3.32a, b, c) in subsection 5.3.2.

$$\rho\frac{\partial\epsilon}{\partial t} + \rho\epsilon_{|i}V^i - \sigma^{ij}V_{j|i} - q^i_{|i} - \rho h = 0 \tag{5.3.15}$$

represent the equations of continuity, momentum, and energy, respectively, in the curvilinear coordinates. Equations (5.3.13)–(5.3.15) are known as the Navier–Stokes system of equations. The momentum equations alone are often called the Navier–Stokes equations, corresponding to the Navier equations for solid mechanics. Notice that Eq. (5.3.13) was derived as the expression for the conservation of mass in Eq. (2.2.28) for the Cartesian coordinates. In subsection 5.3.2 we prove that the conservation of mass is the prerequisite to the conservation of momentum. At the same time, the conservation of both mass and momentum is the prerequisite to the conservation of energy, as will be demonstrated in subsection 5.3.2.

For Cartesian coordinates, the covariant derivatives are replaced by the partial derivatives, leading to the Navier–Stokes system of equations in Cartesian coordinates.

Continuity

$$\frac{\partial\rho}{\partial t} + (\rho V_i)_{,i} = 0. \tag{5.3.16}$$

Momentum

$$\rho\frac{\partial V_j}{\partial t} + \rho V_{j,i}V_i - \rho F_j - \sigma_{ij,i} = 0. \tag{5.3.17}$$

Energy

$$\rho\frac{\partial\epsilon}{\partial t} + \rho\epsilon_{,i}V_i - \sigma_{ij}V_{j,i} - q_{i,i} - \rho h = 0, \tag{5.3.18}$$

where the heat flux q_i consists of conductive and radiative parts,

$$q_i = q_i^{(c)} + q_i^{(R)}, \tag{5.3.19}$$

in which the conductive heat flux $q_i^{(c)}$ may be given by the Fourier heat conduction law, as in Eq. (4.4.36), and the radiative heat flux $q_i^{(R)}$ is a complicated integral expression given in terms of various radiative parameters,

$$q_i^{(R)} = \int_0^\infty q_{\lambda i}^{(R)}\,\mathrm{d}\lambda = \int_0^\infty\int_{4\pi} I_\lambda n_i\,\mathrm{d}\omega\,\mathrm{d}\lambda, \tag{5.3.20}$$

where λ is the wave length, I_λ is the spectral radiation intensity, n_i is the direction cosine, and ω is the *solid angle*. For further details on radiative heat transfer, see Sparrow and Cess (1966) and Chung and Kim (1984).

Substituting Eq. (5.3.4b) into Eq. (5.3.18) yields

$$\rho c_p \frac{\partial T}{\partial t} + \rho c_p T_{,i} V_i - \frac{\partial p}{\partial t} - p_{,i} V_i - V_{i,i} p - \sigma_{ij} V_{j,i} - q_{i,i} - \rho h = 0. \tag{5.3.21}$$

Note that in Eq. (5.3.21) the continuity equation (5.3.16) is utilized to arrive at

$$\frac{1}{\rho}\left(\frac{\partial \rho}{\partial t} + \rho_{,i} V_i\right) = -V_{i,i}.$$

We have now obtained three differential equations (continuity, momentum, and energy), but there are four dependent variables (velocity, pressure, density, and temperature) to be determined. To this end, we require an additional equation, called the equation of state:

$$f(\rho, p, T) = 0, \tag{5.3.22}$$

For an ideal (perfect) gas, the equation of state assumes the form given in Eq. (5.1.2):

$$p = \rho RT. \tag{5.3.23}$$

Equations (5.3.16)–(5.3.23) represent the most general Cartesian form of governing equations in fluid mechanics. The specific forms of the stress tensor σ_{ij} and the equation of state will distinguish one type of *fluid* from another (i.e., viscous or inviscid, compressible or incompressible,[4] ideal or real gases). Furthermore, we must distinguish one type of *flow* from another, based on flow speeds (subsonic, transonic, or supersonic), flow patterns (laminar or turbulent, rotational or irrotational, with or without boundary layers, oscillatory or nonoscillatory), and the influence of body forces (natural or forced convection in heat transfer). Thus, the subject of fluid mechanics is divided into various specialized areas to deal with the various types of 'fluids' and 'flows', which will require each specialty to employ its own version of the governing equations. In the sequel, we demonstrate how the equations introduced in this section are modified to serve each special case.

For the purpose of comparing the foregoing derivation of the

[4] Compressibility and incompressibility may also be the flow properties instead of fluid properties. For example, the air is treated as incompressible for a Mach number lower than approximately 0.3, but compressible otherwise.

Navier–Stokes system of equations (5.3.12) with other approaches, the First Law of Thermodynamics approach presented in this section is referred to as the FLT equation. The next section examines other approaches, called CNS and CVS equations, respectively, referred to as the Conservation form of the Navier–Stokes system of equations and the Control Volume-Surface components of the Navier–Stokes system of equations.

5.3.2 The Conservation Form of the Navier–Stokes System of Equations, Control Volumes, and Control Surfaces

The Navier–Stokes system of equations derived from the first law of thermodynamics in subsection 5.3.1 may be recast in a *conservation form*. For simplicity, let us consider a two-dimensional case as follows.

Mass

$$\frac{\partial \rho}{\partial t} + \frac{\partial}{\partial x}(\rho u) + \frac{\partial}{\partial y}(\rho v) = 0. \tag{5.3.24}$$

Momentum

$$\frac{\partial}{\partial t}(\rho u) + \frac{\partial}{\partial x}(\rho u^2) + \frac{\partial}{\partial y}(\rho v u) = -\frac{\partial p}{\partial x} + \frac{\partial \tau_{xx}}{\partial x} + \frac{\partial \tau_{yx}}{\partial y} + \rho F_x, \tag{5.3.25a}$$

$$\frac{\partial}{\partial t}(\rho v) + \frac{\partial}{\partial x}(\rho u v) + \frac{\partial}{\partial y}(\rho v^2) = -\frac{\partial p}{\partial y} + \frac{\partial \tau_{xy}}{\partial x} + \frac{\partial \tau_{yy}}{\partial y} + \rho F_y. \tag{5.3.25b}$$

Energy

$$\begin{aligned}\frac{\partial}{\partial t}(\rho E) + \frac{\partial}{\partial x}(\rho E u) + \frac{\partial}{\partial y}(\rho E v) = \\ -\frac{\partial}{\partial x}(p u) - \frac{\partial}{\partial y}(p v) \\ + \frac{\partial}{\partial x}(\tau_{xx} u + \tau_{xy} v) + \frac{\partial}{\partial y}(\tau_{yx} u + \tau_{yy} v) \\ + \frac{\partial q_x}{\partial x} + \frac{\partial q_y}{\partial y} + \rho h + \rho F_x u + \rho F_y v.\end{aligned} \tag{5.3.26}$$

These results are obtained as a consequence of the conservation of mass, momentum, and energy in the control volumes and on the control surfaces as shown in Fig. 5.3.1, by summing all components that enter and exit through the inflow and outflow boundaries.

It is now a simple matter to combine Eqs. (5.3.24)–(5.3.26) into a compact form:

$$\frac{\partial \mathbf{U}}{\partial t} + \frac{\partial \mathbf{F}_i}{\partial x_i} + \frac{\partial \mathbf{G}_i}{\partial x_i} = \mathbf{B}, \tag{5.3.27}$$

which is referred to as the CNS equation. Here $\mathbf{U}$, $\mathbf{F}_i$, $\mathbf{G}_i$, and $\mathbf{B}$ are the conservation flow variables, convection flux variables, diffusion flux variables, and source terms, respectively, with

$$\mathbf{U} = \begin{bmatrix} \rho \\ \rho V_j \\ \rho E \end{bmatrix}, \quad \mathbf{F}_i = \begin{bmatrix} \rho V_i \\ \rho V_i V_j + p\delta_{ij} \\ \rho E V_i + p V_i \end{bmatrix}, \tag{5.3.28a–d}$$

$$\mathbf{G}_i = \begin{bmatrix} 0 \\ -\tau_{ij} \\ -\tau_{ij} V_j - q_i \end{bmatrix}, \quad \mathbf{B} = \begin{bmatrix} 0 \\ \rho F_j \\ \rho h + \rho F_j V_j \end{bmatrix},$$

$$\tau_{ij} = \mu(V_{i,j} + V_{j,i} - \tfrac{2}{3} V_{k,k}\delta_{ij}), \tag{5.3.29}$$

$$p = (\gamma - 1)\rho(E - \tfrac{1}{2} V_j V_j), \tag{5.3.30a}$$

$$T = \frac{1}{c_v}(E - \tfrac{1}{2} V_j V_j). \tag{5.3.30b}$$

The Navier–Stokes system of equations without the diffusion flux terms ($\mathbf{G}_i$) is known as the *Euler equations*.

$$\frac{\partial \mathbf{U}}{\partial t} + \frac{\partial \mathbf{F}_i}{\partial x_i} = \mathbf{B}. \tag{5.3.31}$$

We substitute Eqs. (5.3.28a, b, c, d) into Eq. (5.3.27) and perform the required differentiation; we obtain three separate equations:

$$\frac{\partial \rho}{\partial t} + (\rho V_i)_{,i} = 0, \tag{5.3.32a}$$

$$V_j\left[\frac{\partial \rho}{\partial t} + (\rho V_i)_{,i}\right] + \rho\frac{\partial V_j}{\partial t} + \rho V_i V_{j,i} + p_{,j} - \tau_{ij,i} - \rho F_j = 0, \tag{5.3.32b}$$

and

$$E\left[\frac{\partial \rho}{\partial t} + (\rho V_i)_{,i}\right] + V_j\left(\rho\frac{\partial V_j}{\partial t} + \rho V_i V_{j,i} + p_{,j} - \tau_{ij,i} - \rho F_j\right) + \rho\frac{\partial \epsilon}{\partial t} + \rho\epsilon_{,i} V_i + p V_{i,i} - \tau_{ij} V_{j,i} - q_{i,i} - \rho h = 0. \tag{5.3.32c}$$

A glance at Eq. (5.3.32b) indicates that conservation of mass is a prerequisite to the conservation of momentum, as given in Eq. (5.3.17). Similarly, Eq. (5.3.32c) states that conservation of both mass

$$\frac{\partial \rho}{\partial t} + \frac{\partial}{\partial x}(\rho u) + \frac{\partial}{\partial y}(\rho v) = 0$$

x-momentum $\dfrac{\partial}{\partial t}(\rho u) + \dfrac{\partial}{\partial x}(\rho u^2) + \dfrac{\partial}{\partial y}(\rho vu) = -\dfrac{\partial P}{\partial x} + \dfrac{\partial \tau_{xx}}{\partial x} + \dfrac{\partial \tau_{yx}}{\partial y} + \rho F_x$

y-momentum $\dfrac{\partial}{\partial t}(\rho v) + \dfrac{\partial}{\partial x}(\rho uv) + \dfrac{\partial}{\partial y}(\rho v^2) = -\dfrac{\partial P}{\partial y} + \dfrac{\partial \tau_{xy}}{\partial x} + \dfrac{\partial \tau_{yy}}{\partial y} + \rho F_y$

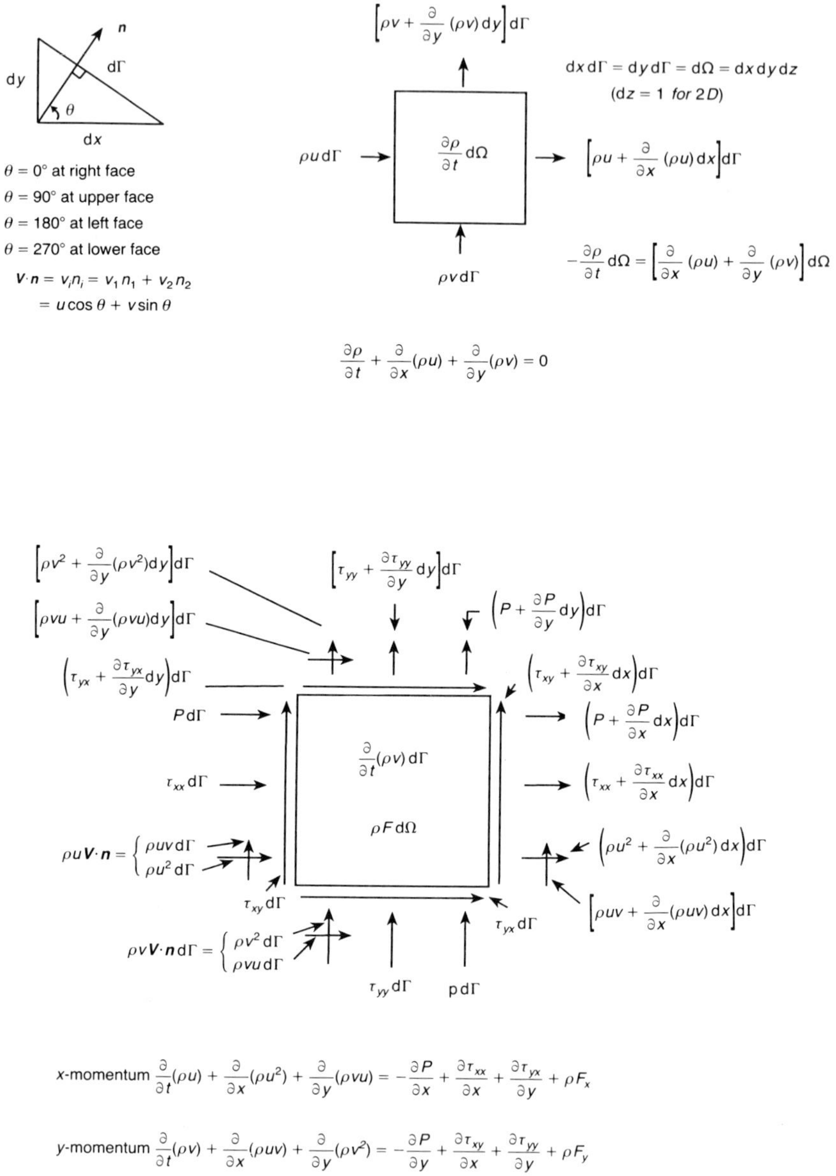

Figure 5.3.1 Free body diagrams for conservation of mass (top), momentum (bottom), and energy–CVS equations (facing page). Here the arrows are placed in the positive x and y directions and do not imply directions of the forces in balance. Also, the signs are determined by direction cosines, and pressure forces are opposite in sign relative to the viscous shear forces.

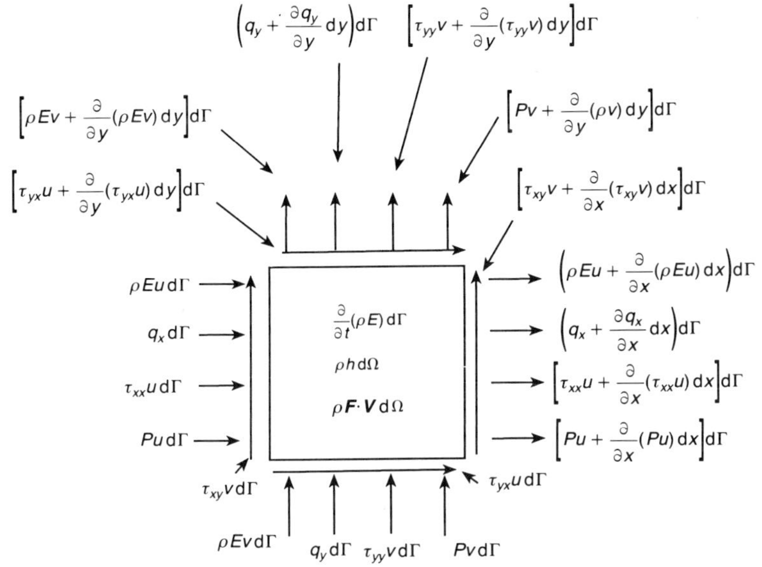

$$\frac{\partial}{\partial t}(\rho E) + \frac{\partial}{\partial x}(\rho Eu) + \frac{\partial}{\partial y}(\rho Ev) = -\frac{\partial}{\partial x}(Pu) - \frac{\partial}{\partial y}(Pv) + \frac{\partial}{\partial x} + \frac{\partial}{\partial x}(\tau_{xx}u + \tau_{xy}v)$$

$$+ \frac{\partial}{\partial y}(\tau_{yx}u + \tau_{yy}v) + \frac{\partial q_x}{\partial x} + \frac{\partial q_y}{\partial y} + \rho h + \rho \mathbf{F}\cdot\mathbf{V}$$

Figure 5.3.1 continued.

and momentum is a prerequisite to the conservation of energy, as given in Eq. (5.3.18). Thus, the conservation form Eq. (5.3.27) is equivalent to the first law of thermodynamics upon differentiation, resulting in the nonconservation form of the Navier–Stokes system of equations given by Eqs. (5.3.16)–(5.3.18). Indeed, the third equation, (5.3.32c), is identical to the integrands of Eq. (5.3.12) in Cartesian coordinates, as derived from the first law of thermodynamics.

If the conservation form (Eq. 5.3.27) is integrated over the domain,

$$\int_\Omega \left(\frac{\partial \mathbf{U}}{\partial t} + \frac{\partial \mathbf{F}_i}{\partial x_i} + \frac{\partial \mathbf{G}_i}{\partial x_i} - \mathbf{B} \right) d\Omega = 0 \tag{5.3.33}$$

we obtain

$$\int_\Omega \left(\frac{\partial \mathbf{U}}{\partial t} - \mathbf{B} \right) d\Omega + \int_\Gamma (\mathbf{F}_i + \mathbf{G}_i) n_i \, d\Gamma = 0 \tag{5.3.34a}$$

or in a discrete form for control volumes (CV) and control surfaces (CS),

$$\sum_{cv}\left(\frac{\partial \mathbf{U}}{\partial t} - \mathbf{B}\right)d\Omega + \sum_{cs}(\mathbf{F}_i + \mathbf{G}_i)n_i\,d\Gamma = 0, \qquad (5.3.34b)$$

which are referred to as the CVS equations. Notice that these equations lend themselves to a convenient numerical solution scheme because the boundary surface integrals in Eq. (5.3.34) can be evaluated numerically on boundaries. They can be computed along the exterior boundary surfaces, and along the interior control surfaces of a control volume for the computational grid points, such as in *finite volume methods*.

Note that for one-dimensional problems the surface integrals in Eq. (5.3.34a) play a major role in evaluating the simple physical phenomena governing the conservation of mass, momentum, and energy. Furthermore, both the volume and surface integrals in Eq. (5.3.34a) can be decomposed into all relevant components on free body diagrams, as shown in Fig. 5.3.1 for two dimensions. To see this the surface integrals (Eq. 5.3.34a) and the corresponding discrete forms (Eq. 5.3.34b) are written as follows.

Continuity

$$\begin{aligned} &\int_\Gamma \rho V_i n_i\,d\Gamma = a \\ &\sum_{cs}\rho V_i n_i\,d\Gamma = a. \end{aligned} \qquad (5.3.35a)$$

Momentum

$$\begin{aligned} &\int_\Gamma (\rho V_i V_j + p\delta_{ij} - \tau_{ij})n_i\,d\Gamma = b_j, \\ &\sum_{cs}(\rho V_i V_j + p\delta_{ij} - \tau_{ij})n_i\,d\Gamma = b_j. \end{aligned} \qquad (5.3.35b)$$

Energy

$$\begin{aligned} &\int_\Gamma (\rho E V_i + pV_i - \tau_{ij}V_j - q_i)n_i\,d\Gamma = c, \\ &\sum_{cs}(\rho E V_i + pV_i - \tau_{ij}V_j - q_i)n_i\,d\Gamma = c. \end{aligned} \qquad (5.3.35c)$$

Expansion of these equations for a two-dimensional domain ($i = 1, 2$)

represents all components that appear on the left and bottom control surfaces (upstream sides) in Fig. 5.3.1. The changes (differentials) of these quantities across the segments dx and dy, which are represented by the integrands of Eq. (5.3.33), appear on the right and top control surfaces (downstream sides). The time derivative and source terms are shown inside the square. Although all of these components may be intuitively predicted, as shown in traditional textbooks, it is seen here that they actually arise from the integration of the conservation form of the Navier–Stokes system of equations, as verified by the first law of thermodynamics. Notice that the direction cosines n_i ($n_1 = \cos\theta$, $n_2 = \sin\theta$, with θ measured from the x axis counterclockwise to the vector normal to the control surfaces) are used in determining the signs of all components in Fig. 5.3.1 and Eqs. (5.3.24)–(5.3.26). We emphasize that the balance of mass, momentum, and energy as these components move across the control volume is maintained mathematically with correct signs through appropriate direction cosines, as they appear in Eq. (5.3.35). Thus, the arrows shown in Fig. 5.3.1 merely indicate the direction of flow and are not indicative of signs to be used in determining the balance or equilibrium. The collection of equations that arise from Eq. (5.3.34b) in line with all components in Fig. 5.3.1 constitutes the CVS equations.

The importance of a thorough understanding of the relationships between the conservation variables in Fig. 5.3.1 (Eq. (5.3.34), CVS equations) and the conservation form of Navier–Stokes system of equations (Eq. (5.3.27), CNS equations) along with the consequence of the first law of thermodynamics (Eq. (5.3.12) or Eqs. (5.3.16)–(5.3.18), FLT equations) cannot be overemphasized. Modifications and simplifications of these equations will be discussed in the remainder of this chapter, as we relate them to various types of fluid flows.

5.3.3 Initial and Boundary Conditions

The basic approach in determining the boundary conditions is to classify the partial differential equations into elliptic, parabolic, and hyperbolic forms. Each form dictates a specific type of initial and boundary conditions.

For simplicity, let us consider the elliptic, parabolic, and hyperbolic equations in the form

$$\phi_{,ii} = 0 \qquad \textit{Elliptic}, \tag{5.3.36}$$

$$\frac{\partial \phi}{\partial t} - \phi_{,ii} = 0 \quad \textit{Parabolic}, \tag{5.3.37}$$

$$\frac{\partial^2 \phi}{\partial t^2} - \phi_{,ii} = 0 \quad \textit{Hyperbolic}. \tag{5.3.38}$$

In general, initial and boundary conditions are written as

$$\alpha(x_i, t) + \beta(x_i, t)\phi_{,i}n_i = \gamma(x_i, t). \tag{5.3.39}$$

Elliptic equations are associated with the Dirichlet condition ($\beta = 0$) or Neumann boundary conditions ($\alpha = 0$) in a closed region. On the other hand, parabolic equations are associated with the Dirichlet condition ($\beta = 0$) or the Neumann condition ($\alpha = 0$) in an open region. For hyperbolic equations, however, either Dirichlet or Neumann conditions with $\partial\phi/\partial t$ should be specified in an open region. For the case of the first-order hyperbolic (Euler) equation the outflow boundary conditions are unspecified. For mixed elliptic and parabolic equations it is often required to provide the so-called Cauchy (or Robin) condition ($\alpha \neq 0$, $\beta \neq 0$), such as in convective heat transfer boundary conditions.

The Navier–Stokes system of equations, in general, is considered as the mixed elliptic, parabolic, and hyperbolic equations if all terms in Eq. (5.3.33) are included. The Neumann boundary conditions are given by

$$\int_\Gamma (\mathbf{F}_i + \mathbf{G}_i)n_i \, d\Gamma = \mathbf{g}, \tag{5.3.40}$$

and explicit components are shown in Eq. (5.3.35).

Correct specifications of Dirichlet (essential) boundary conditions and Neumann (natural) boundary conditions render the solution of the Navier–Stokes system of equations 'well-posed'. Further details are found in Strikweda (1976), Gustafsson and Sundström (1978), and Dutt (1988).

5.3.4 The Second Law of Thermodynamics

Recall that we used the second law of thermodynamics to derive the heat conduction equation for solids in Section 4.4. This is because the only way the internal energy density can be characterized in thermomechanically loaded solids is to use the Helmholtz free energy in terms of entropy, which calls for the second law of thermodynamics. On the other hand, in fluid mechanics, the internal energy can be expressed by enthalpy rather than by entropy. For this reason, the second law of thermodynamics is not required for heat transfer involved in the energy equation. However, in acoustic oscillations in which thermal expansion of fluids occurs, such as in sound emission and absorption, it is necessary that the second law of thermodynamics be invoked. Toward this end, a brief introduction to thermodynamic relations in gases is in order.

The first law of thermodynamics in gases states that the heat added to the system δQ in an infinitesimal process is

$$\delta Q = \mathrm{d}\epsilon + p\,\mathrm{d}V, \tag{5.3.41}$$

where V is the specific volume $V = 1/\rho$ in nonreacting gases. Here δ implies different processes for Q. This equation can also be written in terms of the enthalpy H per unit mass,

$$H = \epsilon + pV. \tag{5.3.42}$$

Thus, in view of Eqs. (5.3.41) and (5.3.42),

$$\delta Q = \mathrm{d}H - V\,\mathrm{d}p. \tag{5.3.43}$$

The second law of thermodynamics in gases is stated in a different form from the case of the thermodynamics of solids, discussed in Section 4.4,

$$\mathrm{d}\eta \geqslant \frac{\delta Q}{T}, \tag{5.3.44}$$

where η is the specific entropy previously defined in the thermodynamics of solids (the more commonly used notation for entropy in thermodynamics in traditional texts is S) and the equality and inequality signs denote reversible and irreversible processes, respectively. For the reversible process, from Eq. (5.3.43) or (5.3.41) we have,

$$T\,\mathrm{d}\eta = \delta Q = \mathrm{d}H - V\,\mathrm{d}p \tag{5.3.45a}$$

or

$$T\,\mathrm{d}\eta = \delta Q = \mathrm{d}\epsilon + p\,\mathrm{d}V. \tag{5.3.45b}$$

It is interesting to note that these results confirm the existence of *Gibbs energy* in the form

$$G = H - T\eta,$$

which is related to the enthalpy and entropy.

To compute entropy changes in a variety of processes, it is necessary to have equations in which the changes are expressed in terms of measurable quantities. This requires the use of some relationships between partial derivatives of thermodynamic properties. These relationships are known as Maxwell's thermodynamic relations. To this end, we begin by expanding $\eta = \eta(T, p)$,

$$d\eta = \left(\frac{\partial \eta}{\partial T}\right)_p dT + \left(\frac{\partial \eta}{\partial p}\right)_T dp, \tag{5.3.46}$$

where the subscripts p and T indicate that pressure and temperature are, respectively, held constant for the appropriate partial derivatives. The following Maxwell's equations are obtained from Eqs. (5.3.41)–(5.3.46):

$$\left(\frac{\partial \eta}{\partial p}\right)_T = -\left(\frac{\partial V}{\partial T}\right)_p, \tag{5.3.47}$$

$$\left(\frac{\partial \eta}{\partial V}\right)_T = \left(\frac{\partial p}{\partial T}\right)_V, \tag{5.3.48}$$

$$\left(\frac{\partial \eta}{\partial V}\right)_p = \left(\frac{\partial p}{\partial T}\right)_\eta, \tag{5.3.49}$$

and

$$\left(\frac{\partial \eta}{\partial p}\right)_V = -\left(\frac{\partial V}{\partial T}\right)_p. \tag{5.3.50}$$

Additional properties associated with Maxwell's equations are

$$\left(\frac{\partial \eta}{\partial T}\right)_p = \frac{c_p}{T}, \tag{5.3.51}$$

$$c_p = c_v + p\left(\frac{\partial V}{\partial T}\right)_p = c_v + R, \tag{5.3.52}$$

where c_p and c_v are the specific heats at constant pressure and constant volume, respectively,

$$c_p = \left(\frac{\partial H}{\partial T}\right)_p, \quad c_v = \left(\frac{\partial \epsilon}{\partial T}\right)_v.$$

In view of Eqs. (5.3.46), (5.3.47), and (5.3.51), we obtain

$$d\eta = \frac{c_p}{T} dT - \left(\frac{\partial V}{\partial T}\right)_p dp, \tag{5.3.53}$$

or

$$d\eta = \frac{c_p}{T} dT - V\alpha\, dp, \tag{5.3.54}$$

where α is the coefficient of thermal expansion defined as

$$\alpha = \frac{1}{V}\left(\frac{\partial V}{\partial T}\right)_p = -\frac{1}{\rho}\left(\frac{\partial \rho}{\partial T}\right)_p. \tag{5.3.55}$$

Similarly, using $\eta = \eta(T, V)$, we obtain

$$T\,\mathrm{d}\eta = c_v\,\mathrm{d}T + T\left(\frac{\partial p}{\partial T}\right)_v \mathrm{d}V. \tag{5.3.56}$$

Let us now return to the energy equation (5.3.18) derived in subsection 5.3.1,

$$\rho\frac{D\epsilon}{Dt} = \sigma_{ij}V_{j,i} + q_{i,i} + \rho h. \tag{5.3.57}$$

Also, rewrite Eq. (5.3.45b) in the form

$$\begin{aligned} T\frac{D\eta}{Dt} &= \frac{D\epsilon}{Dt} + p\frac{D}{Dt}\left(\frac{1}{\rho}\right) \\ &= \frac{D\epsilon}{Dt} + \frac{p}{\rho}V_{i,i} \end{aligned} \tag{5.3.58}$$

with

$$p\frac{D}{Dt}\left(\frac{1}{\rho}\right) = -\frac{p}{\rho^2}\left[\frac{\partial\rho}{\partial t} + (\mathbf{V}\cdot\boldsymbol{\nabla})\rho\right] = \frac{p}{\rho}\boldsymbol{\nabla}\cdot\mathbf{V}. \tag{5.3.59}$$

It follows from Eq. (5.3.54) that

$$T\frac{D\eta}{Dt} = c_p\frac{DT}{Dt} - \frac{\alpha T}{\rho}\frac{Dp}{Dt}. \tag{5.3.60}$$

Combining Eqs. (5.3.57), (5.3.58), and (5.3.59) yields

$$\rho c_p\frac{DT}{Dt} - \alpha T\frac{Dp}{Dt} - pV_{i,i} - \sigma_{ij}V_{j,i} - \boldsymbol{\nabla}\cdot\mathbf{q} - \rho h = 0. \tag{5.3.61}$$

The final form of the energy equation is

$$\rho c_p\frac{\partial T}{\partial t} + \rho c_p T_{,i}V_i - \alpha T\frac{\partial p}{\partial t} - \alpha T p_{,i}V_i - \tau_{ij}V_{j,i} - q_{i,i} - \rho h = 0, \tag{5.3.62}$$

where τ_{ij} is given by Eq. (5.2.30). Here, we notice the interesting similarity of the term

$$-\alpha T\frac{\partial p}{\partial t},$$

associated with the thermal expansion coefficient α, to the counterpart

$$\alpha T_0(3\lambda + 2\mu)\frac{\partial}{\partial t}(u_{i,i})$$

in Eq. (4.4.40b) for solid mechanics. The difference in sign stems from the fact that, in solids, thermal stresses act against the mechanical resistance, whereas in fluids such restraint is absent. Applications of Eq. (5.3.62) can be made for the problems of sound wave emission or radiation and absorption from fluctuating forces in acoustics.

The second law of thermodynamics is also invoked when shock waves are curved, in which enthalpy and entropy gradients normal to the streamlines are nonzero. The flow is rotational behind the shock-waves. This subject is discussed in Section 5.10.

If specific heats c_p and c_v are not constant but are dependent on temperature, then such gas is known as *thermally perfect*. A *real gas* arises when it is dependent on both temperature and pressure. These topics are beyond the scope of this text.

5.4 Compressible Viscous Flow

5.4.1 Cartesian Coordinates

Having established the stress tensor for Newtonian fluids in Eq. (5.2.29) and the conservation equations for mass, momentum, and energy in Eqs. (5.3.16)–(5.3.18), we now proceed to the governing equations for compressible viscous fluids.

Continuity

$$\frac{\partial \rho}{\partial t} + \boldsymbol{\nabla} \cdot (\rho \mathbf{V}) = 0. \qquad (5.4.1)$$

Momentum

$$\rho \frac{\partial \mathbf{V}}{\partial t} + \rho(\mathbf{V} \cdot \boldsymbol{\nabla})\mathbf{V} - \rho \mathbf{F} + \boldsymbol{\nabla} p - \mu[\nabla^2 \mathbf{V} + \tfrac{1}{3}\boldsymbol{\nabla}(\boldsymbol{\nabla} \cdot \mathbf{V})] = 0. \qquad (5.4.2)$$

Energy

$$\rho \frac{\partial \epsilon}{\partial t} + \rho(\mathbf{V} \cdot \boldsymbol{\nabla})\epsilon + p\boldsymbol{\nabla} \cdot \mathbf{V} + \mu(\tfrac{2}{3}V_{i,i}V_{j,j} - V_{i,j}V_{j,i} - V_{i,j}V_{i,j}) - k\nabla^2 T - \rho h = 0, \qquad (5.4.3)$$

with the equation of state for perfect gas given in Eq. (5.1.2).

The momentum and energy equations may be written more explicitly in terms of index notation.

Momentum

$$\rho \frac{\partial V_i}{\partial t} + \rho V_{i,j}V_j - \rho F_i + p_{,i} - \mu(V_{i,jj} + \tfrac{1}{3}V_{j,ji}) = 0. \qquad (5.4.4)$$

Energy

$$\rho c_v \frac{\partial T}{\partial t} + \rho c_v T_{,i} V_i + \rho V_{i,i} + \mu(\tfrac{2}{3} V_{i,i} V_{j,j} - V_{i,j} V_{j,i} - V_{j,i} V_{j,i}) - k T_{,ii} - \rho h = 0, \quad (5.4.5a)$$

or

$$\rho c_p \frac{\partial T}{\partial t} + \rho c_p T_{,i} V_i - \frac{\partial p}{\partial t} - p_{,i} V_i + \mu(\tfrac{2}{3} V_{i,i} V_{j,j} - V_{i,j} V_{j,i} - V_{j,i} V_{j,i}) - k T_{,ii} - \rho h = 0. \quad (5.4.5b)$$

Here, the viscosity constant or dynamic viscosity μ and the coefficient of thermal conductivity k are, in general, given as a function of temperature. For two-dimensional flow ($x_1 = x$, $x_2 = y$, $V_1 = u$, $V_2 = v$), we write the following equations.

Continuity

$$\frac{\partial \rho}{\partial t} + \rho\left(\frac{\partial u}{\partial x} + \frac{\partial v}{\partial y}\right) + u\frac{\partial \rho}{\partial x} + v\frac{\partial \rho}{\partial y} = 0. \quad (5.4.6)$$

Momentum

$$\rho\frac{\partial u}{\partial t} + \rho\left(u\frac{\partial u}{\partial x} + v\frac{\partial u}{\partial y}\right) - \rho F_x + \frac{\partial p}{\partial x} - \mu\left(\frac{4}{3}\frac{\partial^2 u}{\partial x^2} + \frac{\partial^2 u}{\partial y^2} + \frac{1}{3}\frac{\partial^2 v}{\partial x \partial y}\right) = 0, \quad (5.4.7a)$$

$$\rho\frac{\partial v}{\partial t} + \rho\left(u\frac{\partial v}{\partial x} + v\frac{\partial v}{\partial y}\right) - \rho F_y + \frac{\partial p}{\partial y} - \mu\left(\frac{\partial^2 v}{\partial x^2} + \frac{4}{3}\frac{\partial^2 v}{\partial y^2} + \frac{1}{3}\frac{\partial^2 u}{\partial y \partial x}\right) = 0. \quad (5.4.7b)$$

Energy

$$\rho c_v\frac{\partial T}{\partial t} + \rho c_v\left(u\frac{\partial T}{\partial x} + v\frac{\partial T}{\partial y}\right) + p\left(\frac{\partial u}{\partial x} + \frac{\partial v}{\partial y}\right) - \mu\left\{\frac{4}{3}\left[\left(\frac{\partial u}{\partial x}\right)^2 + \left(\frac{\partial v}{\partial y}\right)^2 - \frac{\partial u}{\partial x}\frac{\partial v}{\partial y}\right] + 2\frac{\partial u}{\partial y}\frac{\partial v}{\partial x} + \left(\frac{\partial u}{\partial y}\right)^2 + \left(\frac{\partial v}{\partial x}\right)^2\right\} - k\left(\frac{\partial^2 T}{\partial x^2} + \frac{\partial^2 T}{\partial y^2}\right) - \rho h = 0, \quad (5.4.8a)$$

or

$$\rho c_p \frac{\partial T}{\partial t} + \rho c_p \left(u \frac{\partial T}{\partial x} + v \frac{\partial T}{\partial y} \right) - \frac{\partial p}{\partial t} - u \frac{\partial p}{\partial x} - v \frac{\partial p}{\partial y}$$
$$- \mu \left\{ \frac{4}{3} \left[\left(\frac{\partial u}{\partial x} \right)^2 + \left(\frac{\partial v}{\partial y} \right)^2 - \frac{\partial u}{\partial x} \frac{\partial v}{\partial y} \right] + 2 \frac{\partial u}{\partial y} \frac{\partial v}{\partial x} \right. \tag{5.4.8b}$$
$$\left. + \left(\frac{\partial u}{\partial y} \right)^2 + \left(\frac{\partial v}{\partial x} \right)^2 \right\} - k \left(\frac{\partial^2 T}{\partial x^2} + \frac{\partial^2 T}{\partial y^2} \right) - \rho h = 0,$$

where the dependent variables (u, v, ρ, p, T) are called primitive.

A glance at these equations indicates that there are nonlinear terms in the momentum and energy equations. They arise from $(\mathbf{V} \cdot \nabla)\mathbf{V}$ and $(\mathbf{V} \cdot \nabla)T$, which are known as *convective terms*. The presence of these terms renders the solution difficult both analytically and numerically. These terms are also needed to describe complicated physical phenomena, such as shock waves, turbulence, and convective heat transfer. Furthermore, the terms generated from $\sigma_{ij} V_{j,i}$, called *thermoviscous dissipation*, are also nonlinear. The nonlinearity of these terms is not as troublesome as that of the convective terms, but it is still the source of difficulty in analytical and numerical solutions.

5.4.2 Curvilinear Coordinates

From the derivations given in Section 5.3, we list the equations of continuity, momentum, and energy in terms of curvilinear coordinates as follows.

$$\frac{\partial \rho}{\partial t} + \rho_{|i} V^i + \rho V^i_{|i} = 0, \tag{5.4.9}$$

$$\rho \frac{\partial V^j}{\partial t} + \rho V^j_{|i} V^i - \rho F^j - \sigma^{ij}_{|i} = 0, \tag{5.4.10}$$

$$\rho \frac{\partial \epsilon}{\partial t} + \rho \epsilon_{|i} V^i - \sigma^{ij} V_{j|i} - q^i_{|i} - \rho h = 0, \tag{5.4.11}$$

where, for isotropic fluids,

$$\sigma^{ij} = -p g^{ij} + \tau^{ij}, \tag{5.4.12}$$

$$\begin{aligned} \tau^{ij} &= E^{ijkm} d_{km} = \lambda g^{ij} g^{km} d_{km} + \mu (g^{ik} g^{jm} + g^{im} g^{jk}) d_{km} \\ &= \lambda d^k_k g^{ij} + 2\mu d^{ij} = -\tfrac{2}{3}\mu d^k_k g^{ij} + 2\mu d^{ij} \\ &= \mu (g^{ik} V^j_{|k} + g^{jk} V^i_{|k} - \tfrac{2}{3} g^{ij} V^k_{|k}). \end{aligned} \tag{5.4.13}$$

Recall that Kronecker deltas were replaced by metric tensors (for the deformed state) in Eqs. (4.4.33a, b, and c). Similarly here, metric tensors for curvilinear coordinates are employed in place of Kronecker deltas. The rate-of-deformation tensor is of the form

$$d_{ij} = \tfrac{1}{2}(V_{i|j} + V_{j|i}). \tag{5.4.14}$$

Multiply Eq. (5.4.10) by $\mathbf{g}_j$ to obtain

$$\left(\rho\frac{\partial V^j}{\partial t} + \rho V^j_{|i}V^i - \rho F^j - \sigma^{ij}_{|i}\right)\mathbf{g}_j = 0, \tag{5.4.15}$$

or

$$\rho\frac{\partial \mathbf{V}}{\partial t} + \rho(\mathbf{V}\cdot\nabla)\mathbf{V} - \rho\mathbf{F} - \sigma^{ij}_{|i}\mathbf{g}_j = 0. \tag{5.4.16}$$

In view of Eqs. (5.4.12)–(5.4.14), it follows that

$$\begin{aligned} \sigma^{ij}_{|i}\mathbf{g}_j &= [(-pg^{ij})_{|i} + \tau^{ij}_{,i} + \Gamma^j_{mi}\tau^{mi} + \Gamma^i_{mi}\tau^{jm}]\mathbf{g}_j \\ &= \{-p_{,i}g^{ij} + \mu[(V^i_{|m}g^{jm} + V^j_{|m}g^{im} - \tfrac{2}{3}V^k_{|k}g^{ij})_{,i} \\ &\quad + \Gamma^j_{mi}(V^m_{|k}g^{ik} + V^i_{|k}g^{mk} - \tfrac{2}{3}V^k_{|k}g^{mi}) \\ &\quad + \Gamma^i_{mi}(V^j_{|k}g^{mk} + V^m_{|k}g^{jk} - \tfrac{2}{3}V^k_{|k}g^{jm})]\}\mathbf{g}_j. \end{aligned} \tag{5.4.17}$$

Be cautioned that the tangent vector $\mathbf{g}_j$ must be included in the algebra so that, when all tensor quantities are converted into physical components, they will be based on the unit vectors in the final form.

Instead of using Eq. (5.4.17), one may work with vector notation in the form

$$\sigma^{ij}_{|i}\mathbf{g}_j = -\nabla p + \mu[\nabla^2\mathbf{V} + \tfrac{1}{3}\nabla(\nabla\cdot\mathbf{V})], \tag{5.4.18}$$

where

$$\nabla p = \mathbf{g}^i\frac{\partial p}{\partial \xi_i} = p_{,i}\mathbf{g}^i = p_{,i}g^{ij}\mathbf{g}_j \tag{5.4.19}$$

with

$$\nabla^2\mathbf{V} = (\nabla\cdot\nabla)\mathbf{V} = \left(\mathbf{g}^k\frac{\partial}{\partial\xi_k}\cdot\mathbf{g}^j\frac{\partial}{\partial\xi_j}\right)V^i\mathbf{g}_i = g^{jk}V^i_{|jk}\mathbf{g}_i, \tag{5.4.20}$$

$$\begin{aligned} V^i_{|jk} &= -\Gamma^n_{kj}V^i_{,n} - \Gamma^m_{kj}\Gamma^i_{nm}V^n + V^i_{,jk} + 2V^m_{,k}\Gamma^i_{mj} + V^m(\Gamma^i_{jm})_{,k} \\ &\quad + V^m\Gamma^n_{mj}\Gamma^i_{nk}, \end{aligned}$$

and

$$\nabla(\nabla\cdot\mathbf{V}) = \mathbf{g}^k\frac{\partial}{\partial\xi_k}\left(\mathbf{g}^j\frac{\partial}{\partial\xi_j}\cdot V^i\mathbf{g}_i\right) = V^i_{|ik}\mathbf{g}^k = V^j_{|jk}g^{ki}\mathbf{g}_i \tag{5.4.21}$$

with

$$V^j_{|jk} = V^j_{,jk} + V^j_{,k}\Gamma^m_{mj} + V^j(\Gamma^m_{mj})_{,k}.$$

The reader may find the approach of Eq. (5.4.18) more straightforward than that of Eq. (5.4.17).

The equations of continuity and energy in curvilinear coordinates can be obtained similarly. Thus, the governing equations for compressible viscous fluids in cylindrical coordinates are written as follows.

Continuity

$$\frac{\partial \rho}{\partial t} + V_r \frac{\partial \rho}{\partial r} + \frac{V_\theta}{r}\frac{\partial \rho}{\partial \theta} + V_z \frac{\partial \rho}{\partial z} + \rho\left(\frac{\partial V_r}{\partial r} + \frac{1}{r}\frac{\partial V_\theta}{\partial \theta} + \frac{\partial V_z}{\partial z} + \frac{V_r}{r}\right) = 0. \quad (5.4.22)$$

Momentum

Momentum in the r direction:

$$\rho\frac{\partial V_r}{\partial t} + \rho\left(\frac{\partial V_r}{\partial r}V_r + \frac{1}{r}\frac{\partial V_r}{\partial \theta}V_\theta + \frac{\partial V_r}{\partial z}V_z - \frac{V_\theta^2}{r}\right) - \rho F_r + \frac{\partial p}{\partial r} - \mu\left[\frac{\partial^2 V_r}{\partial r^2} + \frac{1}{r}\frac{\partial V_r}{\partial r} + \frac{1}{r^2}\frac{\partial^2 V_r}{\partial \theta^2} - \frac{2}{r^2}\frac{\partial V_\theta}{\partial \theta} - \frac{V_r}{r^2} + \frac{\partial^2 V_r}{\partial z^2} + \frac{1}{3}\left(\frac{\partial^2 V_r}{\partial r^2} + \frac{1}{r}\frac{\partial^2 V_\theta}{\partial \theta \partial r} + \frac{\partial^2 V_z}{\partial r \partial z} - \frac{1}{r^2}\frac{\partial V_\theta}{\partial \theta} + \frac{1}{r}\frac{\partial V_r}{\partial r} - \frac{V_r}{r^2}\right)\right] = 0. \quad (5.4.23a)$$

Momentum in the θ direction:

$$\rho\frac{\partial V_\theta}{\partial t} + \rho\left(\frac{\partial V_\theta}{\partial r}V_r + \frac{1}{r}\frac{\partial V_\theta}{\partial \theta}V_\theta + \frac{\partial V_\theta}{\partial z}V_z + \frac{V_r V_\theta}{r}\right) - \rho F_\theta + \frac{1}{r}\frac{\partial p}{\partial \theta} - \mu\left[\frac{\partial^2 V_\theta}{\partial r^2} + \frac{1}{r}\frac{\partial V_\theta}{\partial r} + \frac{1}{r^2}\frac{\partial^2 V_\theta}{\partial \theta^2} + \frac{2}{r^2}\frac{\partial V_r}{\partial \theta} + \frac{\partial^2 V_\theta}{\partial z^2} - \frac{V_\theta}{r^2} + \frac{1}{3}\left(\frac{1}{r}\frac{\partial^2 V_r}{\partial r \partial \theta} + \frac{1}{r^2}\frac{\partial^2 V_\theta}{\partial \theta^2} + \frac{1}{r}\frac{\partial^2 V_z}{\partial z \partial \theta} + \frac{1}{r^2}\frac{\partial V_r}{\partial \theta}\right)\right] = 0. \quad (5.4.23b)$$

Momentum in the z direction:

$$\rho\frac{\partial V_z}{\partial t} + \rho\left(\frac{\partial V_z}{\partial r}V_r + \frac{1}{r}\frac{\partial V_z}{\partial \theta}V_\theta + \frac{\partial V_z}{\partial z}V_z\right) - \rho F_z + \frac{\partial p}{\partial z} - \mu\left[\frac{\partial^2 V_z}{\partial r^2} + \frac{1}{r}\frac{\partial V_2}{\partial r} + \frac{1}{r^2}\frac{\partial^2 V_z}{\partial \theta^2} + \frac{\partial^2 V_z}{\partial z^2} + \frac{1}{3}\left(\frac{\partial^2 V_r}{\partial r \partial z} + \frac{1}{r}\frac{\partial^2 V_\theta}{\partial \theta \partial z} + \frac{\partial^2 V_z}{\partial z^2} + \frac{1}{r}\frac{\partial V_r}{\partial z}\right)\right] = 0. \quad (5.4.23c)$$

Energy Equation

$$\rho c_p \frac{\partial T}{\partial t} + \rho c_p(\mathbf{V}\cdot\boldsymbol{\nabla})T - \frac{\partial p}{\partial t} - (\mathbf{V}\cdot\boldsymbol{\nabla})p - p(\boldsymbol{\nabla}\cdot\mathbf{V}) - \sigma^{ij}V_{j|i} - \boldsymbol{\nabla}\cdot\mathbf{q} - \rho h = 0$$

or

$$\begin{aligned}\rho c_p \frac{\partial T}{\partial t} &+ \rho c_p\left(\frac{\partial T}{\partial r}V_r + \frac{V_\theta}{r}\frac{\partial T}{\partial \theta} + \frac{\partial T}{\partial z}V_z\right) - \frac{\partial p}{\partial t} - \frac{\partial p}{\partial r}V_r \\ &- \frac{V_\theta}{r}\frac{\partial p}{\partial \theta} - \frac{\partial p}{\partial z}V_z \\ &- \tau^{ij}V_{j|i} - k\left(\frac{\partial^2 T}{\partial r^2} + \frac{1}{r^2}\frac{\partial^2 T}{\partial \theta^2} + \frac{\partial^2 T}{\partial z^2} + \frac{1}{r}\frac{\partial T}{\partial r}\right) - \rho h = 0,\end{aligned} \tag{5.4.24}$$

where

$$\begin{aligned}\tau^{ij}V_{j|i} &= \mu(g^{ik}V^j_{\ |k}V_{j|i} + V^i_{\ |k}V^k_{\ |i} - \tfrac{2}{3}V^i_{\ |i}V^k_{\ |k}) \\ &= \mu\left\{2\left(\frac{\partial V_r}{\partial r}\right)^2 + \frac{2}{r^2}\left(\frac{\partial V_\theta}{\partial \theta} + V_r\right)^2 + 2\left(\frac{\partial V_z}{\partial z}\right)^2\right. \\ &\quad + \left(\frac{\partial V_\theta}{\partial r}\right)^2 + \left(\frac{\partial V_z}{\partial r}\right)^2 + \frac{1}{r^2}\left(\frac{\partial V_r}{\partial \theta} - V_\theta\right)^2 \\ &\quad + \frac{1}{r^2}\left(\frac{\partial V_z}{\partial \theta}\right)^2 + \left(\frac{\partial V_r}{\partial z}\right)^2 + \left(\frac{\partial V_\theta}{\partial z}\right)^2 \\ &\quad + \frac{2}{r}\frac{\partial V_\theta}{\partial r}\left(\frac{\partial V_r}{\partial \theta} - V_\theta\right) + \frac{2}{r}\frac{\partial V_\theta}{\partial z}\frac{\partial V_z}{\partial \theta} + 2\frac{\partial V_z}{\partial r}\frac{\partial V_r}{\partial z} \\ &\quad \left. - \frac{2}{3}\left[\frac{\partial V_r}{\partial r} + \frac{1}{r}\left(\frac{\partial V_\theta}{\partial \theta} + V_r\right) + \frac{\partial V_z}{\partial z}\right]^2\right\}.\end{aligned} \tag{5.4.25}$$

Equations (5.4.22), (5.4.23), and (5.4.24) and the equation of state, (5.3.25), are the governing equations for compressible viscous flow in cylindrical coordinates.

The governing equations in spherical coordinates are derived in a similar manner. The results are as follows.

Continuity

$$\begin{aligned}\frac{\partial \rho}{\partial t} &+ V_R\frac{\partial \rho}{\partial R} + \frac{V_\alpha}{R}\frac{\partial \rho}{\partial \alpha} + \frac{V_\theta}{R\sin\alpha}\frac{\partial \rho}{\partial \theta} \\ &+ \rho\left(\frac{\partial V_R}{\partial R} + \frac{1}{R}\frac{\partial V_\alpha}{\partial \alpha} + \frac{1}{R\sin\alpha}\frac{\partial V_\theta}{\partial \theta} + \frac{2V_R}{R} + \frac{V_\alpha}{R}\cot\alpha\right) = 0.\end{aligned} \tag{5.4.26}$$

R Momentum

$$\rho\frac{\partial V_R}{\partial t} + \rho\left(V_R\frac{\partial V_R}{\partial R} + \frac{V_\alpha}{R}\frac{\partial V_R}{\partial \alpha} - \frac{V_\alpha^2}{R} + \frac{V_\theta}{R\sin\alpha}\frac{\partial V_R}{\partial \theta} - \frac{V_\theta^2}{R}\right)$$
$$- \rho F_R + \frac{\partial p}{\partial R} - \mu\left[\frac{\partial^2 V_R}{\partial R^2} + \frac{1}{R^2}\frac{\partial^2 V_R}{\partial \alpha^2} + \frac{1}{R^2\sin^2\alpha}\frac{\partial^2 V_R}{\partial \theta^2}\right.$$
$$- \frac{2}{R^2}\frac{\partial V_\alpha}{\partial \alpha} - \frac{2}{R^2\sin\alpha}\frac{\partial V_\theta}{\partial \theta} - \frac{2V_R}{R^2} - \frac{2V_\alpha\cot\alpha}{R^2} + \frac{2}{R}\frac{\partial V_R}{\partial R}$$
$$+ \frac{\cot\alpha}{R^2}\frac{\partial V_R}{\partial \alpha} + \frac{1}{3}\left(\frac{\partial^2 V_R}{\partial R^2} - \frac{1}{R^2}\frac{\partial V_R}{\partial \alpha} + \frac{1}{R}\frac{\partial^2 V_\alpha}{\partial \alpha \partial R} - \frac{1}{R^2\sin\alpha}\frac{\partial V_\theta}{\partial \theta}\right.$$
$$\left.\left. + \frac{1}{R\sin\alpha}\frac{\partial^2 V_\theta}{\partial \theta \partial R} - \frac{2V_R}{R^2} + \frac{2}{R}\frac{\partial V_R}{\partial R} - \frac{V_\alpha\cot\alpha}{R^2} + \frac{\cot\alpha}{R}\frac{\partial V_\alpha}{\partial R}\right)\right] = 0. \tag{5.4.27a}$$

α Momentum

$$\rho\frac{\partial V_\alpha}{\partial t} + \rho\left(V_R\frac{\partial V_R}{\partial R} + \frac{V_\alpha}{R}\frac{\partial V_\alpha}{\partial \alpha} + \frac{V_\alpha V_R}{R} + \frac{V_\theta}{R\sin\alpha}\frac{\partial V_\alpha}{\partial \theta} - \frac{V_\theta^2\cot\alpha}{R}\right)$$
$$- \rho F_\alpha + \frac{1}{R\sin^2\alpha}\frac{\partial p}{\partial \alpha} - \mu\left[\frac{2}{R}\frac{\partial V_\alpha}{\partial R} + \frac{\partial^2 V_\alpha}{\partial R^2} + \frac{1}{R^2}\frac{\partial^2 V_\alpha}{\partial \alpha^2}\right.$$
$$+ \frac{1}{R^2\sin^2\alpha}\frac{\partial^2 V_\alpha}{\partial \theta^2} + \frac{2}{R^2}\frac{\partial V_R}{\partial \alpha} - \frac{2\cos\alpha}{R^2\sin^2\alpha}\frac{\partial V_\theta}{\partial \theta} - \frac{V_\alpha}{R^2\sin^2\alpha}$$
$$+ \frac{\cot\alpha}{R^2}\frac{\partial V_\alpha}{\partial \alpha} + \frac{1}{3}\left(\frac{1}{R}\frac{\partial^2 V_R}{\partial R \partial \alpha} + \frac{1}{R^2}\frac{\partial^2 V_\alpha}{\partial \alpha^2} - \frac{\cos\alpha}{R^2\sin^2\alpha}\frac{\partial V_\theta}{\partial \theta}\right.$$
$$\left.\left. + \frac{1}{R^2\sin\alpha}\frac{\partial^2 V_\theta}{\partial \alpha \partial \theta} + \frac{2}{R^2}\frac{\partial V_R}{\partial \alpha} + \frac{\cot\alpha}{R^2}\frac{\partial V_\alpha}{\partial \alpha} - \frac{V_\alpha}{R_2} - \frac{V_\alpha\cot^2\alpha}{R^2}\right)\right] = 0. \tag{5.4.27b}$$

θ Momentum

$$\rho\frac{\partial V_\theta}{\partial t} + \rho\left(V_R\frac{\partial V\theta}{\partial R} + \frac{V_\alpha}{R}\frac{\partial V_\theta}{\partial \alpha} + \frac{V_\theta}{R\sin\alpha}\frac{\partial V_\theta}{\partial \theta} + \frac{V_\theta V_R}{R}\right.$$
$$\left. + \frac{V_\alpha V_\theta}{R}\cot\alpha\right)$$
$$-\rho F_\theta + \frac{1}{R\sin\alpha}\frac{\partial p}{\partial \theta} - \mu\left[\frac{2}{R}\frac{\partial V_\theta}{\partial R} + \frac{\partial^2 V_\theta}{\partial R^2} + \frac{\cot\alpha}{R^2}\frac{\partial V_\theta}{\partial \alpha} + \frac{1}{R^2}\frac{\partial^2 V_\theta}{\partial \alpha^2}\right.$$
$$\frac{1}{R^2\sin^2\alpha}\frac{\partial^2 V_\theta}{\partial \theta^2} - \frac{V_\theta}{R^2\sin^2\alpha} + \frac{2}{R^2\sin\alpha}\frac{\partial V_R}{\partial \theta} + \frac{2\cos\alpha}{R^2\sin^2\alpha}\frac{\partial V_\alpha}{\partial \theta}$$

$$+ \frac{1}{3R\sin\alpha}\left(\frac{\partial^2 V_R}{\partial R\partial\theta} + \frac{1}{R}\frac{\partial^2 V_\alpha}{\partial\alpha\partial\theta} + \frac{1}{R\sin\alpha}\frac{\partial^2 V_\theta}{\partial\theta^2}\right.$$
$$\left.\left. + \frac{2}{R}\frac{\partial V_R}{\partial\theta} + \frac{\cot\alpha}{R}\frac{\partial V_\alpha}{\partial\theta}\right)\right] = 0. \qquad (5.4.27c)$$

Energy

$$\rho c_p\frac{\partial T}{\partial t} - \frac{\partial p}{\partial t} + \rho c_p\left(V_R\frac{\partial T}{\partial R} + \frac{V_\alpha}{R}\frac{\partial T}{\partial\alpha} + \frac{V_\theta}{R\sin\alpha}\frac{\partial T}{\partial\theta}\right)$$
$$- V_R\frac{\partial p}{\partial R} - \frac{V_\alpha}{R}\frac{\partial p}{\partial\alpha} - \frac{V_\theta}{R\sin\alpha}\frac{\partial p}{\partial\theta}$$
$$- \mu\left\{\frac{4}{3}\left(\frac{\partial V_R}{\partial R}\right)^2 + \frac{2}{R}\frac{\partial V_\alpha}{\partial R}\frac{\partial V_R}{\partial\alpha} - \frac{2V_\alpha}{R}\frac{\partial V_\alpha}{\partial R}\right.$$
$$+ \frac{1}{R\sin\alpha}\frac{\partial V_\theta}{\partial R}\left(\frac{\partial V_R}{\partial\theta} - V_\theta\sin\alpha\right)$$
$$+ \left(\frac{1}{R}\frac{\partial V_\alpha}{\partial\alpha} + \frac{V_R}{R}\right)^2 + \frac{2}{R\sin\alpha}\frac{\partial V_\theta}{\partial\alpha}\left(\frac{1}{R}\frac{\partial V_\alpha}{\partial\theta} - \frac{V_\theta\cos\alpha}{R}\right)$$
$$+ \left(\frac{1}{R\sin\alpha}\frac{\partial V_\theta}{\partial\theta} + \frac{V_\alpha}{R}\cot\alpha + \frac{V_R}{R}\right)^2 + \left(\frac{\partial V_\alpha}{\partial R}\right)^2 + \left(\frac{\partial V_\theta}{\partial R}\right)^2$$
$$+ \frac{1}{R^2}\left[\left(\frac{\partial V_R}{\partial\alpha} - V_\alpha\right)^2 + \left(\frac{\partial V_\alpha}{\partial\alpha} + V_R\right)^2 + \left(\frac{\partial V_\theta}{\partial\alpha}\right)^2\right]$$
$$+ \frac{1}{R^2\sin^2\alpha}\left[\left(\frac{\partial V_R}{\partial\theta} - V_\theta\sin\alpha\right)^2 + \left(\frac{\partial V_\alpha}{\partial\theta} - V_\theta\cos\alpha\right)^2\right.$$
$$+ \left(\frac{1}{R\sin\alpha}\frac{\partial V_\theta}{\partial\theta} + \frac{V_\alpha}{R}\cot\alpha + \frac{V_R}{R}\right)$$
$$\left.\times\left(R\sin\alpha\frac{\partial V_\theta}{\partial\theta} + RV_R\sin^2\alpha + RV_R\sin\alpha\cos\alpha\right)\right]$$
$$- \frac{2}{3}\left[\frac{2}{R}\frac{\partial V_R}{\partial R}\left(\frac{\partial V_\alpha}{\partial\alpha} + V_R\right) + \frac{2}{R}\frac{\partial V_R}{\partial R}\left(\frac{1}{\sin\alpha}\frac{\partial V_\theta}{\partial\theta} + V_\alpha\cot\alpha + V_R\right)\right.$$
$$+ \frac{1}{R^2}\left(\frac{\partial V_\alpha}{\partial\alpha} + V_R\right)^2 + \frac{2}{R^2}\left(\frac{\partial V_\alpha}{\partial\alpha} + V_R\right)\left(\frac{1}{\sin\alpha}\frac{\partial V_\theta}{\partial\theta} + V_\alpha\cot\alpha + V_R\right)$$
$$\left.\left. + \frac{1}{R^2}\left(\frac{1}{\sin\alpha}\frac{\partial V_\theta}{\partial\theta} + V_\alpha\cot\alpha + V_R\right)^2\right]\right\}$$
$$- k\left(\frac{\partial^2 T}{\partial R^2} + \frac{1}{R^2}\frac{\partial^2 T}{\partial\alpha^2} + \frac{1}{R^2\sin^2\alpha}\frac{\partial^2 T}{\partial\theta^2} + \frac{2}{R}\frac{\partial T}{\partial R} + \frac{\cot\alpha}{R^2}\frac{\partial T}{\partial\alpha}\right)$$
$$- \rho h = 0. \qquad (5.4.28)$$

In summary, we have discussed a complete set of governing equations for compressible viscous flow in both Cartesian and curvilinear coordinates. Equations for other types of flow may be obtained by simplifying and modifying the equations for compressible viscous flow. We discuss these topics in the sections that follow.

5.5 Ideal Flow

5.5.1 General

An ideal flow is characterized by the vanishing of the divergence and curl of the velocity vector:

$$\nabla \cdot \mathbf{V} = 0, \tag{5.5.1}$$

$$\nabla \times \mathbf{V} = 0, \tag{5.5.2}$$

in which the fluid is assumed to be inviscid.

Recall that $\nabla \cdot \mathbf{V} = 0$ indicates the conservation of mass, which is also a requirement for incompressible flow. Notice that density must remain constant in space and time in this case. On the other hand, $\boldsymbol{\omega} = \nabla \times \mathbf{V} = 0$ implies irrotationality, or the vorticity vector $\boldsymbol{\omega}$ is zero.

The solution of Eqs. (5.5.1) and (5.5.2) is facilitated by the scalar functions called *velocity potential function* ϕ and *stream function* ψ such that, for two dimensions,

$$\mathbf{V} = \nabla\phi = \phi_{,i}\mathbf{i}_i \quad (i = 1, 2), \tag{5.5.3}$$

or

$$\mathbf{V} = \epsilon_{ij}\psi_{,j}\mathbf{i}_i \quad (i, j = 1, 2), \tag{5.5.4}$$

where ϵ_{ij} is the second-order permutation symbol defined by $\epsilon_{12} = 1$, $\epsilon_{21} = -1$, and all other ϵ_{ij} equal to zero. We prove the relation (5.5.4) in subsection 5.5.2. Substituting Eqs. (5.5.3) and (5.5.4) into Eqs. (5.5.1) and (5.5.2), respectively, we have

$$\nabla \cdot \mathbf{V} = \nabla^2\phi = \phi_{,ii} = 0 \quad (i = 1, 2), \tag{5.5.5}$$

and

$$\begin{aligned} \nabla \times \mathbf{V} &= V_{j,i}\epsilon_{ijk}\mathbf{i}_k = \psi_{,mi}\epsilon_{jm}\epsilon_{ijk}\mathbf{i}_k = (V_{2,1} - V_{1,2})\mathbf{i}_3 \\ &= -(\psi_{,11} + \psi_{,22})\mathbf{i}_3 = 0 \quad (i, j = 1, 2). \end{aligned}$$

Thus,

$$\nabla^2\psi = \psi_{,ii} = 0 \quad (i = 1, 2). \tag{5.5.6}$$

Equations (5.5.5) and (5.5.6) are the Laplace equations written as

$$\nabla^2 \phi = \frac{\partial^2 \phi}{\partial x^2} + \frac{\partial^2 \phi}{\partial y^2} = 0,$$

$$\nabla^2 \psi = \frac{\partial^2 \psi}{\partial x^2} + \frac{\partial^2 \psi}{\partial y^2} = 0.$$

The solution of Laplace equations $\nabla^2 \phi = 0$ and $\nabla^2 \psi = 0$ will provide data for potential lines and streamlines, respectively. The two must intersect orthogonally, i.e., $\nabla \phi \cdot \nabla \psi = 0$, everywhere in the domain; proof of this is left to the reader. An example of ideal flow over a cylinder is shown in Fig. 5.5.1 with points A and B indicating the positions of highest velocity and zero velocity (stagnation), respectively. In the absence of viscosity (friction) and rotation, the flow field is symmetrical everywhere, with ψ and ϕ lines forming perfect squares, known as 'flow net', as $\Delta x \rightarrow 0$ and $\Delta y \rightarrow 0$. The region of high velocity is indicated by smaller squares. The symmetric flow pattern will no longer be maintained if viscosity becomes significant.

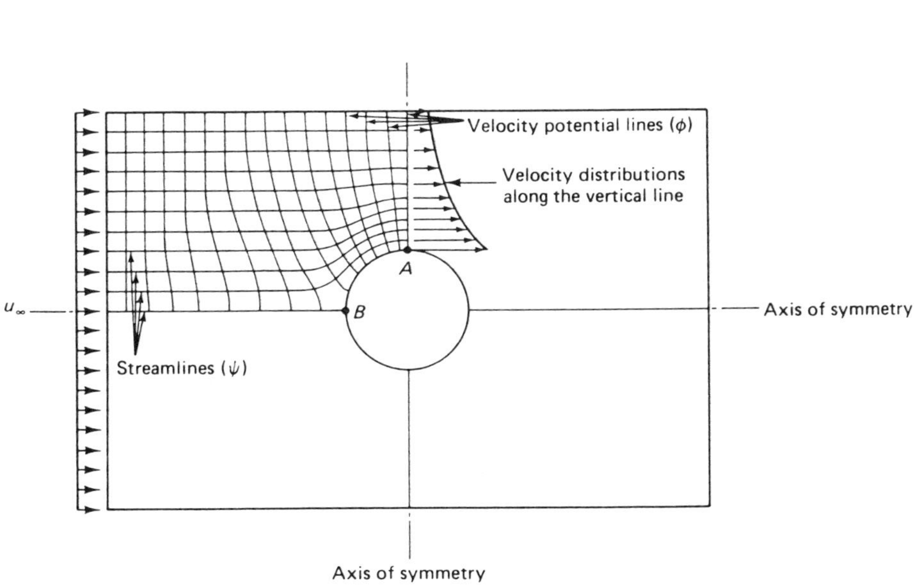

Figure 5.5.1 Two-dimensional ideal flow around a cylinder between plates. Streamlines and velocity potential lines intersect orthogonally so that in the limit they form perfect squares everywhere. The smaller the square, the higher the velocity. The smallest square occurs at A giving rise to the highest velocity; the largest square occurs at B corresponding to the lowest velocity. A square can be formed at B only when infinitely many ϕ and ψ lines are drawn.

5.5.2 Existence of the Stream Function in Two Dimensions

The relation between the velocity and stream function given by Eq. (5.5.4) is proved in this section. To this end, let $\mathbf{dR} \equiv \mathbf{dx}$ be an element of the streamline passing through any point and $\mathbf{V}$ denote the velocity vector at this point (see Fig. 5.5.2) such that $\mathbf{dx}$ is parallel to $\mathbf{V}$:

$$\mathrm{d}\mathbf{x} \times \mathbf{V} = 0, \tag{5.5.7}$$

or

$$\mathrm{d}x_i \mathbf{i}_i \times V_j \mathbf{i}_j = \mathrm{d}x_i V_j \epsilon_{ijk} \mathbf{i}_k = 0.$$

Upon expansion, we obtain

$$V_3 \, \mathrm{d}x_2 - V_2 \, \mathrm{d}x_3 = 0, \tag{5.5.8a}$$

$$V_1 \, \mathrm{d}x_3 - V_3 \, \mathrm{d}x_1 = 0, \tag{5.5.8b}$$

$$V_2 \, \mathrm{d}x_1 - V_1 \, \mathrm{d}x_2 = 0, \tag{5.5.8c}$$

which may be written as

$$\frac{\mathrm{d}x_1}{V_1} = \frac{\mathrm{d}x_2}{V_2} = \frac{\mathrm{d}x_3}{V_3}. \tag{5.5.9}$$

Equation (5.5.8) or (5.5.9) is the Cartesian form of the set of differential equations for a streamline. Since only two of the equations can be independent in equation set (5.5.8), we may write

$$C_{mj}(x_k) \, \mathrm{d}x_j = 0 \quad (m = 1, 2, j = 1, 2, 3). \tag{5.5.10}$$

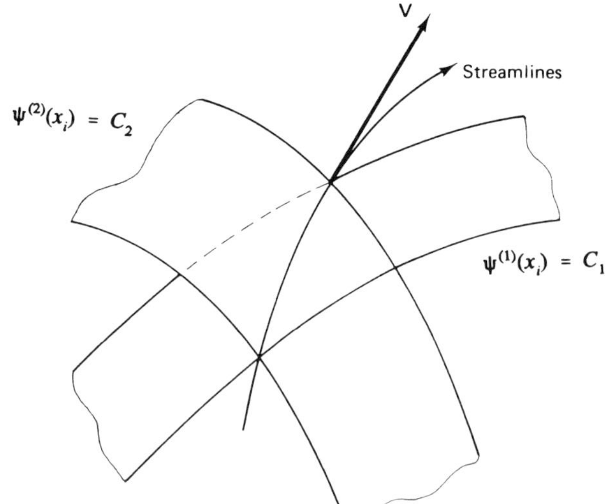

Figure 5.5.2 Stream functions ($\psi^{(1)}$, $\psi^{(2)}$) on surfaces C_1 and C_2 and streamlines parallel to $\mathbf{V}$.

Equations of this kind, (5.5.8)–(5.5.10), are known as *Pfaffian differential equations* (Sneddon 1957), and represent a streamline (see Eq. 2.2.32). The solution of Eq. (5.5.10) is given by two independent functions,

$$\psi^{(1)}(x_i) = c_1,$$
$$\psi^{(2)}(x_i) = c_2.$$

This indicates that a line in space, such as a streamline, may be described as the curve of the intersection of two surfaces such that

$$\mathbf{V} \cdot \nabla \psi^{(1)} = 0, \tag{5.5.11a}$$
$$\mathbf{V} \cdot \nabla \psi^{(2)} = 0, \tag{5.5.11b}$$

or

$$\mathbf{V} = \nabla \psi^{(1)} \times \nabla \psi^{(2)}. \tag{5.5.12}$$

Equation (5.5.12) implies that $\mathbf{V}$ is parallel to the cross-product $\nabla \psi^{(1)} \times \nabla \psi^{(2)}$ (see Fig. 5.5.2).

For two-dimensional problems ($V_3 = 0$), the streamline equation takes the form

$$\frac{\mathrm{d}x_1}{V_1} = \frac{\mathrm{d}x_2}{V_2} = \frac{\mathrm{d}x_3}{0}. \tag{5.5.13}$$

This requires that $\mathrm{d}x_3 = 0$ and $x_3 =$ constant, suggesting that

$$\psi^{(1)} = \psi(x_1, x_2), \tag{5.5.14a}$$
$$\psi^{(2)} = x_3. \tag{5.5.14b}$$

Expand Eq. (5.5.12) to get

$$\begin{aligned} \mathbf{V} &= \psi^{(1)}_{,i} \psi^{(2)}_{,j} \epsilon_{ijk} \mathbf{i}_k \\ &= (\psi^{(1)}_{,2} \psi^{(2)}_{,3} - \psi^{(1)}_{,3} \psi^{(2)}_{,2}) \mathbf{i}_1 + (\psi^{(1)}_{,3} \psi^{(2)}_{,1} - \psi^{(1)}_{,1} \psi^{(2)}_{,3}) \mathbf{i}_2 \\ &\quad + (\psi^{(1)}_{,1} \psi^{(2)}_{,2} - \psi^{(1)}_{,2} \psi^{(2)}_{,1}) \mathbf{i}_3. \end{aligned} \tag{5.5.15}$$

Then substitute Eq. (5.5.14) into Eq. (5.5.15), which gives

$$\mathbf{V} = V_1 \mathbf{i}_1 + V_2 \mathbf{i}_2 = \psi^{(1)}_{,2} \psi^{(2)}_{,3} \mathbf{i}_1 - \psi^{(1)}_{,1} \psi^{(2)}_{,3} \mathbf{i}_2$$

or, in view of Eq. (5.5.14),

$$V_1 \mathbf{i}_1 + V_2 \mathbf{i}_2 = \psi_{,2} \mathbf{i}_1 - \psi_{,1} \mathbf{i}_2 = \epsilon_{ij} \psi_{,j} \mathbf{i}_i \quad (i, j = 1, 2),$$

which is in agreement with Eq. (5.5.4). Using the standard notation, we obtain

$$V_x = \frac{\partial \psi}{\partial y}, \quad V_y = -\frac{\partial \psi}{\partial x}. \tag{5.5.16}$$

This result satisfies Eq. (5.5.1):

$$\nabla \cdot \mathbf{V} = (\epsilon_{ij} \psi_{,j})_{,i} = 0.$$

For axisymmetric cylindrical motion ($V_\theta = 0$) (see Fig. 2.3.4a), we have $x_1 = r$, $x_2 = \theta$, $x_3 = z$, $V^1 = V_r$, $V^2 = V_\theta/r$, and $V^3 = V_z$. Thus, it follows that

$$\frac{\mathrm{d}z}{V_z(z, r)} = \frac{\mathrm{d}r}{V_r(z, r)} = \frac{r\,\mathrm{d}\theta}{0}.$$

Since we must have $\mathrm{d}\theta = 0$ and thus $\theta =$ constant, we define

$$\psi^{(1)} = \psi, \quad \psi^{(2)} = \theta.$$

These conditions lead to

$$\begin{aligned}
\mathbf{V} = \nabla\psi^{(1)} \times \nabla\psi^{(2)} &= \frac{\partial\psi^{(1)}}{\partial\xi_i}\frac{\partial\psi^{(2)}}{\partial\xi_j}\mathbf{g}^i \times \mathbf{g}^j \\
&= \psi^{(1)}_{,i}\psi^{(2)}_{,j}\frac{1}{\sqrt{g}}\epsilon^{ijk}\mathbf{g}_k \\
&= \frac{1}{\sqrt{g}}[(\psi^{(1)}_{,2}\psi^{(2)}_{,3} - \psi^{(1)}_{,3}\psi^{(2)}_{,2})\mathbf{g}_1 + (\psi^{(1)}_{,3}\psi^{(2)}_{,1} - \psi^{(1)}_{,1}\psi^{(2)}_{,3})\mathbf{g}_2 \\
&\quad + (\psi^{(1)}_{,1}\psi^{(2)}_{,2} - \psi^{(1)}_{,2}\psi^{(2)}_{,1})\mathbf{g}_3] \\
&= \frac{1}{r}(-\psi^{(1)}_{,3}\psi^{(2)}_{,2}\mathbf{g}_1 + \psi^{(1)}_{,1}\psi^{(2)}_{,2}\mathbf{g}_3) \\
&= -\frac{1}{r}\frac{\partial\psi}{\partial z}\mathbf{e}_r + \frac{1}{r}\frac{\partial\psi}{\partial r}\mathbf{e}_z.
\end{aligned}$$

Since $\mathbf{V} = V_r\mathbf{e}_r + V_z\mathbf{e}_z$, the velocity-stream function relations are expressed by:

$$V_r = -\frac{1}{r}\frac{\partial\psi}{\partial z}, \quad V_z = \frac{1}{r}\frac{\partial\psi}{\partial r}. \tag{5.5.17}$$

Note that, once again, Eq. (5.5.1) is satisfied. Here streamlines are turned into stream tubes in the three-dimensional domain.

Similarly, for axisymmetric spherical motion (see Fig. 2.3.4b), we have $x_1 = R$, $x_2 = \alpha$, $x_3 = \theta$, $V^1 = V_R$, $V^2 = V_\alpha/R$, and $V^3 = v_\theta/(R\sin\alpha)$. Thus,

$$\frac{\mathrm{d}R}{V_R(R, \alpha)} = \frac{R\,\mathrm{d}\alpha}{V_\alpha(R, \alpha)} = \frac{R\sin\alpha\,\mathrm{d}\theta}{0},$$

which requires

$$\mathrm{d}\theta = 0, \quad \theta = \textit{constant}, \quad \psi^{(1)} = \psi(R, \alpha), \quad \psi^{(2)} = \theta$$

and

$$\mathbf{V} = \nabla\psi^{(1)} \times \nabla\psi^{(2)} = \frac{\partial\psi^{(1)}}{\partial\xi_i}\frac{\partial\psi^{(2)}}{\partial\xi_j}\mathbf{g}^i \times \mathbf{g}^j$$

$$= \psi_{,i}^{(1)}\psi_{,j}^{(2)}\frac{1}{\sqrt{g}}\epsilon^{ijk}\mathbf{g}_k$$

$$= \frac{1}{\sqrt{g}}[(\psi_{,2}^{(1)}\psi_{,3}^{(2)} - \psi_{,3}^{(1)}\psi_{,2}^{(2)})\mathbf{g}_1 + (\psi_{,3}^{(1)}\psi_{,1}^{(2)} - \psi_{,1}^{(1)}\psi_{,3}^{(2)})\mathbf{g}_2 + (\psi_{,1}^{(1)}\psi_{,2}^{(2)} - \psi_{,2}^{(1)}\psi_{,1}^{(2)})\mathbf{g}_3]$$

$$= \frac{1}{R^2 \sin\alpha}[\psi_{,2}^{(1)}\psi_{,3}^{(2)}\mathbf{g}_1 - \psi_{,1}^{(1)}\psi_{,3}^{(2)}\mathbf{g}_2]$$

$$= \frac{1}{R^2 \sin\alpha}\frac{\partial\psi}{\partial\alpha}\mathbf{e}_R - \frac{1}{R\sin\alpha}\frac{\partial\psi}{\partial R}\mathbf{e}_\alpha.$$

This gives

$$V_R = \frac{1}{R^2 \sin\alpha}\frac{\partial\psi}{\partial\alpha}, \quad V_\alpha = -\frac{1}{R\sin\alpha}\frac{\partial\psi}{\partial R}, \tag{5.5.18}$$

and Eq. (5.5.1) is also satisfied.

5.5.3 Existence of the Stream Function in Three Dimensions

The most general approach to derive the relationship between the velocity and the stream function in three-dimensional flow is to consider that the velocity $\mathbf{V}$ is parallel or tangent to the streamline identified by a stream function ψ (thus $\nabla\psi$ is perpendicular to $\mathbf{V}$), and that $\mathbf{V}$ is equal to the curl of $\nabla\psi$ and the unit normal vector $\hat{\mathbf{n}}$ (perpendicular to the plane of $\nabla\psi$ and $\mathbf{V}$), as shown in Fig. 5.5.3a.

$$\mathbf{V} = \nabla\psi \times \hat{\mathbf{n}} = \epsilon_{ijk}\psi_{,j}\hat{n}_k\mathbf{i}_i \quad (i, j, k = 1, 2, 3) \tag{5.5.19a}$$

with each component given by

$$V_1 = \frac{\partial\psi}{\partial y}\hat{n}_3 - \frac{\partial\psi}{\partial z}\hat{n}_2, \quad V_2 = \frac{\partial\psi}{\partial z}\hat{n}_1 - \frac{\partial\psi}{\partial x}\hat{n}_3, \quad V_3 = \frac{\partial\psi}{\partial x}\hat{n}_2 - \frac{\partial\psi}{\partial y}\hat{n}_1. \tag{5.5.19b}$$

We may identify the stream function described on each surface designated by $\hat{n}_i$ as (see Fig. 5.5.3b)

$$\Psi_1 = \psi\hat{n}_1, \quad \Psi_2 = \psi\hat{n}_2, \quad \Psi_3 = \psi\hat{n}_3. \tag{5.5.19c}$$

These definitions suggest that there exist three independent stream function components

$$\boldsymbol{\Psi} = \Psi_i\mathbf{i}_i \quad (i = 1, 2, 3)$$

so that

$$\mathbf{V} = \nabla \times \boldsymbol{\Psi} = \epsilon_{ijk}\Psi_{k,j}\mathbf{i}_i \tag{5.5.20}$$

with $\Psi_k = \psi\hat{n}_k$, as given by Eq. (5.5.19c). Equations (5.5.19a) and

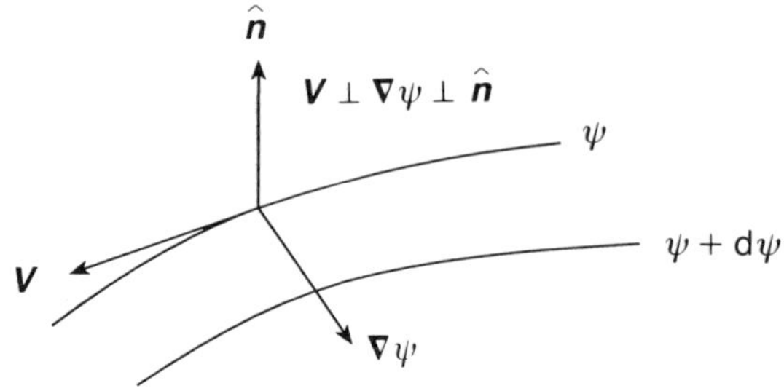

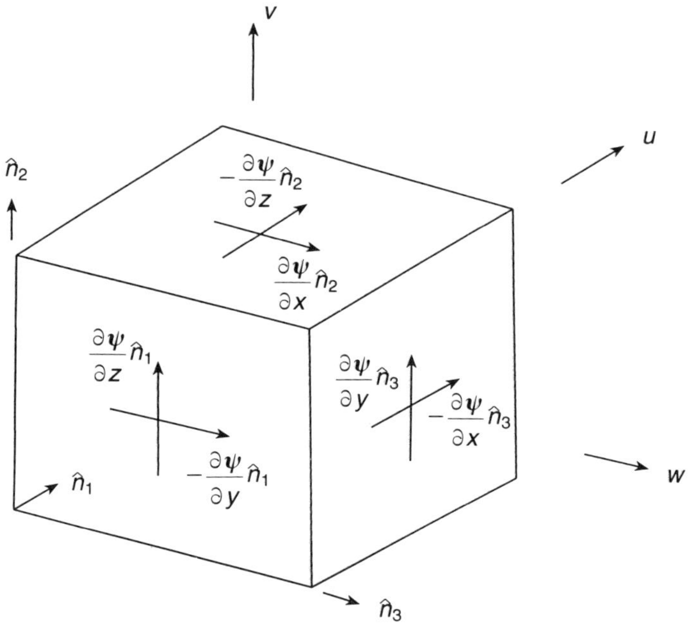

Figure 5.5.3 General descriptions of three-dimensional stream functions.

(5.5.20) satisfy the conservation of mass in steady state and reduce automatically to the two-dimensional incompressible flow in terms of a single stream function $\psi = \Psi_3$.

$$u = \frac{\partial \psi}{\partial y}\hat{n}_3 = \frac{\partial \psi}{\partial y} \quad (\hat{n}_3 = 1 \; on \; \hat{n}_3 \; surface)$$

$$v = -\frac{\partial \psi}{\partial x}\hat{n}_3 = -\frac{\partial \psi}{\partial x} \quad (\hat{n}_3 = 1 \; on \; \hat{n}_3 \; surface),$$

which represents the two-dimensional definition given by Eq. (5.5.4). The deduction from Eqs. (5.5.19a)–(5.5.20), as depicted in Fig. 5.5.3b, has an important physical significance, in which a scalar stream function ψ is extended to the three-dimensional stream function vector $\mathbf{\Psi}$.

As a result of the above assessments, the vorticity vector assumes the form

$$\boldsymbol{\omega} = \nabla \times \mathbf{V} = \nabla \times (\nabla \times \mathbf{\Psi}) = \nabla(\nabla \cdot \mathbf{\Psi}) - \nabla^2 \mathbf{\Psi} = -\nabla^2 \mathbf{\Psi}, \quad (5.5.21)$$

where $\nabla \cdot \mathbf{\Psi} = 0$ arises simply from the geometrical property, $\nabla \psi \cdot \hat{\mathbf{n}} = 0$ in Fig. (5.5.3a). Thus, an irrotational ideal flow in three dimensions can be given by $\nabla^2 \mathbf{\Psi} = 0$ in terms of the three-dimensional stream function vector.

The foregoing analyses allow one to formulate three-dimensional analyses in terms of three stream function components through the vorticity transport equation discussed in Section 5.6.

The notion of the three-dimensional stream function vector described is in contrast to the so-called three-dimensional vector potential concept in which the velocity vector is assumed to be the sum of the irrotational part and the rotational (solenoidal) parts,

$$\mathbf{V} = \nabla \phi + \nabla \times \mathbf{A},$$

where $\mathbf{A}$ is the vector potential, different from the three-dimensional stream function vector $\mathbf{\Psi}$. Unfortunately, the vector potential approach to deal with three-dimensional flow encounters unnecessary complications in the solution of the resulting differential equations, particularly with regard to the specification of boundary conditions.

5.5.4 The Bernoulli Equation

An ideal flow may also be treated by means of the *Bernoulli equation*. To derive this equation we may begin with momentum equations for incompressible flow ($\nabla \cdot \mathbf{V} = 0$),

$$\frac{\partial \mathbf{V}}{\partial t} + (\mathbf{V} \cdot \nabla)\mathbf{V} + \frac{1}{\rho}(\nabla p - \rho \mathbf{F}) - \nu \nabla^2 \mathbf{V} = 0, \quad (5.5.22a)$$

where $\nu = \mu/\rho$ is the kinematic viscosity. It can be shown that

$$\rho \mathbf{F} = -\nabla(\rho g h) \quad (5.5.22b)$$

$$\nabla^2 \mathbf{V} = \nabla(\nabla \cdot \mathbf{V}) - \nabla \times (\nabla \times \mathbf{V}) = -\nabla \times \boldsymbol{\omega}. \quad (5.5.22c)$$

Here g is the gravitational acceleration, h is the height, and $\nabla^2 \mathbf{V} = 0$ for irrotation flow ($\boldsymbol{\omega} = 0$), indicating that viscosity plays no role in irrotational flow. Thus, Eq. (5.5.22a) results in the form known as the

Euler equation, given by

$$\frac{\partial \mathbf{V}}{\partial t} + (\mathbf{V} \cdot \nabla)\mathbf{V} = -\nabla\left(\frac{p}{\rho} + gh\right).$$

Recall that the vector identity (Problem 1.7)

$$(\mathbf{V} \cdot \nabla)\mathbf{V} = \nabla\left(\frac{V^2}{2}\right) - \mathbf{V} \times \boldsymbol{\omega},$$

where

$$V^2 = \mathbf{V} \cdot \mathbf{V}.$$

It follows that for steady motion,

$$\mathbf{V} \times \boldsymbol{\omega} = \nabla\left(\frac{p}{\rho} + \frac{V^2}{2} + gh\right).$$

For irrotational flow, or $\boldsymbol{\omega} = 0$, we have

$$\nabla\left(\frac{p}{\rho} + \frac{V^2}{2} + gh\right) = 0.$$

Integrate this equation to obtain

$$\frac{p}{\rho} + \frac{V^2}{2} + gh = C, \tag{5.5.23}$$

where C is the constant of integration. This is known as the Bernoulli equation, which may be generalized to include an unsteady motion in the form

$$\frac{\partial \mathbf{V}}{\partial t} + \nabla\left(\frac{p}{\rho} + \frac{V^2}{2} + gh\right) = 0.$$

From the definition in Eq. (5.5.3), $\mathbf{V} = \nabla\phi$, we obtain

$$\nabla\left(\frac{\partial \phi}{\partial t} + \frac{p}{\rho} + \frac{V^2}{2} + gh\right) = 0,$$

or

$$\frac{\partial \phi}{\partial t} + \frac{p}{\rho} + \frac{V^2}{2} + gh = C(t). \tag{5.5.24}$$

The Bernoulli equation relates the total velocity of flow to the local pressure. Thus, once we have obtained the distribution of velocity in the flow field from the velocity potential, the Bernoulli equation allows us to obtain the pressure distribution in the flow field.

A similar procedure may be followed to derive the Bernoulli equation for energy with compressibility, viscosity, and diffusion. Such an equation is useful for the analysis of, for example, turbomachine problems. In this case, the flow is no longer ideal.

5.6 Rotational Flow

5.6.1 General

For fluids in rotational motion, the curl of the velocity produces the vorticity vector $\boldsymbol{\omega}$:

$$\boldsymbol{\omega} = 2\mathbf{w} = \boldsymbol{\nabla} \times \mathbf{V} = \epsilon_{ijk} V_{j,i} \mathbf{i}_k, \tag{5.6.1}$$

where $\mathbf{w}$ is the rotation vector. For two-dimensional flow, we have

$$\omega_3 = V_{2,1} - V_{1,2}, \tag{5.6.2}$$

or we may write

$$\omega = \omega_3 = \frac{\partial v}{\partial x} - \frac{\partial u}{\partial y}.$$

This defines the vorticity in two-dimensional flow (in the xy plane) whose direction is oriented toward the z axis.

Substituting the definition of the stream function, Eq. (5.5.16), we obtain the Poisson equation of the form

$$\frac{\partial^2 \psi}{\partial x^2} + \frac{\partial^2 \psi}{\partial y^2} = -\omega. \tag{5.6.3}$$

The vorticity normal to a surface element ds, $\boldsymbol{\omega} \cdot \mathbf{n}\, \mathrm{d}s$, is called the *circulation*. The circulation around any closed curve that contains a given set of fluid particles remains constant with time as long as the external force field remains conservative (or constant). This phenomenon is known as the *Kelvin circulation theorem*. It follows from the Kelvin theorem that a vortex line can neither begin nor end in the fluid; hence, it appears as a closed loop or ends on a boundary.

Vortex motions cannot be formed in an ideal fluid, but regions of localized vorticity are often produced in portions of real fluids, thus affecting the fluid motion. The action of viscosity is, in general, a prerequisite to the appearance of vorticity. The regions in which rotational motion and circulation exist are relatively small, and the effect of vortices on fluid motion is felt through the velocities which they induce throughout the rest of the flow field.

In certain real fluid flows, we encounter flow patterns which may be idealized as perfect fluids that contain infinite or semi-infinite rows of vortices, called *vortex streets*. The occurrence of vortex streets is usually associated with the formation of wakes behind bodies, which would lead to a kinematic instability of a form that might be conducive to possible rearrangements of vortices. In this regard, Kármán showed that only one particular arrangement of alternating vortices is stable; this is known as the *Kármán vortex street*.

In what follows we discuss the vorticity transport equation coupled with either the velocity or stream function and the associated boundary conditions.

5.6.2 The Vorticity Transport Equation

To obtain the governing equations for incompressible vortex flow, we take the curl of the momentum equation, (5.4.2):

$$\boldsymbol{\nabla} \times \left[\frac{\partial \mathbf{V}}{\partial t} + (\mathbf{V} \cdot \boldsymbol{\nabla})\mathbf{V} - \mathbf{F} + \frac{1}{\rho}\boldsymbol{\nabla} p - \nu \nabla^2 \mathbf{V}\right] = 0, \qquad (5.6.4)$$

where $\nu = \mu/\rho$ is the kinematic viscosity and the term containing $\boldsymbol{\nabla} \cdot \mathbf{V}$ is eliminated, which is a requirement for incompressible flow. We recall that

$$(\mathbf{V} \cdot \boldsymbol{\nabla})\mathbf{V} = \boldsymbol{\nabla}(\tfrac{1}{2}\mathbf{V} \cdot \mathbf{V}) - \mathbf{V} \times \boldsymbol{\omega}$$

and from Eq. (1.2.23)

$$\boldsymbol{\nabla} \times (\mathbf{V} \times \boldsymbol{\omega}) = (V_{i,j}\omega_j + V_i\omega_{j,j} - V_{j,j}\omega_i - V_j\omega_{i,j})\mathbf{i}_i,$$

where

$$V_{i,j}\omega_j\mathbf{i}_i = (\boldsymbol{\omega} \cdot \boldsymbol{\nabla})\mathbf{V} = (\boldsymbol{\nabla} \times \mathbf{V} \cdot \boldsymbol{\nabla})\mathbf{V} = \epsilon_{ijk} V_{j,i} V_{m,k}\mathbf{i}_m = 0$$

for two-dimensional flow. Furthermore, $V_{j,j}\omega_i = 0$ for incompressible flow, and because the velocity is the continuous function, we obtain

$$V_i\omega_{j,j}\mathbf{i}_i = \mathbf{V}(\boldsymbol{\nabla} \cdot \boldsymbol{\omega}) = \mathbf{V}[\boldsymbol{\nabla} \cdot (\boldsymbol{\nabla} \times \mathbf{V})] = \mathbf{V}(\epsilon_{ijk} V_{j,ik}) = 0,$$

where $\partial^2 V_1/\partial x_1 \partial x_2 = \partial^2 V_1/\partial x_2 \partial x_1$, etc. Note also that $\boldsymbol{\nabla} \times \mathbf{F} = 0$ for the conservative field or $\boldsymbol{\nabla} \times \boldsymbol{\nabla}(gh) = 0$ and $\boldsymbol{\nabla} \times \boldsymbol{\nabla} p = 0$ from the definition of vector algebra. With these observations, it follows from Eq. (5.6.4) that

$$\frac{\partial \omega}{\partial t} + (\mathbf{V} \cdot \boldsymbol{\nabla})\omega - \nu \nabla^2 \omega = 0. \qquad (5.6.5)$$

This is known as the *vorticity transport equation* for two-dimensional incompressible flow.

In terms of stream functions, the vorticity transport equation for two dimensions becomes

$$\frac{\partial \omega}{\partial t} + \omega_{,i}\epsilon_{ij}\psi_{,j} - \nu\omega_{,jj} = 0 \quad (i, j = 1, 2). \qquad (5.6.6)$$

Since there are two unknowns (ω, ψ), we require an additional equation, namely, the stream function Poisson equation, (5.6.3). Equations (5.6.6) and (5.6.3) should be solved simultaneously for ω and ψ.

To calculate the pressure, we take the divergence of the momentum equation,

$$\nabla \cdot \left[\frac{\partial \mathbf{V}}{\partial t} + (\mathbf{V} \cdot \nabla)\mathbf{V} - \mathbf{F} + \frac{1}{\rho}\nabla p - \nu \nabla^2 \mathbf{V} \right] = 0$$

in which the first and last terms vanish because of incompressibility, and $\nabla \cdot \mathbf{F} = 0$ for the conservative field. Thus, we obtain

$$\nabla^2 p = -B, \tag{5.6.7}$$

where

$$B = \rho \nabla \cdot [(\mathbf{V} \cdot \nabla)\mathbf{V}] = \rho (V_{i,j} V_j)_{,i}. \tag{5.6.8}$$

Then expand Eq. (5.6.8) and use the incompressibility condition to obtain

$$\nabla^2 p = 2\rho \left(\frac{\partial u}{\partial x}\frac{\partial v}{\partial y} - \frac{\partial u}{\partial y}\frac{\partial v}{\partial x} \right). \tag{5.6.9}$$

Once the velocity field is calculated, then Eq. (5.6.9) can be used to determine the pressure.

An alternate approach to rotational flow analysis is to rewrite Eq. (5.6.6) by eliminating the vorticity through Eq. (5.5.4), to arrive at

$$\nabla^4 \psi = R, \tag{5.6.10}$$

where

$$R = \frac{1}{\nu}\left[\frac{\partial^3 \psi}{\partial x \partial y^2}\frac{\partial \psi}{\partial y} - \frac{\partial^3 \psi}{\partial x^2 \partial y}\frac{\partial \psi}{\partial x} + \frac{\partial^3 \psi}{\partial x^3}\frac{\partial \psi}{\partial y} - \frac{\partial^3 \psi}{\partial y^3}\frac{\partial \psi}{\partial x} \right],$$

in which steady state is assumed. This approach has the advantage over Eq. (5.6.6) of involving only one variable.

The standard boundary conditions to be specified for Eqs. (5.6.6) and (5.6.3) include $\psi = \hat{\psi}$ on Γ_1, $\omega = \hat{\omega}$ on Γ_2, $\psi_{,i} n_i = 0$ on Γ_3, and $\omega_i n_i = 0$ on Γ_4. These boundary conditions are to be chosen, depending on the governing equations used for the solution.

For three-dimensional incompressible flows the two-dimensional vorticity transport equation must be modified using the definition of three stream functions, as shown in Eq. (5.5.20). This leads to

$$\frac{\partial \boldsymbol{\omega}}{\partial t} + (\mathbf{V} \cdot \nabla)\boldsymbol{\omega} - (\boldsymbol{\omega} \cdot \nabla)\mathbf{V} = \nu \nabla^2 \boldsymbol{\omega}, \tag{5.6.11a}$$

or

$$\frac{\partial}{\partial t}\nabla^2 \boldsymbol{\Psi} + (\nabla \times \boldsymbol{\Psi} \cdot \nabla)\nabla^2 \boldsymbol{\Psi} - (\nabla^2 \boldsymbol{\Psi} \cdot \nabla)(\nabla \times \boldsymbol{\Psi}) = \nu \nabla^4 \boldsymbol{\Psi}. \tag{5.6.11b}$$

Once the components of $\boldsymbol{\Psi}$ are calculated, then the velocity and vorticity components and pressure are determined from Eqs. (5.5.20), (5.5.21), and (5.6.7), respectively.

5.7 Turbulence

5.7.1 Time and Ensemble Averages

It is well known that the motion of fluids may occur in irregular fluctuations, resulting in *mixing*, and/or *eddying*. Such motions are called *turbulent flows*. When the Reynolds number ($Re = u_\infty L/\nu$, u_∞ = free-stream velocity, L = characteristic length) is increased, internal flows and boundary layers around solid bodies change from laminar to turbulent. Such a transition is influenced by geometries, pressure gradients, suction, compressibility, and heat transfer. The exact mathematical treatment of turbulence is hopelessly complex. In turbulence, the motion appears as if the viscosity were increased tremendously. At large Reynolds numbers, there exists a continuous transport of energy from the main flow into large eddies. The velocity and pressure at a fixed point in space do not remain constant with time, but undergo large, irregular fluctuations of high frequency. The flow is then assumed to consist of a mean motion and an eddying motion. Therefore, in the analysis of turbulent flow, we assume that the instantaneous value of any variable f for the motion is represented as the sum of the time-averaged mean part $\bar{f}$ and the eddying (fluctuating) part f':

$$f = \bar{f} + f'. \tag{5.7.1}$$

This relationship is shown in Fig. 5.7.1. The *time average* is given by

$$\bar{f} = \lim_{\Delta t \to \infty} \frac{1}{\Delta t} \int_{t_0}^{t_0 + \Delta t} f \, \mathrm{d}t, \tag{5.7.2}$$

where Δt is a very small time interval, but large in comparison with the *time scale* of the turbulence and sufficiently small in comparison with the period of slow variations of the averaged quantities in the flow field. It is apparent from the definition in Eq. (5.7.2) that the time average of the fluctuation is zero,

$$\bar{f}' = \lim_{\Delta t \to \infty} \frac{1}{\Delta t} \int_{t_0}^{t_0 + \Delta t} f' \, \mathrm{d}t = 0. \tag{5.7.3}$$

On the other hand, the time average of the product of two fluctuations is not zero. For example, the time average of the product of f' and

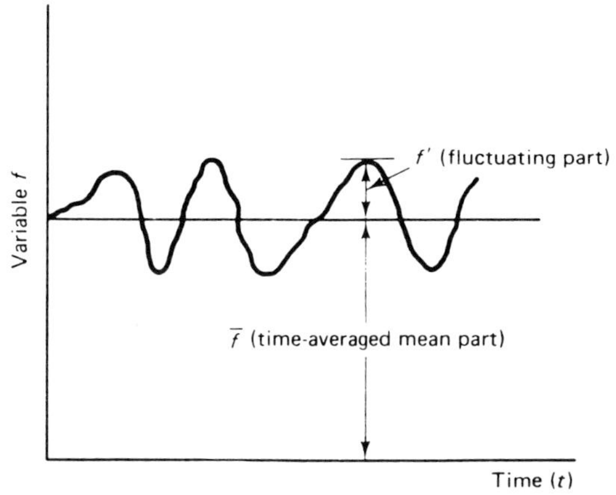

Figure 5.7.1 Fluctuations of any variable f in turbulent flow.

another variable g' is written as

$$\lim_{\Delta t \to \infty} \frac{1}{\Delta t} \int_{t_0}^{t_0+\Delta t} f' g' \, \mathrm{d}t = \overline{f' g'}, \tag{5.7.4}$$

where the bar denotes the time average. This implies that the time average of the product of two different variable fluctuations does not cancel out, but tends to grow with respect to time, known as correlations.

Another means of averaging, known as the *ensemble average*, is given by

$$\bar{f} = \frac{1}{N} \sum_{k=1}^{N} f_k, \tag{5.7.5}$$

where f_k denotes the values of N identical measurements. For quasi-steady systems, i.e., for $\partial f/\partial t = 0$ and $\partial \bar{f}/\partial t = 0$, it may be assumed that the two averaging systems lead to the same result.

5.7.2 Correlations

For compressible viscous flow, the variables involved in turbulence are the velocity, pressure, temperature, and density, consisting of the mean values and fluctuations as follows

$$V_i = \bar{V}_i + V'_i, \tag{5.7.6a}$$

$$p = \bar{p} + p', \tag{5.7.6b}$$

$$T = \bar{T} + T', \tag{5.7.6c}$$

$$\rho = \bar{\rho} + \rho'. \tag{5.7.6d}$$

For simplicity, consider an isothermal flow governed by the conservation form of the Navier–Stokes system of equations as follows.

Continuity

$$\frac{\partial \rho}{\partial t} + (\rho V_i)_{,i} = 0. \tag{5.7.7}$$

Momentum

$$\frac{\partial}{\partial t}(\rho V_i) + (\rho V_i V_j)_{,j} = -p_{,i} + \left[\mu\left(V_{i,j} + \frac{1}{3}V_{j,i}\right)\right]_{,j}. \tag{5.7.8}$$

Substituting Eq. (5.7.6) into Eqs. (5.7.7) and (5.7.8), taking time averages, and satisfying the conservation of mass lead to

$$\frac{\partial \bar{\rho}}{\partial t} + (\bar{\rho}\bar{V}_i + \overline{\rho' V_i'})_{,i} = 0, \tag{5.7.9}$$

$$\bar{\rho}\frac{\partial \bar{V}_i}{\partial t} + \bar{\rho}\bar{V}_{i,j}\bar{V}_j = -\bar{p}_{,i} + \mu\left(\bar{V}_{i,jj} + \frac{1}{3}\bar{V}_{j,ji}\right) + \Phi_i, \tag{5.7.10}$$

where

$$\Phi_i = -\left[\frac{\partial}{\partial t}(\overline{\rho' V_i'}) + (\bar{\rho}\overline{V_i'V_j'})_{,j} + (\overline{\rho' V_j'}\bar{V}_i)_{,j} + (\overline{\rho' V_i'}\bar{V}_j)_{,j} + (\overline{\rho' V_i' V_j'})_{,j}\right] \tag{5.7.11}$$

in which the last four terms of the variables within the brackets in Eq. (5.7.11) are known as turbulent (or Reynolds) stresses. These additional terms in Eq. (5.7.10), which originated from the convective terms in Eq. (5.7.8), are responsible for turbulent flow. For incompressible flow, Eq. (5.7.10) is simplified to read

$$\rho\left(\frac{\partial \bar{V}_i}{\partial t} + \bar{V}_{i,j}\bar{V}_j\right) = -\bar{p}_{,i} + (\mu\bar{V}_{i,j} - \rho\overline{V_i'V_j'})_{,j}. \tag{5.7.12}$$

The total stress tensor may be written as

$$\sigma_{ij} = -\bar{p}\delta_{ij} + 2\mu\bar{d}_{ij} - \rho\overline{V_i'V_j'}. \tag{5.7.13}$$

The turbulent (Reynolds) stress,

$$-\rho\overline{V_i'V_j'}$$

comes from the convective terms, and it is only logical that the convective motion results in turbulence. It is important to realize that the momentum equations for turbulent flows with the turbulent stress appearing as shown in Eq. (5.7.12) arise only through the conservation form, not through the nonconservation form. In 1877 Boussinesq proposed the concept of *eddy viscosity* $\hat{\mu}$, such that

$$\sigma_{ij} = -\bar{p}\delta_{ij} + 2(\mu + \hat{\mu})\,\bar{d}_{ij}. \tag{5.7.14}$$

By equating this with Eq. (5.7.13), we get

$$-\rho\overline{V_i'V_j'} = 2\hat{\mu}\bar{d}_{ij} = \hat{\mu}(\bar{V}_{i,j} + \bar{V}_{j,i}). \tag{5.7.15}$$

If we assume that $i = j$ in Eq. (5.7.15), then

$$-\rho\overline{V_i'V_i'} = 2\hat{\mu}\bar{V}_{i,i}. \tag{5.7.16}$$

The right-hand side vanishes for incompressible flow, whereas the left-hand side is not zero unless turbulence is absent. This is clearly a contradiction.

Another possibility is to assume that the eddy viscosity can be expressed in terms of the second- or fourth-order tensor:

$$-\rho\overline{V_i'V_j'} = \hat{\mu}_{ik}\bar{d}_{kj}, \tag{5.7.17}$$

or

$$-\rho\overline{V_i'V_j'} = \hat{\mu}_{ijkm}\bar{d}_{km}. \tag{5.7.18}$$

However, these relations are not invariant with respect to a rotation of the coordinate system with a constant angular velocity, and so they violate the principle of material objectivity (see subsection 4.4.4).

In searching for acceptable forms of representation for eddy viscosity, let us examine the equation

$$-\rho\overline{V_i'V_j'} = \hat{\mu}_0\delta_{ij} + \hat{\mu}_1\bar{d}_{ij} + \hat{\mu}_2\bar{d}_{ik}\bar{d}_{kj}, \tag{5.7.19}$$

where the scalars $\hat{\mu}_0$, $\hat{\mu}_1$, and $\hat{\mu}_2$, are functions of the principal invariants of $\bar{d}_{ij}$, much as the scalars were defined for the constitutive equation of Reiner–Rivlin fluids (see Eqs. [5.2.7] and [5.2.8]). Unfortunately, the model suggested in Eq. (5.7.19) has not been supported by experimental evidence, despite the fact that Eq. (5.7.19) meets the requirement for material objectivity (frame indifference).

The turbulent process may be thought of as a memory effect of the sort that occurs in viscoelasticity, giving rise to an integral of the form

$$-\rho\overline{V_iV_j} = -\bar{p}_t\delta_{ij} + A\bar{p}_t\int_0^{\infty} M(\tau)\left[\frac{\partial V_i^*}{\partial x_j}(t-\tau) + \frac{\partial V_j^*}{\partial x_i}(t-\tau)\right]d\tau, \tag{5.7.20}$$

where $\bar{p}_t$ is the average turbulence pressure, defined as

$$\bar{p}_t = \tfrac{1}{3}\rho\bar{V}'_k\bar{V}'_k \tag{5.7.21}$$

and $M(\tau)$ is an exponential-type memory function characteristic of turbulent flow (Hinze 1975). Although this approach is appealing from the theoretical viewpoint, it is not likely to contribute to practical applications.

The turbulence correlation for the energy equation can be obtained in a similar manner:

$$\frac{\partial \bar{T}}{\partial t} + \bar{V}_i \bar{T}_{,i} = (\alpha^* \bar{T}_{,i} - \overline{V_i' T'})_{,i}, \qquad (5.7.22)$$

where $\alpha^* = k/\rho c_p$ is the thermal diffusivity. Also, the second-order correlation

$$-\overline{V_i' T'}$$

between the turbulent fluctuations of the velocity and the temperature may be given by

$$-\overline{V_i' T'} = (\hat{\alpha} \bar{T})_{,i}, \qquad (5.7.23)$$

where $\hat{\alpha}$ is the eddy-transport coefficient. If this coefficient is chosen to be of second order, then

$$-\overline{V_i T'} = (\hat{\alpha}_{ij} \bar{T})_{,j}. \qquad (5.7.24)$$

Along with these correlation data, we need to examine the notion of isotropic and homogeneous properties, and kinetic and dissipative energies in turbulence, which we discuss in the sections that follow.

5.7.3 Isotropic and Homogeneous Turbulence

The turbulence shear stress

$$\tau_{ij}^* = \rho \overline{V_i' V_j'}$$

has a normal component ($i = j$) and a shear component ($i \neq j$). Let us consider a turbulence stress in a direction normal to an arbitrary surface of a fluid element in incompressible flows,

$$\tau_{(i)}^* = \tau_{ij}^* n_j. \qquad (5.7.25)$$

The energy required for this fluid element in the normal direction is

$$E_{(n)} = \tau_{ij}^* n_i n_j = \rho \overline{V_i' V_j'} n_i n_j \qquad (5.7.26)$$

since

$$n_i n_j = \tfrac{1}{3}\delta_{ij}$$

$$E_{(n)} = \tfrac{1}{3}\rho \overline{V_i' V_i'} = \tfrac{2}{3} K^*, \qquad (5.7.27)$$

where K^* is the turbulent kinetic energy:

$$K^* = \tfrac{1}{2}\rho \overline{V_i' V_i'}. \qquad (5.7.28)$$

If the turbulence does not show a "preference" for any particular direction, i.e., if $V_1'^2 = V_2'^2 = V_3'^2 = u'^2$, then

$$K^* = \tfrac{3}{2}\rho u'^2 = \tfrac{3}{2}E_{(n)}. \qquad (5.7.29)$$

Equations (5.7.26)–(5.7.29) imply that the surface of this fluid element becomes a sphere and they hold with respect to any coordinate system. They are invariant under rotation of the coordinate system. Such a flow field is said to show *isotropic turbulence*, in which its statistical features have no preference for any direction. Here, no average shear stress can occur, the gradient of the mean velocity is zero, and the mean velocity is constant throughout the field. If the mean velocity has a gradient, then the turbulence is *anisotropic*, as it is, for example, in wall turbulence, where the average shear stress must be taken into account.

In general, viscosity effects result in a conversion of the kinetic energy of the flow into heat. Other effects of viscosity make the turbulence more homogeneous. If the turbulence has the same structure throughout the flow field, it is said to be *homogeneous turbulence*. Otherwise, the turbulence is inhomogeneous. This is the result of the shear stress varying through the flow.

5.7.4 Transport Equations of Reynolds Stresses, Turbulent Kinetic Energy, and Dissipation Rates

The equation of motion for an incompressible fluid with a steady mean flow reads

$$\frac{\partial}{\partial t}(\bar{V}_i + V_i') + (\bar{V}_k + V_k')(\bar{V}_i + V_i')_{,k} =$$
$$-\frac{1}{\rho}(\bar{p} + p')_{,i} + \nu(\bar{V}_i + V_i')_{,kk}. \qquad (5.7.30)$$

The so-called Reynolds stress transport equation is obtained by (1) subtracting Eq. (5.7.12) from Eq. (5.7.30), (2) multiplying the result by V_i', (3) repeating (1) and (2) by reversing i and j above, (4) adding the resulting equations obtained in (2) and (3), and (5) taking the time average of the final equation obtained in (4).

$$\frac{D}{Dt}(\overline{V_i'V_j'}) =$$
$$-\underset{\text{I}}{(\overline{V_i'V_k'}\bar{V}_{j,k} + \overline{V_j'V_k'}\bar{V}_{i,k})} - \underset{\text{II}}{2\nu\overline{V_{i,k}'V_{j,k}'}} + \underset{\text{III}}{\frac{1}{\rho}\overline{p'(V_{i,j}' + V_{j,i}')}} \qquad (5.7.31)$$
$$-\left[\underset{\text{IV}}{(\overline{V_i'V_j'V_k'}) + \frac{1}{\rho}(\overline{p'V_i'\delta_{jk} + p'V_j'\delta_{ik}}) - \nu(\overline{V_i'V_j'})_{,k}}\right]_{,k},$$

where the quantities designated by I, II, III, and IV are identified as the production, destruction, pressure strain, and diffusive transport, respectively.

The differential equation for turbulent kinetic energy per unit mass is obtained by setting $j = i$ in Eq. (5.7.31),

$$\frac{DK}{Dt} = -\overline{V_i'V_j'}\bar{V}_{i,j} - \epsilon - \left(\tfrac{1}{2}\overline{V_i'V_i'V_j'} - \nu K_{,j} + \frac{1}{\rho}\overline{p'V_j'}\right)_{,j} = 0, \tag{5.7.32}$$

where K and ϵ are the turbulent kinetic energy per unit mass and its dissipation rate, respectively:

$$K = \tfrac{1}{2}\overline{V_i'V_i'}, \tag{5.7.33}$$

$$\epsilon = \nu\overline{V_{i,j}'V_{i,j}'}. \tag{5.7.34}$$

The differential equation for the dissipation rates of the turbulent kinetic energy can be derived by differentiating the momentum equation with respect to x_k, multiplying throughout by $2\nu V_{i,k}$, and time-averaging the resulting equation.

$$\begin{aligned}\frac{D\epsilon}{Dt} = &-2\nu(\overline{V_{k,i}'V_{j,k}'} + \overline{V_{k,i}'V_{k,i}'})\bar{V}_{i,j} - 2\nu\overline{V_j'V_{i,k}'}\bar{V}_{i,j} \\ &-2\nu\overline{V_{i,j}'V_{i,k}'V_{j,k}'} - 2\nu\overline{V_{i,jk}'V_{i,jk}'} \\ &-\left(\nu\overline{V_j'V_{i,k}'V_{i,k}'} + \frac{2\nu}{\rho}\overline{V_{j,k}'p_{,k}'} - \nu\epsilon_{,j}\right)_{,j}\end{aligned} \tag{5.7.35}$$

From Eqs. (5.7.32) and (5.7.35), the following observations are made. (1) The turbulent shear stress

$$\overline{V_i'V_j'} \quad (i \neq j)$$

is produced only if the main flow is not uniform; (2) the pressure-velocity-gradient correlations tend to decrease the nonisotropy by equalizing the three turbulence velocity components and by decreasing the turbulent shear stress; (3) this tendency to isotropy is greater in the smaller-scale range of turbulence; and (4) the viscosity effect, through dissipation, increases with increasing intensity of turbulence, and contributes to the damping of the greater-intensity components at a higher rate than the smaller, the effect being greater in the higher wave number range, although such a process due to viscosity is rather slow. Notice that the transport equations for the Reynolds stress, turbulent kinetic energy, and dissipation rates contain many unknowns of multiple velocity products. They must be modeled empirically in terms of known quantities.

Discussions carried out so far indicate that we have introduced additional unknowns, namely, the product terms with overbars in Eq. (5.7.11). To solve the Navier–Stokes system, we will then require additional equations, with the number of equations matching the number of unknowns. To this end, we are able to use the transport equations of the Reynolds stress (Eq. 5.3.31), turbulent kinetic energy (Eq. 5.7.32), and dissipation rates (Eq. 5.7.35). Such a process is known as *turbulence closure*. Although we do not delve into this subject, it is informative to examine the earliest and simplest method of turbulence closure in the next subsection.

5.7.5 Phenomenological Theories

In earlier theories, such as the Prandtl mixing-length theory, the diffusive action of turbulence is considered to result in a constant, known as the eddy viscosity or eddy heat conductivity. This implies that the local gradient of the mean property may be used together with such constants. To this end, in a way that was analogous to the reasoning in the kinetic theory of gases, Prandtl (1904) assumed that the eddy viscosity was equal to the product of the *mixing length* and some suitable velocity. If we consider a plane flow along with its mean velocity in the x direction and negligible x-direction velocity gradients (boundary layer), then the Reynolds stress originally proposed by Boussinesq reads

$$-\rho\overline{u'v'} = \hat{\mu}\frac{\mathrm{d}\bar{u}}{\mathrm{d}y}, \tag{5.7.36}$$

where $\hat{\mu}$ is the dynamic eddy viscosity:

$$\hat{\mu} = \rho L^2 \frac{\mathrm{d}\bar{u}}{\mathrm{d}y} \tag{5.7.37}$$

with L being the Prandtl mixing length:

$$L = \kappa y, \tag{5.7.38}$$

where $\kappa = 0.4$ is an empirical dimensionless constant, as proposed by Kármán. This is considered the simplest phenomenological model in which the u velocity is zero at the wall and parabolic u velocity distributions are assumed along the y direction known as the boundary layer, as discussed in Section 5.8. Here, we require no additional equation because no additional variable is introduced, and so the model is known as a *zero-equation model*.

One of the most popular *two-equation models* is known as the $K - \epsilon$ model, in which the incompressible flow transport equations for K and ϵ are given by

$$\frac{\partial}{\partial t}(\bar{\rho} K) + (\bar{\rho} K V_i')_{,i} = (\mu_k K_{,i})_{,i} + (\bar{\tau}_{ij} V_j')_{,i} - \bar{\rho}\epsilon, \tag{5.7.39a}$$

$$\frac{\partial}{\partial t}(\bar{\rho}\epsilon) + (\bar{\rho}\epsilon V_i')_{,i} = (\mu_\epsilon \epsilon_{,i})_{,i} + c_{\epsilon 1}(\bar{\tau}_{ij} V_j')_{,i} - c_{\epsilon 2}\bar{\rho}\frac{\epsilon^2}{K}, \tag{5.7.39b}$$

with

$$\bar{\tau}_{ij} = -\overline{\rho V_i' V_j'} = \hat{\mu}(\bar{V}_{i,j} + \bar{V}_{j,i} - \tfrac{2}{3}\bar{V}_{k,k}\delta_{ij}) - \tfrac{2}{3}\bar{\rho} K \delta_{ij},$$

$$\hat{\mu} = \bar{\rho} c_\mu \frac{K^2}{\epsilon}, \quad \mu_k = \mu + \frac{\hat{\mu}}{\sigma_k}, \quad \mu_\epsilon = \mu + \frac{\hat{\mu}}{\sigma_\epsilon},$$

$$c_\mu = 0.09, \quad C_{\epsilon 1} = 1.45\text{~}1.55, \quad C_{\epsilon 2} = 1.92\text{~}2.0, \quad \sigma_k = 1, \quad \sigma_\epsilon = 1.3,$$

as reported in Launder and Spalding (1972).

Very complicated models can be developed that take into account many variables. Various empirical assumptions have been advanced, but no phenomenological models developed to date are considered perfect. It is not likely that we will ever achieve a complete understanding of the mechanism of turbulence through phenomenological models, because its physical nature is extremely complicated.

For further details concerning the subject of turbulence, the reader should consult specialized books such as by Hinze (1975) for theory and Hanjalic and Launder (1976) for numerical modeling, where the unknowns in Eqs. (5.7.31), (5.7.32), and (5.7.35) are replaced by explicit empirical models. Details of the various turbulence models are beyond the scope of this book.

5.7.6 Other Approaches to Turbulence

In turbulence, large eddies interact strongly with the mean flow. Small eddies are created mainly by nonlinear interactions among large eddies. Most transport of mass, momentum, energy, and concentration is due to large eddies. Small eddies dissipate fluctuations of these quantities, but affect the mean properties only slightly. Large eddies are anisotropic, whereas small eddies are nearly isotropic.

In view of these properties, large structures in the flow are computed explicitly and small ones are modeled. This approach is known as large eddy simulation (Ferziger 1977). In this case, variables are space-averaged rather than time-averaged, so that the large-scale field is defined as

$$\bar{f}(\mathbf{x}) = \int_{\Omega} G(\mathbf{x}, \mathbf{x}') f(\mathbf{x}') \, d\Omega, \tag{5.7.40}$$

where $G(\mathbf{x}, \mathbf{x}')$ is a filter function. To this end, one defines a subgrid scale velocity V_i' by

$$V_i' = V_i - \bar{V}_i. \tag{5.7.41}$$

Thus, the filtered continuity and momentum equations can be written as

$$\frac{\partial \bar{V}_i}{\partial x_i} = 0, \tag{5.7.41a}$$

$$\frac{\partial \bar{V}_i}{\partial t} + \frac{\partial}{\partial x_j}(\overline{V_i V_j}) = -\frac{1}{\rho}\frac{\partial \bar{p}}{\partial x_i} + \nu \frac{\partial^2 \bar{V}_i}{\partial x_j \partial x_j}, \tag{5.7.41b}$$

where the overbar means a space-averaged quantity, and

$$\overline{V_i V_j} = \overline{\bar{V}_i \bar{V}_j} + \overline{V_i' \bar{V}_j} + \overline{\bar{V}_i V_j'} + \overline{V_i' V_j'}. \tag{5.7.42}$$

The last three terms contain the subgrid scale velocity and therefore must be treated by modeling. The first term depends only on the large-scale velocities, $\bar{V}_i$, and can be calculated explicitly. This is compatible with use of the Fourier transform in view of the space average given by (5.7.40).

Turbulence may be computed without resorting to any turbulence models, an approach known as direct numerical simulation (Ferziger 1983, Huser and Biringen 1993, among others). Here, turbulence microscales are resolved, using refined discretization of the domain.

5.8 The Boundary Layer

5.8.1 General

When a fluid is in contact with the surface of a solid body, the flow experiences a marked velocity change in a region from the immediate vicinity of the wall to some distance from the wall, forming a so-called *boundary layer*, an idea first proposed by Prandtl in 1904. Although viscosity plays a major role within the boundary layer, it has little influence in the flow outside the boundary layer, as pictured in Fig. 5.8.1. Boundary layers prevail for both compressible and incompressible flows.

In the study of the boundary layer, we consider a very thin layer in the immediate neighborhood of the body, where the velocity gradient normal to the wall is large, contributing to high shear stresses. No such

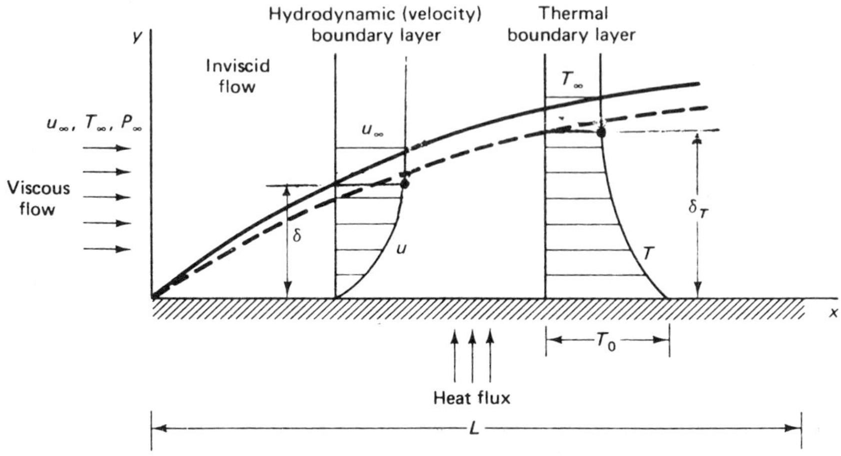

Figure 5.8.1 Laminar hydrodynamic (velocity) and thermal boundary layers along a flat wall: $\delta_T = \delta$ for $Pr = 1$; $\delta_T < \delta$ for $Pr > 1$; $\delta_T > \delta$ for $Pr < 1$.

large velocity gradients occur away from the surface of the body or channel, and the flow becomes frictionless (known as potential). This is the basic assumption for the boundary layer theory. The boundary-layer thickness is proportional to the square root of the kinematic viscosity of the fluid. More specifically, for the boundary layer on a flat plate, using the momentum equation, we obtain

$$\mu \frac{u_\infty}{\delta^2} \sim \rho \frac{u_\infty^2}{L}, \tag{5.8.1}$$

where u_∞ is the free-stream velocity and L is the plate length. This gives

$$\delta \sim \sqrt{\frac{\mu L}{\rho u_\infty}} = \sqrt{\frac{\nu L}{u_\infty}}. \tag{5.8.2}$$

Blasius (1908) showed that for laminar flow,

$$\delta = 5 \sqrt{\frac{\nu L}{u_\infty}}$$

or

$$\delta = \frac{5L}{\sqrt{\text{Re}}}, \tag{5.8.3}$$

where $\text{Re} = u_\infty L/\nu$ is the Reynolds number. It is seen that as the

Reynolds number becomes infinity, the boundary-layer thickness vanishes. In general, the boundary-layer thickness is represented by

$$\Delta^2 = \left(\frac{\delta}{L}\right)^2 \sim \frac{1}{\mathrm{Re}} \ll 1. \tag{5.8.4}$$

Similarly, for the thermal boundary layer δ_T (see Fig. 5.8.1), using the energy equation, we write

$$\Delta_T^2 = \left(\frac{\delta_T}{L}\right)^2 \sim \frac{1}{\mathrm{Re\,Pr}} \ll 1, \tag{5.8.5}$$

in which Pr is the Prandtl number, defined as

$$\mathrm{Pr} = \frac{c_p \mu}{k}. \tag{5.8.6}$$

It is clear from these definitions that $\delta_T = \delta$ for $\mathrm{Pr} = 1$, $\delta_T < \delta$ for $\mathrm{Pr} > 1$, and $\delta_T > \delta$ for $\mathrm{Pr} < 1$.

5.8.2 Laminar and Turbulent Boundary Layers

We consider a flow over a flat plate. The flow will be laminar as it first emerges on the left side in Fig. 5.8.1. The continuity, momentum, and energy equations for steady state may be written in conservation form using the nondimensional quantities:

$$X = \frac{x}{L}, \quad Y = \frac{y}{L}, \quad U = \frac{u}{U_\infty}, \quad V = \frac{v}{U_\infty}, \quad P = \frac{p}{\rho U_\infty^2},$$
$$\theta = \frac{T}{T_s - T_\infty}.$$

Continuity

$$\frac{\partial U}{\partial X} + \frac{\partial V}{\partial Y} = 0. \tag{5.8.7}$$

X Momentum

$$\frac{\partial U^2}{\partial X} + \frac{\partial}{\partial Y}(UV) = -\frac{\partial P}{\partial X} + \frac{1}{\mathrm{Re}}\left(\frac{\partial^2 U}{\partial X^2} + \frac{\partial^2 U}{\partial Y^2}\right). \tag{5.8.8}$$

Y Momentum

$$\frac{\partial}{\partial X}(VU) + \frac{\partial V^2}{\partial Y} = -\frac{\partial P}{\partial Y} + \frac{1}{\mathrm{Re}}\left(\frac{\partial^2 V}{\partial X^2} + \frac{\partial^2 V}{\partial Y^2}\right). \tag{5.8.9}$$

Energy

$$\frac{\partial}{\partial X}(U\theta) + \frac{\partial}{\partial Y}(V\theta) = \frac{1}{\mathrm{Re}\,\mathrm{Pr}}\left(\frac{\partial^2\theta}{\partial X^2} + \frac{\partial^2\theta}{\partial Y^2}\right) + \frac{\mathrm{Ec}}{\mathrm{Re}}\left[2\left(\frac{\partial U}{\partial X}\right)^2 + 2\left(\frac{\partial V}{\partial Y}\right)^2 + \left(\frac{\partial V}{\partial X} + \frac{\partial U}{\partial Y}\right)^2\right], \tag{5.8.10}$$

where Ec is the Eckert number, defined as

$$\mathrm{Ec} = \frac{U_\infty^2}{c_p(T_s - T_\infty)} \tag{5.8.11}$$

and where T_s is the surface temperature.

In view of the definitions given in Eqs. (5.8.4) and (5.8.5), the orders of magnitude of the quantities in Eqs. (5.8.7)–(5.8.10) can be characterized in terms of 1 and Δ or $\Delta_T \simeq \Delta$, such that $\Delta \ll 1$. Specifically, we set

$$\delta \ll L, \quad \frac{\partial U}{\partial X} \sim \frac{U}{L}, \quad \frac{\partial V}{\partial Y} \sim \frac{U}{L},$$

$$V \sim \frac{U\delta}{L}, \quad \frac{\partial V}{\partial X} \sim \frac{U\delta}{L^2}, \quad \frac{\partial U}{\partial Y} \sim \frac{U}{\delta},$$

where the third relation is due to the conservation of mass. These results are then applied to determine the orders of magnitude of the various terms in Eqs. (5.8.7)–(5.8.10). They are simplified by neglecting the terms of the order of Δ in comparison with those of order unity. As a result, the Y momentum vanishes and the simplified equations are as follows.

Continuity

$$\frac{\partial U}{\partial X} + \frac{\partial V}{\partial Y} = 0. \tag{5.8.12}$$

X Momentum

$$U\frac{\partial U}{\partial X} + V\frac{\partial U}{\partial Y} = -\frac{\partial P}{\partial X} + \frac{1}{\mathrm{Re}}\frac{\partial^2 U}{\partial Y^2}. \tag{5.8.13}$$

Energy

$$U\frac{\partial \theta}{\partial X} + V\frac{\partial \theta}{\partial Y} = -\frac{1}{\mathrm{Re}\,\mathrm{Pr}}\frac{\partial^2\theta}{\partial Y^2} + \frac{\mathrm{Ec}}{\mathrm{Re}}\left(\frac{\partial U}{\partial Y}\right)^2. \tag{5.8.14}$$

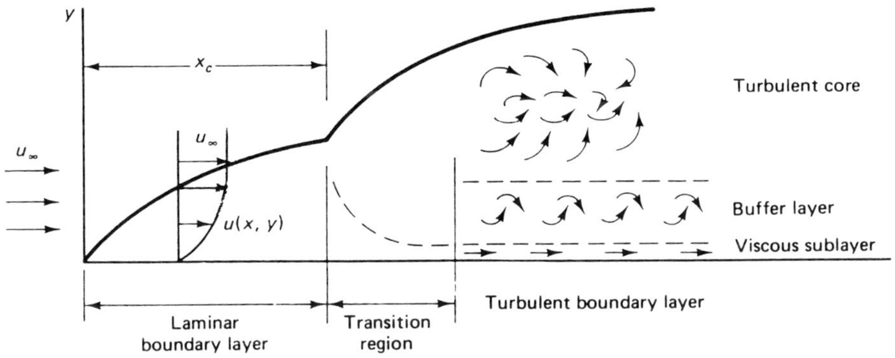

Figure 5.8.2 Turbulent boundary layers for flow over a flat plate.

Note that for gases, Pr is of the order unity; hence, the two boundary layers are almost the same thickness. For liquids, the Prandtl number ranges between 10 and 1,000; thus, the thermal boundary-layer thickness is smaller than the thickness of the velocity or hydrodynamic boundary layer. For liquid metals, Pr varies from about 0.003 to 0.03, and the thermal boundary layer is much thicker than the velocity boundary layer.

As shown in Fig. 5.8.2, starting from the leading edge of the plate, the laminar boundary layer continues to develop until some critical distance x_c is reached, beyond which small disturbances start and grow inside the boundary layer and the transition from laminar to turbulent boundary layer takes place, with the transition region characterized by unstable fluid motions. This critical distance is determined from the relation $x_c = \mu \mathrm{Re}/\rho u_\infty$. Usually, turbulence may occur at $5 \times 10^5 <$ $\mathrm{Re} < 5 \times 10^6$ for a flat plate, but as low as 3,000 for a circular tube. In the boundary layer next to the wall, the flow is characterized by a very thin viscous-flow region called the *viscous sublayer*. Adjacent to the sublayer is a highly turbulent region known as the *buffer layer*, in which the mean axial velocity increases rapidly with the distance from the wall. The buffer layer is then followed by the *turbulence core*, in which there is relatively lower intensity and larger-scale turbulence, with the velocity changing slightly with the distance from the wall.

The governing equations for two-dimensional, turbulent, thermal-hydrodynamic boundary-layer flow can be obtained by substituting Eqs. (5.7.6a, b, and c) into Eqs. (5.8.7)–(5.8.10), taking time averages, and using the boundary layer assumptions as follows.

Continuity

$$\frac{\partial \bar{u}}{\partial x} + \frac{\partial \bar{v}}{\partial y} = 0. \tag{5.8.15}$$

X Momentum

$$\bar{u}\frac{\partial \bar{u}}{\partial x} + \bar{v}\frac{\partial \bar{u}}{\partial y} = -\frac{1}{\rho}\frac{\mathrm{d}\bar{p}}{\mathrm{d}x} + \frac{\partial}{\partial y}\left(\nu\frac{\partial \bar{u}}{\partial y} - \overline{u'v'}\right). \tag{5.8.16}$$

Energy

$$\bar{u}\frac{\partial \bar{T}}{\partial x} + \bar{v}\frac{\partial \bar{T}}{\partial y} = \frac{\partial}{\partial y}\left(\alpha^*\frac{\partial \bar{T}}{\partial y} - \overline{T'v'}\right), \tag{5.8.17}$$

where $\alpha^* = k/\rho c_p$ is the diffusivity, and all variables are in dimensional quantity. Once again the Reynolds stress and the Reynolds heat flux

$$\overline{u'v'} \quad \text{and} \quad \overline{T'v'}$$

require suitable correlations, as discussed in Section 5.7. Also note that the turbulent boundary layer equations can be obtained only through the conservation form of the Navier–Stokes system of equations (5.8.7)–(5.8.10), not through the nonconservation form. For a large Reynolds number, flow turbulence is essentially a three-dimensional structure and the simplified two-dimensional analysis is not acceptable.

5.9 Convective Heat Transfer

5.9.1 General

The subject of heat transfer is often treated as an independent area of study in engineering. However, the governing equations derived in Section 5.3 have already suggested that heat transfer arises as a part of the first and/or second laws of thermodynamics. To see this, we return to Eqs. (5.3.16)–(5.3.18), and note that heat transfer consists of conduction, convection, and radiation. The simplest form of heat transfer is through conduction, usually characterized by Fourier's law – Eq. (4.4.33a). It is unaffected by the flow of fluids. Radiation, the most complicated mode of heat transfer, is an additional source of heat flux, as shown in Eqs. (5.3.19) and (5.3.20), which may or may not participate with the flow media. Convection is the study of heat transport processes directly affected by the flow of fluids.

Convection is either of two types: natural (free) or forced. Natural convection is due to the differences in temperature in space. In a fluid this activates motion. On the other hand, forced convection arises when external disturbances are introduced into the domain by force. We discuss appropriate governing equations for natural and forced convection in the next section.

5.9.2 Natural and Forced Convection

A simple example of natural convection is a heated body immersed in a cold reservoir; it acts as a heat engine that drives the fluid (see Fig. 5.9.1). Thus, in natural convection the heat engine that drives the flow is built into the flow itself, whereas in forced convection the heat engine that drives the flow is external.

Consider the governing equations of incompressible viscous flow:

$$V_{i,i} = 0, \tag{5.9.1}$$

$$\rho V_{i,j} V_j = \rho F_i - p_{,i} + \mu V_{i,jj}, \tag{5.9.2}$$

$$T_{,i} V_i = \alpha^* T_{,ii}. \tag{5.9.3}$$

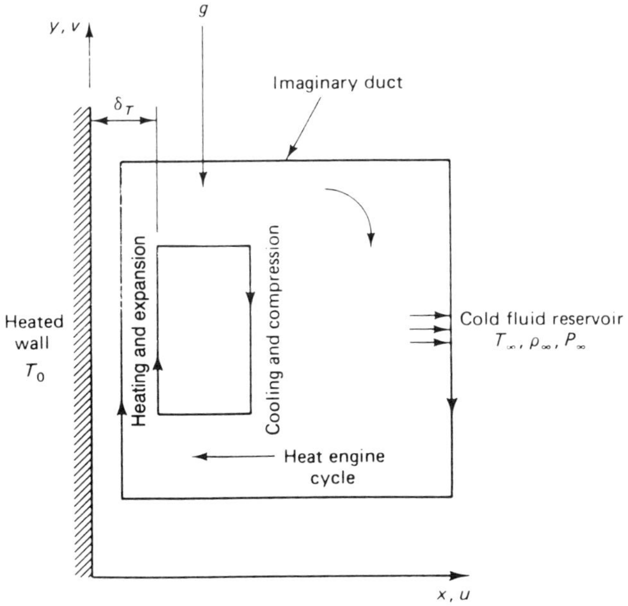

Figure 5.9.1 Natural convection along a vertical wall and analogy of heat engine responsible for driving the flow.

The boundary-layer equations for the flow system in Fig. 5.9.1 are

$$\rho\left(u\frac{\partial v}{\partial x} + v\frac{\partial v}{\partial y}\right) = -\rho g - \frac{\partial p}{\partial y} + \mu\frac{\partial^2 v}{\partial x^2}, \tag{5.9.4}$$

$$u\frac{\partial T}{\partial x} + v\frac{\partial T}{\partial y} = \alpha^*\frac{\partial^2 T}{\partial x^2}, \tag{5.9.5}$$

where $F_2 = F_y$ in Eq. (5.9.2) is set equal to gravity g. Note that $\partial p/\partial y = \mathrm{d}p_\infty/\mathrm{d}y$ is equivalent to the gravity force $-\rho_\infty g$ in the reservoir to obtain

$$\rho\left(u\frac{\partial v}{\partial x} + v\frac{\partial v}{\partial y}\right) = \mu\frac{\partial^2 v}{\partial x^2} + (\rho_\infty - \rho)g. \tag{5.9.6}$$

Define the thermal expansion coefficient as

$$\alpha = -\frac{1}{\rho}\left(\frac{\partial \rho}{\partial T}\right)_p \tag{5.9.7}$$

and invoke the Boussinesq (1877) approximation

$$\alpha \simeq -\frac{1}{\rho}\frac{\rho - \rho_\infty}{T - T_\infty}, \tag{5.9.8}$$

so we now write the momentum equation in the form

$$u\frac{\partial v}{\partial x} + v\frac{\partial v}{\partial y} = \nu\frac{\partial^2 v}{\partial x^2} + \alpha g(T - T_\infty). \tag{5.9.9}$$

Notice that the term $\alpha g(T - T_\infty)$ represents the driving force that is due to the body force and the pressure gradient in Eq. (5.9.4). This implies that we must add another pressure gradient term that accounts for the contribution from additional driving forces applied externally to the boundaries, if forced convection is combined with natural convection. Thus, the momentum equation (5.9.9) is revised to read

$$u\frac{\partial v}{\partial x} + v\frac{\partial v}{\partial y} = -\frac{1}{\rho}\frac{\partial p}{\partial x} + \nu\frac{\partial^2 v}{\partial x^2} + \alpha g(T - T_\infty). \tag{5.9.10}$$

Notice that if the source of natural convection no longer exists, Eq. (5.9.10) assumes the form of the standard boundary-layer equation, (5.9.4), with forced convection only.

5.10 High-Speed Aerodynamics

5.10.1 General

In high-speed flows a convenient measure of speed is a quantity called the Mach number, M, defined as the ratio of the flow velocity to the speed of sound. When a Mach number is less than approximately 0.3 the flow is considered, in general, to be incompressible (see Problem 5.18). However, for $M \geqslant 0.3$ the flow becomes compressible, and density changes are significant. For Mach numbers near unity (transonic: $0.8 \leqslant M \leqslant 1.2$) shock waves may develop and persist throughout supersonic and hypersonic ranges.

In this section, we explore the nature of the governing equation for transonic flows, its mathematical implications, and the computational difficulties closely associated with the complicated physical phenomenon of shock waves.

5.10.2 Full Potential Equation

Consider the steady flow governed by the Euler equations:

$$(\rho V_i)_{,i} = 0 \tag{5.10.1}$$

and

$$V_{i,j}V_j + \frac{1}{\rho}p_{,i} = 0. \tag{5.10.2}$$

Defining the speed of sound as $a = (\partial p/\partial \rho)^{1/2}$, we obtain

$$p_{,i} = a^2 \rho_{,i}. \tag{5.10.3}$$

Multiplying Eq. (5.10.2) by V_i and using Eq. (5.10.3) yields

$$V_i V_{i,j} V_j + \frac{a^2}{\rho} V_i \rho_{,i} = 0. \tag{5.10.4}$$

From the relation expressed in Eq. (5.10.1), we rewrite Eq. (5.10.4) in the form

$$V_i V_j V_{i,j} - a^2 V_{i,i} = 0. \tag{5.10.5}$$

The expansion of Eq. (5.10.5) in two dimensions gives

$$(1 - M_1^2)V_{1,1} + (1 - M_2^2)V_{2,2} - \frac{V_1 V_2}{a^2}(V_{1,2} + V_{2,1}) = 0, \tag{5.10.6}$$

where $M_1 = V_1/a$ and $M_2 = V_2/a$ are known as local Mach numbers in the x_1 and x_2 directions, respectively. Rearrange Eq. (5.10.6), which leads to

$$(1 - M_1^2)V_{1,1} + (1 - M_2^2)V_{2,2} - \frac{2V_1 V_2}{a^2}V_{1,2} = E \tag{5.10.7}$$

with

$$E = (V_{2,1} - V_{1,2})\frac{V_1 V_2}{a^2}. \tag{5.10.8}$$

For $E = 0$, Eq. (5.10.7) is known as the full potential equation for irrotational flow. If $E \neq 0$, then Eq. (5.10.7) represents nonvanishing vorticity. For a general case, E must be evaluated from the thermodynamic point of view. To this end, we begin with the definitions of total enthalpy $\hat{H}$ and entropy η as follows.

$$\hat{H} = H + \tfrac{1}{2}V_j V_j, \tag{5.10.9}$$

$$\eta = c_v \ln \frac{p}{\rho^\gamma}. \tag{5.10.10}$$

Along the streamline, the velocity of sound is

$$a = \left(\sqrt{\frac{\partial p}{\partial \rho}}\right)_{\eta = constant} = \sqrt{\frac{\gamma p}{\rho}} = \sqrt{\gamma R T}. \tag{5.10.11}$$

For an ideal gas, it follows that

$$\frac{p}{\rho^\gamma} = \frac{p_0}{\rho_0^\gamma} \exp \frac{(\eta - \eta_0)}{c_v} = C \exp \frac{\eta}{c_v} \tag{5.10.12}$$

with the subscript 0 indicating the initial condition, while the function C remains constant along each streamline. By virtue of Eqs. (5.10.10)–(5.10.12) and from $p = f(\rho, \eta)$, we obtain

$$p_{,i} = a^2 \rho_{,i} + \frac{\rho a^2}{c_p} \eta_{,i}. \tag{5.10.13}$$

Recall that Eq. (5.10.2) may be rewritten in the form

$$\tfrac{1}{2}(V_j V_j)_{,i} - \epsilon_{ijk} \epsilon_{kmn} V_{n,m} V_j + \frac{1}{\rho} p_{,i} = 0. \tag{5.10.14}$$

Use the definition of enthalpy

$$H = c_p T = \frac{\gamma}{\gamma - 1} \frac{p}{\rho}, \tag{5.10.15}$$

and in view of Eq. (5.10.13), we rewrite Eq. (5.10.14) as

$$T \eta_{,i} + \epsilon_{ijk} \epsilon_{kmn} V_{n,m} V_j - (H + \tfrac{1}{2} V_j V_j)_{,i} = 0 \tag{5.10.16a}$$

or

$$\mathbf{V} \times (\nabla \times \mathbf{V}) = \mathbf{V} \times \omega = \nabla \hat{H} - T \nabla \eta, \tag{5.10.16b}$$

with $\hat{H}$ being the total enthalpy. This is known as Crocco's equation. It implies that, if entropy and enthalpy gradients are nonvanishing across the streamlines, then there exists a nonvanishing vorticity, a condition of nonisentropic and nonadiabatic flow. Note that Eq. (5.10.16) also can be derived by combining the reversible process of the second law of thermodynamics and the momentum equation, respectively,

$$T\nabla\eta = \nabla H - \frac{1}{\rho}\nabla p$$

and

$$\nabla(\tfrac{1}{2}\mathbf{V}\cdot\mathbf{V}) - \mathbf{V}\times\boldsymbol{\omega} = -\frac{1}{\rho}\nabla p,$$

from which Eq. (5.10.16) can be obtained.

For the direction normal to the streamline, Eq. (5.10.16) takes the form

$$T\eta_{,i}n_i + \epsilon_{ijk}\epsilon_{kmn}V_{n,m}V_jn_i - \hat{H}_{,i}n_i = 0 \tag{5.10.17}$$

or

$$T\frac{\partial\eta}{\partial n} - \frac{\partial\hat{H}}{\partial n} + \mathbf{V}\times(\nabla\times\mathbf{V})\cdot\mathbf{n} = 0, \tag{5.10.18}$$

where, in two dimensions,

$$\mathbf{V}\times(\nabla\times\mathbf{V})\cdot\mathbf{n} = \epsilon_{ijk}\epsilon_{kmn}V_{n,m}V_jn_i = V^*(V_{2,1} - V_{1,2})$$

with

$$V^* = V_2n_1 - V_1n_2. \tag{5.10.19}$$

Hence,

$$E = \frac{1}{V^*}\left(\frac{\partial\hat{H}}{\partial n} - \frac{a^2}{\gamma R}\frac{\partial\eta}{\partial n}\right)\frac{V_1V_2}{a^2}. \tag{5.10.20}$$

Behind the interacting shock waves of different strengths a slipstream (contact surface, vortex sheet) develops, across which (normal to the slipstream) entropy is discontinuous, resulting in rotational flow with $E \neq 0$. In terms of the velocity potential function, from Eq. (5.10.5) we obtain

$$\phi_{,ii} - \frac{1}{a^2}\phi_{,i}\phi_{,j}\phi_{,ij} = 0, \tag{5.10.21a}$$

or

$$(1 - M_x^2)\frac{\partial^2\phi}{\partial x^2} + (1 - M_y^2)\frac{\partial^2\phi}{\partial y^2} - \frac{2}{a^2}\frac{\partial\phi}{\partial x}\frac{\partial\phi}{\partial y}\frac{\partial^2\phi}{\partial x\partial y} = 0. \tag{5.10.21b}$$

It is important to note that the same result is obtained from Eq. (5.10.7) with $E = 0$, implying that the velocity potential representation leads to irrotational flow.

The boundary conditions on the surface of the body are

$$\phi = g(x, y) \quad \text{on } \Gamma_1 \tag{5.10.22}$$

and

$$\phi_{,i}n_i = 0 \quad \text{on } \Gamma_2. \tag{5.10.23}$$

Since the solution of the full potential equation, (5.10.21), represents a major computational effort, simplifications known as *small-perturbation* approximations are made in many engineering applications. We assume that a body such as an airfoil is so thin that disturbances (in the y direction) are limited to the immediate vicinity of the body and, thus, do not propagate far from it. In this case, Eq. (5.10.21) is simplified to the small perturbation potential equation,

$$(1 - M_\infty^2)\frac{\partial^2 \phi'}{\partial x^2} + \frac{\partial^2 \phi'}{\partial y^2} = M_\infty^2\left(\frac{1+\gamma}{u_\infty}\right)\frac{\partial \phi'}{\partial x}\frac{\partial^2 \phi'}{\partial x^2}, \tag{5.10.24}$$

where M_∞ and u_∞ refer, respectively, to the Mach number and the velocity in the free-stream region, and ϕ' is the perturbation velocity potential defined as

$$\phi' = \phi - u_\infty x.$$

A linearized, small perturbation equation arises when the right-hand side of Eq. (5.10.24) is neglected. Further details are found in Liepmann and Roshko (1957).

5.10.3 Shock Waves

An interesting feature of Eq. (5.10.24) is that, depending on the magnitude of the Mach number, mathematical properties of the governing equation are altered in such a way that the following arise:

For $M_\infty < 1$ (subsonic speed) *elliptic partial differential equation*,

For $M_\infty = 1$ (transonic) *parabolic partial differential equation*,

For $M_\infty > 1$ (supersonic) *hyperbolic partial differential equation*.

Notice that the full potential equation, (5.10.21), is of the mixed type in which the local Mach numbers M_x and M_y vary spatially within the flow field. Shock wave discontinuities usually occur as the Mach number approaches unity. Shock wave structures for various geometries and Mach numbers are illustrated in Fig. 5.10.1.

The occurrence of shock waves is a nonisentropic process. If geometrical changes of flow domain (compression corners, etc.) are encountered in high-speed flows (with the Mach number higher than approximately 0.8), then sudden changes of streamline directions are accompanied by an increase in entropy, causing all flow variables to change abruptly. The conservation of mass, momentum, and energy is

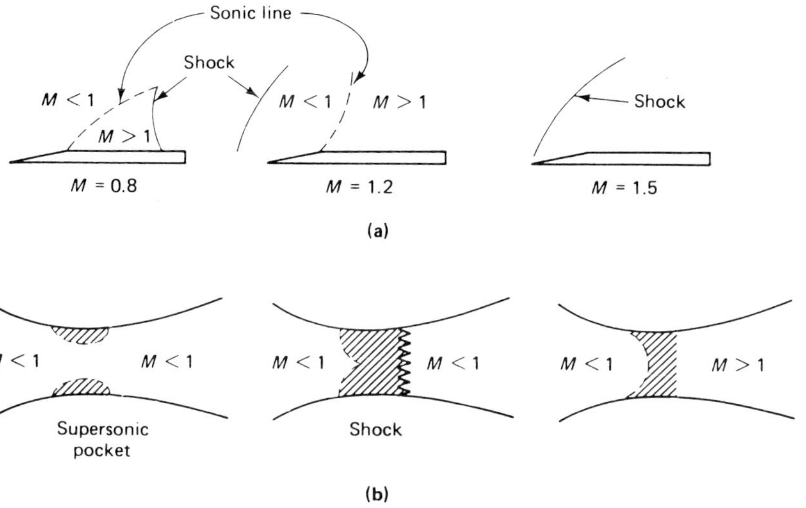

Figure 5.10.1 External and internal flow with various speed ranges. (a) Flow around a wedge. (b) Nozzle flow. Broken lines represent $M = 1$, shaded area supersonic.

enforced in this process by the presence of shock waves, which allows the streamlines to bend nonisentropically.

The pressure discontinuity associated with shock waves is the most critical design information for high-speed vehicles. To this end, the pressure coefficient C_p is a convenient means to measure pressure variations over the body,

$$C_p = \frac{p - p_\infty}{\frac{1}{2}\rho_\infty u_\infty^2}. \tag{5.10.25}$$

For compressible flow, we may modify Eq. (5.10.25) as

$$C_p = \frac{2}{\gamma M_\infty^2}\left(\frac{p}{p_\infty} - 1\right), \tag{5.10.26}$$

whereas, for incompressible flow, Eq. (5.10.25) assumes the form

$$C_p = 1 - \left(\frac{V}{u_\infty}\right)^2 \tag{5.10.27}$$

with $V = (\mathbf{V} \cdot \mathbf{V})^{1/2}$.

A typical pressure coefficient distribution is shown in Fig. 5.10.2. Note that the negative and positive signs imply that $p < p_\infty$ and $p > p_\infty$, respectively. The precise form of C_p depends upon the relationship between p and p_∞ for various types of flows and approximations, such as perturbations and linearization.

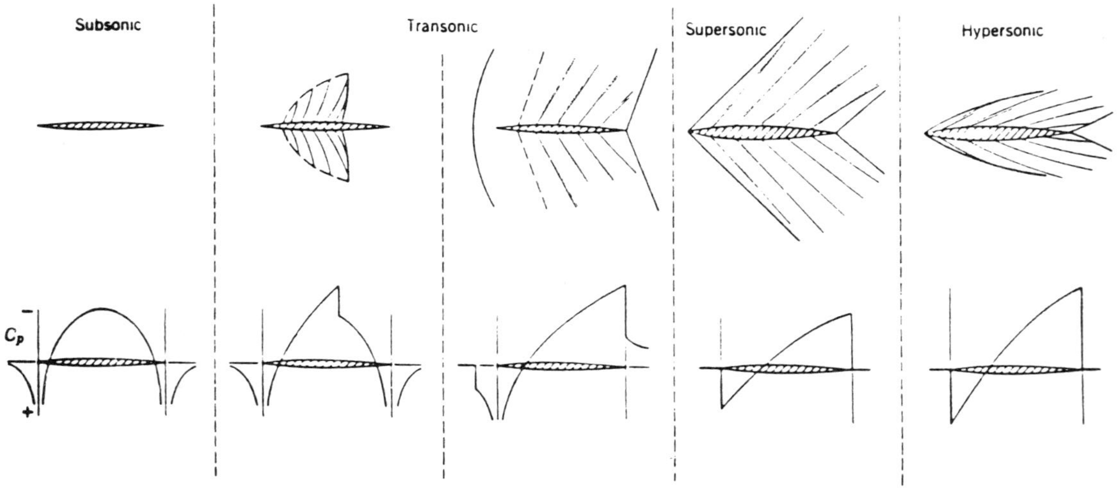

Figure 5.10.2 Surface pressure distributions for a typical airfoil.

The supersonic flow with Mach numbers higher than about 5 is referred to as hypersonic flow. In this case, the inviscid flow theory breaks down because the flow around a body is affected by viscous boundary layers. The flow may even become turbulent, leading to one of the most difficult and complicated fluid mechanics problems – shock-turbulent boundary-layer interactions. The potential equation is no longer valid and we must invoke the Navier–Stokes system of equations (5.3.27), together with additional equations involved in turbulence, as given in Section 5.7.

In summary, the nonlinear terms of the governing equation, (5.10.21), which are responsible for shock discontinuities, originate from the convective terms of Eq. (5.10.2). This is similar in spirit to the turbulent (Reynolds) stress arising from the convective terms of Eq. (5.7.8). Thus, we conclude that the convection of flow leads to both turbulence and shock wave discontinuities. This is in contrast to solid mechanics in which nonlinearity does not arise unless strains are large, which is analogous to non-Newtonian fluids in which stress is nonlinearly proportional to velocity gradients. We have explored the fluid mechanics problems only to the extent that tensor analysis plays an important role in the derivation of various forms of the governing equations in this chapter.

5.11 Acoustics

5.11.1 General

We have discussed the standard problems involved in continuum mechanics, including solid mechanics, fluid mechanics, and heat transfer. The subject of acoustics, however, is regarded as an application area of continuum mechanics. Wave oscillations take place in both solids and fluids; they include sound waves or acoustic waves through solid materials, in a pressurized chamber filled with gases or liquids, or combustion waves in gases. This section focuses on acoustic waves in fluids in order to limit the scope of our discussion. Neither wave propagation in solids nor combustion waves will be included.

The governing equations for acoustics are based on the conservation laws of fluid mechanics. The prerequisite for a sound wave to be transmitted is the presence of a compressible medium. In the absence of viscosity, the following conservation equations are invoked.

Continuity

$$\frac{D\rho}{Dt} + \rho\boldsymbol{\nabla}\cdot\mathbf{V} = 0. \tag{5.11.1}$$

Momentum

$$\rho\frac{D\mathbf{V}}{Dt} + \boldsymbol{\nabla}p = 0. \tag{5.11.2}$$

Energy

$$\frac{D\eta}{Dt} = 0. \tag{5.11.3}$$

For the isentropic process, the changes of pressure are given by

$$\mathrm{d}p = \left(\frac{\partial p}{\partial\rho}\right)_\eta \mathrm{d}\rho + \left(\frac{\partial p}{\partial\eta}\right)_\rho \mathrm{d}\eta = a^2\,\mathrm{d}\rho \tag{5.11.4}$$

with the speed of sound being defined as

$$a = \left(\frac{\partial p}{\partial\rho}\right)_{\eta_0}^{1/2}.$$

Thus, from Eq. (5.11.4) we obtain

$$\frac{Dp}{Dt} - a^2\frac{D\rho}{Dt} = 0.$$

In view of the continuity equation,

$$\frac{Dp}{Dt} + \rho a^2\boldsymbol{\nabla}\cdot\mathbf{V} = 0. \tag{5.11.5}$$

Consider the state of equilibrium slightly disturbed so that

$$\rho = \rho_0(\mathbf{x}) + \rho'(\mathbf{x}, t), \tag{5.11.6a}$$

$$p = p_0(\mathbf{x}) + p'(\mathbf{x}, t), \tag{5.11.6b}$$

where the symbols 0 and prime denote equilibrium and disturbance, respectively. The spatial representation $\mathbf{x}$ refers to the Eulerian description in this section. Substituting Eq. (5.11.6) into Eq. (5.11.5) gives

$$\frac{\partial p'}{\partial t} + \rho_0(\mathbf{x})a_0^2(\mathbf{x})\boldsymbol{\nabla}\cdot\mathbf{V} = 0, \tag{5.11.7}$$

where $a_0^2 = (\partial p/\partial\rho)_{\eta_0,\rho_0}$ and $(\mathbf{V}\cdot\boldsymbol{\nabla})p_0$ is neglected. This can be justified by the fact that, for the characteristic velocity U, length L, and pressure gradient $\boldsymbol{\nabla}p_0 = \rho_0 g$, the ratio

$$|(\mathbf{V} \cdot \boldsymbol{\nabla})p_0|/\rho_0 a_0^2 |\boldsymbol{\nabla} \cdot \mathbf{V}| \simeq \frac{L}{a_0^2/g} \simeq L/12 \text{ km}$$

in air at standard temperature and pressure may be negligibly small, because $L \ll 12$ km in a practical system. Similarly, from the momentum equation, (5.11.2), we have

$$\rho_0 \frac{\partial \mathbf{V}}{\partial t} + \boldsymbol{\nabla} p' = 0. \tag{5.11.8}$$

Now, taking the time derivative of Eq. (5.11.7) and the spatial derivative of Eq. (5.11.8) leads to, respectively,

$$\frac{\partial^2 p'}{\partial t^2} + \rho_0(\mathbf{x}) a_0^2(\mathbf{x}) \boldsymbol{\nabla} \cdot \frac{\partial \mathbf{V}}{\partial t} = 0, \tag{5.11.9a}$$

$$\boldsymbol{\nabla} \cdot \frac{\partial \mathbf{V}}{\partial t} + \frac{1}{\rho_0} \nabla^2 p' = \frac{1}{\rho_0^2} \boldsymbol{\nabla} p' \cdot \boldsymbol{\nabla} \rho_0 \simeq 0. \tag{5.11.9b}$$

Combine Eqs. (5.11.9a) and (5.11.9b), which gives

$$\frac{\partial^2 p'}{\partial t^2} - a_0^2(\mathbf{x}) \nabla^2 p' = 0. \tag{5.11.10}$$

This is the wave equation (or acoustic equation) for a fluid whose equilibrium state is not uniform. It is the same as the classical wave equation, except that the speed of sound a_0 is a function of position.

For many engineering problems, the linearized continuity and momentum equations can be used to derive the wave equations, assuming that the flow is uniform, or that

$$\frac{\partial p'}{\partial t} + \rho_0 a_0^2 \boldsymbol{\nabla} \cdot \mathbf{V} = 0, \tag{5.11.11}$$

$$\rho_0 \frac{\partial \mathbf{V}}{\partial t} + \boldsymbol{\nabla} p' = 0. \tag{5.11.12}$$

We introduce the velocity potential ϕ such that $\mathbf{V} = \boldsymbol{\nabla}\phi$, which is substituted into Eq. (5.11.12), hence we have

$$\boldsymbol{\nabla}\left(\rho_0 \frac{\partial \phi}{\partial t} + p'\right) = 0$$

and

$$\rho_0 \frac{\partial \phi}{\partial t} + p' = c(t),$$

where the constant of integration $c(t)$ may be set equal to zero without loss of generality. Thus,

$$p' = -\rho_0 \frac{\partial \phi}{\partial t}.$$

Now substitute this into Eq. (5.11.11) to get

$$\frac{\partial^2 \phi}{\partial t^2} - a_0^2 \nabla^2 \phi = 0. \tag{5.11.13}$$

Similarly, taking the gradient of Eq. (5.11.11) and substituting in Eq. (5.11.12) yields

$$\frac{\partial^2 \mathbf{V}}{\partial t^2} - a_0^2 \nabla^2 \mathbf{V} = 0, \tag{5.11.14}$$

Equations (5.11.10), (5.11.13), and (5.11.14) are the wave equations in terms of pressure, velocity potential, and velocity vector, respectively. The choice depends on convenience as determined by the available boundary conditions and other aspects of the particular engineering application.

5.11.2 Monochromatic Waves

A *monochromatic wave* is one in which the pressure and velocity depend on time only through periodic functions of time of a single (circular) frequency ω. The velocity potential may be written as the real part ($\mathbb{R}$) of the oscillatory behavior.

$$\phi(\mathbf{x}, t) = \mathbb{R}\{\phi(\mathbf{x})\, \mathrm{e}^{-i\omega t}\}. \tag{5.11.15}$$

Substituting Eq. (5.11.15) into Eq. (5.11.13) gives

$$\nabla^2 \Phi + k^2 \Phi = 0, \tag{5.11.16}$$

where $k = \omega/a_0$ is the wave number. This is known as the Helmholtz equation, which is now independent of time.

Let us consider a *plane wave*, which is defined as a wave that impinges on a plane in a one-, two-, or three-dimensional geometry. For simplicity, we examine a plane wave in one direction.

$$\frac{\mathrm{d}^2 \Phi}{\mathrm{d}x^2} + k^2 \Phi = 0. \tag{5.11.17}$$

The general solution of this equation can be written as

$$\Phi = A\, \mathrm{e}^{ikx} + B\, \mathrm{e}^{-ikx}. \tag{5.11.18}$$

Substitute this into Eq. (5.11.15) to get

$$\Phi(x, t) = A\, \mathrm{e}^{i(kx - \omega t)} + B\, \mathrm{e}^{-i(kx + \omega t)}. \tag{5.11.19}$$

Set $B = 0$, and we arrive at the *positive-going wave*, defined as

$$\phi = \widetilde{A} \cos(kx - \omega t + \beta), \tag{5.11.20}$$

where $A = |A|\, \mathrm{e}^{i\beta}$ and $\widetilde{A} = |A|$ are called the *amplitudes* of the wave and the quantity

$$\psi = kx - \omega t + \beta \tag{5.11.21}$$

is known as the *phase*. Note that the phase is constant for an observer moving with velocity $\mathrm{d}x/\mathrm{d}t$, so that

$$\frac{\mathrm{d}\psi}{\mathrm{d}t} = k\frac{\mathrm{d}x}{\mathrm{d}t} - \omega = 0.$$

Thus,

$$\frac{\mathrm{d}x}{\mathrm{d}t} = \frac{\omega}{k} = c_f,$$

where c_f is the phase velocity and, in this simple case, is equal to the speed of sound a_0. Since the acoustic variables in the wave are sinusoidal functions of time with period T, we have

$$\cos(kx - \omega t + \beta) = \cos[kx - \omega(t + T) + \beta], \tag{5.11.22}$$

which requires that $\omega = 2\pi/T$. Similarly, the spatial period, called the *wave-length* λ, may be introduced in the form

$$\cos kx = \cos[k(x + \lambda)]. \tag{5.11.23}$$

This leads to $k = 2\pi/\lambda$ and $a_0 = \lambda/T = f\lambda$, where $f = 1/T$ is the frequency.

Let us now consider a plane wave in three dimensions propagating along some axis $s(\mathbf{x})$ such that $\mathbf{x} = s\mathbf{n}$, $s = \mathbf{n}\cdot\mathbf{x}$, and

$$\Phi = A\,\mathrm{e}^{ik\mathbf{n}\cdot\mathbf{x}}. \tag{5.11.24}$$

We define a wave vector $\mathbf{k}$ as

$$\mathbf{k} = k\mathbf{n} = \frac{\omega}{a_0}\mathbf{n}.$$

Then,

$$\Phi = A\,\mathrm{e}^{i\mathbf{k}\cdot\mathbf{x}},$$

or

$$\Phi = A\,\mathrm{e}^{ik_jx_j}.$$

Thus,

$$\begin{aligned}\frac{\partial\Phi}{\partial x_m} &= Aik_j\frac{\partial x_j}{\partial x_m}\,\mathrm{e}^{ik_nx_n} = Aik_j\delta_{jm}\,\mathrm{e}^{ik_nx_n}\\ &= Aik_m\,\mathrm{e}^{ik_nx_n} = ik_m\Phi.\end{aligned}$$

This leads to

$$\frac{\partial^2\Phi}{\partial x_m\partial x_m} + k^2\Phi = 0 \tag{5.11.25}$$

in which the following identity has been used:

$$k_j k_j = \mathbf{k} \cdot \mathbf{k} = \frac{\omega}{a_0}\mathbf{n} \cdot \frac{\omega}{a_0}\mathbf{n} = \left(\frac{\omega}{a_0}\right)^2 = k^2.$$

Therefore, we conclude that the plane wave in three dimensions assumes the solution

$$\phi(\mathbf{x}, t) = A\,\mathrm{e}^{i(k_j x_j - \omega t)} + B\,\mathrm{e}^{-i(k_j x_j + \omega t)}. \tag{5.11.26}$$

This solution is useful in problems that deal with plane waves over boundaries which are neither parallel nor perpendicular to the direction of propagation.

The acoustic pressure p' may be written as:

$$p'(\mathbf{x}, t) = \hat{p}'(\mathbf{x})\,\mathrm{e}^{-i\omega t}, \tag{5.11.27}$$

whereas the velocity vector $\mathbf{V}$ is expressed in the form

$$\mathbf{V} = \hat{\mathbf{V}}(\mathbf{x})\,\mathrm{e}^{-i\omega t}. \tag{5.11.28}$$

From the momentum equation,

$$\frac{\partial \mathbf{V}}{\partial t} = -\frac{1}{\rho_0}\boldsymbol{\nabla} p'. \tag{5.11.29}$$

For plane waves,

$$\hat{p}' = p_0\,\mathrm{e}^{ik_j x_j}, \quad \hat{p}'_{,j} = ik_j p_0\,\mathrm{e}^{ik_n x_n} = ik_j \hat{p}'.$$

Thus, from Eq. (5.11.29) together with Eqs. (5.11.27) and (5.11.28), we obtain

$$\hat{\mathbf{V}}'(\mathbf{x}) = \frac{1}{i\rho_0\omega}\boldsymbol{\nabla}\hat{p}'(\mathbf{x}), \tag{5.11.30}$$

or

$$\hat{V}'_j = \frac{1}{i\rho_0\omega}\hat{p}'_{,j} = \frac{k_j\hat{p}'}{\rho_0\omega}. \tag{5.11.31}$$

For the one-dimensional case,

$$\hat{u} = \frac{1}{i\rho_0\omega}\frac{\partial\hat{p}'}{\partial x} = \frac{k\hat{p}'}{\rho_0\omega}. \tag{5.11.32}$$

Let us now examine the incident wave in the positive x direction:

$$p' = \bar{A}\,\mathrm{e}^{-i(kx - \omega t)} \tag{5.11.33}$$

and let the rigid reflector be placed at $x = 0$. The solution of the wave equation for the space in front of the reflector is given by

$$p' = \bar{A}\,\mathrm{e}^{i(\omega t - kx)} + \bar{B}\,\mathrm{e}^{i(\omega t + kx)}. \tag{5.11.34}$$

Thus, the sound field in front of the reflector consists of the incident wave and a second wave that travels from the reflector to $-\infty$.

Obviously, this wave can be interpreted as the wave reflected at the rigid reflector where

$$u'|_{x=0} = 0 = \frac{1}{i\omega\rho_0}\left(\frac{\partial p'}{\partial x}\right)_{x=0} = \frac{1}{ik\rho_0 a_0}(-ik\bar{A} + ik\bar{B})\,e^{i\omega t}$$
$$= \frac{1}{\rho_0 a_0}(\bar{A} - \bar{B})\,e^{i\omega t}, \tag{5.11.35}$$

which suggests that $\bar{B} = \bar{A}$ and

$$\bar{R} = \frac{\bar{B}}{\bar{A}} = 1, \tag{5.11.36}$$

where $\bar{R}$ is the reflection coefficient for the pressure. The resultant pressure,

$$p' = \bar{A}\,e^{i\omega t}(e^{-ikx} + e^{ikx}) = 2\bar{A}\cos(kx)\,e^{i\omega t}, \tag{5.11.37}$$

represents a *standing wave* of amplitude $2\bar{A}\cos(kx)$ and frequency ω. The standing wave is a constant harmonic oscillation at any point x equivalent to the superposition of two *traveling waves*, positive and negative, at any given time, which may grow or decay as a function of time only. The relationship between traveling and standing waves is illustrated in Fig. 5.11.1. The real part of the solution in Eq. (5.11.37) reads

$$p' = 2A\cos(kx)\cos(\omega t + \phi_0). \tag{5.11.38}$$

The complex particle velocity is obtained from Eq. (5.11.35) by setting $\bar{B} = \bar{A}$, so that

$$u' = \frac{1}{\rho_0 a_0}(\bar{A}\,e^{-ikx} - \bar{A}\,e^{ikx})\,e^{i\omega t},$$

or

$$u' = -\frac{2i\bar{A}}{\rho_0 a_0}\sin(kx)\,e^{i\omega t}. \tag{5.11.39}$$

The real part of the solution of the particle velocity is

$$u' = \frac{2\bar{A}}{\rho_0 a_0}\sin(kx)\sin(\omega t + \phi_0)$$
$$= \frac{2\bar{A}}{\rho_0 a_0}\sin(kx)\cos\left(\omega t + \phi_0 - \frac{\pi}{2}\right). \tag{5.11.40}$$

The particle velocity vanishes at the reflector and lags the pressure in phase by 90°. Note that the pressure is reflected without phase change at a rigid surface. Because the reflected wave travels in the direction with the opposite sign of the incident wave at the reflector, the resultant particle velocity at the reflector is zero.

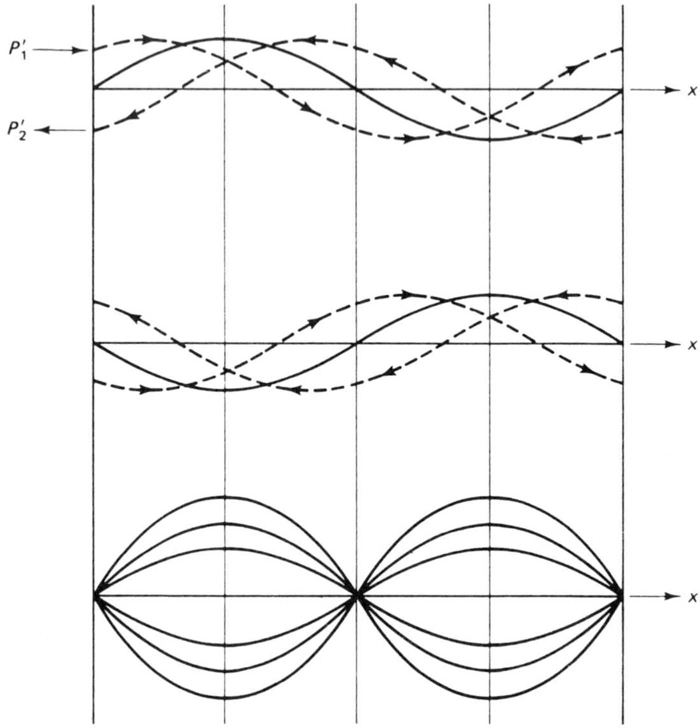

——— Standing Wave

- - - - Traveling Wave

Traveling Wave : $P_1' = A\cos(\omega t - \kappa x)$

$P_2' = A\cos[\omega t + \kappa(x - l)] = A\cos(\omega t + \kappa x)$

Standing Wave : $P' = P_1' + P_2' = 2A\cos\omega t\cos\kappa x$

Figure 5.11.1 Standing wave. Constant harmonic oscillation at any point x equivalent to superposition of two traveling waves, positive and negative, at any given time, may grow or decay as a function of time only.

5.11.3 Cylindrical and Spherical Waves

If the wave fronts form a circular cylinder, we refer to the waves as *cylindrical waves*, in contrast to the plane wave discussed in the previous section. Similarly, another nonplane wave is the *spherical wave*, where the wave fronts are a doubly curved surface.

The Helmholtz equation for cylindrical waves takes the form

$$\nabla^2\phi + k^2\phi = 0, \tag{5.11.41}$$

where

$$\nabla^2\phi = \frac{\partial^2\phi}{\partial r^2} + \frac{1}{r}\frac{\partial\phi}{\partial r} + \frac{1}{r^2}\frac{\partial^2\phi}{\partial\theta^2} + \frac{\partial^2\phi}{\partial x^2} \tag{5.11.42}$$

with the axial, transverse, and tangential velocities defined, respectively, as

$$u = \frac{\partial\phi}{\partial x}, \quad v \doteq \frac{\partial\phi}{\partial r}, \quad w = \frac{1}{r}\frac{\partial\phi}{\partial\theta}$$

and with

$$k = \frac{\omega}{a}, \quad \phi = \hat{\phi}\mathrm{e}^{-i\omega t}.$$

Similarly, the Helmholtz equation for spherical waves is written with $\nabla^2\phi$ replaced as follows.

$$\nabla^2\phi = \frac{\partial^2\phi}{\partial R^2} + \frac{2}{R}\frac{\partial\phi}{\partial R} + \frac{1}{R^2\sin\alpha}\frac{\partial}{\partial\alpha}\left(\sin\alpha\frac{\partial\phi}{\partial\alpha}\right) + \frac{1}{R^2\sin\alpha}\frac{\partial^2\phi}{\partial\theta^2}, \tag{5.11.43}$$

where the radial, meridian, and tangential velocities are defined, respectively, as

$$u_R + \frac{\partial\phi}{\partial R}, \quad u_\alpha = \frac{1}{R}\frac{\partial\phi}{\partial\alpha}, \quad u_\theta = \frac{1}{R\sin\alpha}\frac{\partial\phi}{\partial\theta}.$$

For the centrally symmetric case, the velocity potential is only a function of R. Thus,

$$\nabla^2\phi = \frac{\partial^2\phi}{\partial R^2} + \frac{2}{R}\frac{\partial\phi}{\partial R}. \tag{5.11.44}$$

The solution of cylindrical waves is of interest in many practical engineering problems, such as pressure oscillations in energy systems, missiles, and space shuttle engines. Advanced topics in acoustics include sound emission, sound absorption, and thermal waves. Textbooks specializing in these subjects should be consulted for details (Pierce 1981, Temkin 1981).

5.12 Reacting Flows

5.12.1 General

This section considers the various conservation equations as they would apply in a system of reacting chemical species for which macroscopic viewpoints are assumed to be valid. Thus, our discussion will be limited to a reacting fluid that can be treated as a continuum in which the ideal gas law still holds. In such a situation, the continuum theory presented in the previous sections for Newtonian fluids may be extended to construct the conservation equations for multicomponent reacting systems.

In reacting fluids we are dealing with various species, mass concentrations, molar concentrations, mass fractions, and mole fractions, defined as follows:

1. *Mass concentration* ρ_i: the mass of species i per unit volume of solution;
2. *Molar concentration* $C_i = \rho_i/W_i$: the number of moles of species i per unit volume, where W_i is the molecular weight;
3. *Mass fraction* $Y_i = \rho_i/\rho$: the mass concentration of species i divided by the total mass density of the solution ($\sum_{i=1}^{N} Y_i = 1$, where N is the total number of species);
4. *Mole fraction* $X_i = C_i/C$: the molar concentration of species i divided by the total molar density of the solution.

Here the subscript i represents the species rather than the tensor index. If tensor indices are used, they will be clearly distinguished from the subscripts for the species.

The gaseous state under ordinary conditions (i.e., the air we breathe) is extremely dilute, with the molecules separated by large distances compared with their molecular diameters. Thus, in general, the ideal gas law holds

$$p = \rho RT, \tag{5.12.1}$$

where R is the specific gas constant and has the units $J\,kg^{-1}K^{-1}$. It is also known that at a constant temperature and in the limit as p approaches zero, the ratio p/ρ is equal to the inverse molecular weight ratio

$$\frac{\left(\dfrac{p}{\rho}\right)_A}{\left(\dfrac{p}{\rho}\right)_B} = \frac{R_A}{R_B} = \frac{W_B}{W_A}.$$

Therefore, we may write

$$R^0 = R_A W_A = R_B W_B$$

in which R^0 is the universal gas constant for all substances with the value 8.31434J $\text{mole}^{-1}\,\text{K}^{-1}$. The ideal gas law then becomes

$$p = \frac{1}{V} R^0 T = \frac{\rho}{W} R^0 T,$$

where $V = W/\rho$, the volume of one mole or 6.02283×10^{23} molecules of the gas. The volume of one mole of an ideal gas at $p = 101{,}325$ pascals (or one atmosphere) and $T = 273.15$ K (0 °C) is $0.022414\ \mathrm{m}^3$.

For a mixture of thermally perfect gases, the equation of state is expressed in various ways:

$$p = R^0 T \sum_{i=1}^{N} C_i = \rho R^0 T \sum_{i=1}^{N} \frac{Y_i}{W_i} = R^0 T \frac{\sum_{i=1}^{N} N_i}{V}, \tag{5.12.2}$$

where N_i is the number of moles of species i.

The various chemical species in a diffusing mixture have different velocities $\mathbf{v}_i$ measured with respect to stationary coordinate axes. Thus, for a mixture of N species, the local mass-average velocity $\mathbf{v}$ is defined as:

$$\mathbf{v} = \frac{\sum_{i=1}^{N} \rho_i \mathbf{v}_i}{\sum_{i=1}^{N} \rho_i}. \tag{5.12.3a}$$

The difference between the local mass-average velocity $\mathbf{v}$ and the velocity of the ith species $\mathbf{v}_i$ is known as the diffusion velocity $\mathbf{V}_i$ (Hirschfelder, Curtis, and Bird 1954):

$$\mathbf{V}_i = \mathbf{v}_i - \mathbf{v}. \tag{5.12.3b}$$

A diffusion velocity is zero if $\mathbf{v}_i$ is equal to $\mathbf{v}$, in which case the motion of component i coincides with the local motion of the fluid stream. Here, the lower-case $\mathbf{v}$ is to be taken as the Eulerian mass-average velocity (recall the earlier use of $\mathbf{v}$ for the Lagrangian velocity).

Similarly, a local molar-average velocity $\mathbf{v}^*$ may be defined as:

$$\mathbf{v}^* = \frac{\sum_{i=1}^{N} C_i \mathbf{v}_i}{\sum_{i=1}^{N} C_i}. \tag{5.12.4a}$$

Thus the molar diffusion velocity becomes

$$\mathbf{V}_i^* = \mathbf{v}_i - \mathbf{v}^*. \tag{5.12.4b}$$

The mass flux or molar flux of species i is a vector quantity that denotes the mass or number of moles of species i that passes through a unit arca per unit time relative to stationary coordinates:

$$\dot{\mathbf{m}}_i = \rho_i \mathbf{v}_i \quad (\textit{mass flux}), \tag{5.12.5a}$$

$$\dot{\mathbf{n}}_i = C_i \mathbf{v}_i \quad (\textit{molar flux}). \tag{5.12.5b}$$

The relative mass flux and relative molar flux are

$$\mathbf{J}_i = \rho_i(\mathbf{v}_i - \mathbf{v}) = \rho_i \mathbf{V}_i, \tag{5.12.6a}$$

$$\mathbf{J}_i^* = C_i(\mathbf{v}_i - \mathbf{v}^*) = C_i \mathbf{V}_i^*. \tag{5.12.6b}$$

The mass diffusivity $D_{AB} = D_{BA}$ (cm^2/sec) in a binary system (a system with species A and B) is related to these quantities by *Fick's first law*:

$$\mathbf{J}_A = \rho_A \mathbf{V}_A = -\rho D_{AB} \nabla Y_A \quad (\textit{mass diffusion flux}), \tag{5.12.7a}$$

$$\mathbf{J}_A^* = C_A \mathbf{V}_A^* = -C D_{AB} \nabla X_A \quad (\textit{molar diffusion flux}). \tag{5.12.7b}$$

In general, the reaction mechanism for multistep reversible reactions is written

$$\sum_{i=1}^{N} \nu'_{i,k} M_i \rightleftarrows \sum_{i=1}^{N} \nu''_{i,k} M_i \quad (k = 1, 2, \ldots, M), \tag{5.12.8}$$

in which $\nu_{i,k}$ is the stoichiometric coefficient of species i for reaction step k, the prime and double prime represent the reactant and product, respectively, and M_i is a stand-in for the chemical formula for species i. The rate of production of species i for reaction step k is

$$\omega_i = (\nu''_{i,k} - \nu'_{i,k})\omega'_k, \quad (i = 1, 2, \ldots, N;\ k = 1, 2, \ldots, M)$$

with the net forward rate of the kth reaction defined as

$$\omega'_k = K_{f,k} \prod_{i=1}^{N} C_i^{\nu'_{i,k}} - K_{b,k} \prod_{i=1}^{N} C_i^{\nu''_{i,k}} \quad (k = 1, 2, \ldots, M),$$

where $K_{f,k}$ and $K_{b,k}$ denote the specific reaction rate constants for the forward and backward reactions, respectively, and $\prod$ represents a cumulative product.

The specific reaction rate constant is usually evaluated by the Arrhenius law,

$$K_{f,k},\ K_{b,k} = B_k T^{\alpha_k} \exp\left(-\frac{E_k}{R^0 T}\right),$$

where B_k is the frequency factor and α_k is the constant. Thus, the reaction rate is written as:

$$\omega_i = W_i \sum_{k=1}^{M} (\nu''_{i,k} - \nu'_{i,k}) B_k T^{\alpha_k} \exp\left(-\frac{E_k}{R^0 T}\right) \prod_{j=1}^{N} \left(\frac{X_j p}{R^0 T}\right)^{\nu'_{j,k}}. \tag{5.12.9}$$

With these preliminaries, we are now prepared to develop the equations for the conservation of mass, momentum, and energy in reacting flows.

5.12.2 Conservation of Mass for Mixture and Species

Consider the continuity equation for component A in a binary mixture with a chemical reaction at a rate ω_A (kg^{-3}–sec^{-1}), known as the *mass rate of production of species A*:

$$\frac{\partial \rho_A}{\partial t} + \nabla \cdot (\rho_A \mathbf{v}_A) = \omega_A. \tag{5.12.10}$$

Similarly, the equation of continuity for component B is

$$\frac{\partial \rho_B}{\partial t} + \nabla \cdot (\rho_B \mathbf{v}_B) = \omega_B. \tag{5.12.11}$$

Add Eqs. (5.12.10) and (5.12.11) to give

$$\frac{\partial \rho}{\partial t} + \nabla \cdot (\rho \mathbf{v}) = 0. \tag{5.12.12}$$

Equation (5.12.12) results from the law of conservation of mass,

$$\omega_A + \omega_B = 0, \quad \rho_A + \rho_B = \rho, \quad \text{and} \quad \rho_A \mathbf{v}_A + \rho_B \mathbf{v}_B = \rho \mathbf{v}.$$

In terms of molar units, the continuity equation takes the form

$$\frac{\partial C_A}{\partial t} + \nabla \cdot (C_A \mathbf{v}_A) = \bar{\omega}_A, \tag{5.12.13}$$

where $\bar{\omega}_A$ is the molar rate of production of A per unit volume. It follows from Eqs. (5.12.3), (5.12.5a), (5.12.6a), and (5.12.10) that

$$\frac{\partial \rho_A}{\partial t} + \nabla \cdot (\rho_A \mathbf{v}) = \nabla \cdot (\rho D_{AB} \nabla Y_A) + \omega_A. \tag{5.12.14}$$

Similarly, from Eqs. (5.12.4), (5.12.5b), (5.12.6b), and (5.12.13), we obtain

$$\frac{\partial C_A}{\partial t} + \nabla \cdot (C_A \mathbf{v}^*) = \nabla \cdot (C D_{AB} \nabla X_A) + \bar{\omega}_A. \tag{5.12.15}$$

Notice that if chemical reactions are absent and all velocities vanish, then

$$\frac{\partial C_A}{\partial t} = D_{AB} \nabla^2 C_A, \tag{5.12.16}$$

which is called *Fick's second law* of diffusion, and is valid in solids or stationary nonreacting fluids.

In view of Eqs. (5.12.2) and (5.12.10) the continuity equation for a multicomponent system becomes

$$\frac{\partial}{\partial t}(\rho Y_i) + \nabla \cdot [\rho Y_i(\mathbf{v} + \mathbf{V}_i)] = \omega_i, \tag{5.12.17}$$

where we have used the relation $\rho_i = \rho Y_i$. Carrying out the differentiations in Eq. (5.12.17) and using Eq. (5.12.12) leads to

$$\rho\frac{\partial Y_i}{\partial t} + \rho(\mathbf{v} \cdot \nabla)Y_i + \nabla \cdot (\rho Y_i \mathbf{V}_i) = \omega_i \quad (i = 1, 2, \ldots, N), \tag{5.12.18}$$

which indicates the existence of N equations. Thus, the addition of these equations gives the continuity equation for the mixture. It is now obvious that any one of these N equations may be replaced by the continuity equation for the mixture in any given problem, indicating that only $N - 1$ of the Y_i's are independent.

Upon substitution of the relation from Eqs. (5.12.7) and (5.12.8) into Eq. (5.12.18), the conservation of mass equation for Y_i with $D_i = D$ takes the form

$$\rho\frac{\partial Y_i}{\partial t} + \rho(\mathbf{v} \cdot \nabla)Y_i - \nabla \cdot (\rho D \nabla Y_i) = \omega_i, \tag{5.12.19}$$

where the reaction rate ω_i is determined by the phenomenological chemical kinetic expression that will be presented in subsection 5.12.6. Further discussion of diffusion velocities is given in subsection 5.12.5.

5.12.3 Conservation of Momentum

For a reacting fluid mixture that involves a species k ($k = 1, 2, \ldots, N$), the body forces ρF_i acting on species k in direction i will contribute to the rate of change of the momentum:

$$\rho F_i = \rho \sum_{k=1}^{N} Y_k f_{ki}. \tag{5.12.20}$$

These body forces are now added to the forces due to velocity gradients so that, using Eq. (5.3.17), we can write the momentum equations in the form

$$\rho\frac{\partial \mathbf{v}}{\partial t} + \rho(\mathbf{v} \cdot \nabla)\mathbf{v} = -\nabla p + \mu(\nabla^2\mathbf{v} + \tfrac{1}{3}\nabla(\nabla \cdot \mathbf{v})) + \rho\sum_{k=1}^{N} Y_k f_k, \tag{5.12.21}$$

in which the bulk modulus is neglected. If Eq. (5.12.21) is written in terms of tensorial components, then we have

$$\rho\frac{\partial v_i}{\partial t} + \rho\frac{\partial v_i}{\partial x_j}v_j = -\frac{\partial p}{\partial x_i} + \mu\left(\frac{\partial^2 v_i}{\partial x_j \partial x_j} + \frac{1}{3}\frac{\partial^2 v_j}{\partial x_j \partial x_i}\right) + \rho\sum_{k=1}^{N} Y_k f_{ki}, \tag{5.12.22}$$

where the indices i, j refer to the directions in Eulerian coordinates and the subscript $k = 1, 2, \ldots, N$ denotes the species.

5.12.4 Conservation of Energy

In deriving the equation for the conservation of energy for a multicomponent system, we must be aware that the heat flux is contributed by the temperature gradient and radiation and by an additional heat flux due to diffusion velocities given by

$$\rho\sum_{i=1}^{N} H_i Y_i \mathbf{V}_i,$$

where H_i is the average enthalpy (per unit mass) associated with the species i,

$$H_i = H_i^0 + \int_{T^0}^{T} c_{pi}\,\mathrm{d}T \quad (i = 1, \ldots, N), \tag{5.12.23}$$

and where H_i^0 is the standard heat of formation per unit mass for species i at reference temperature T^0. The heat flux is further augmented by the "thermal-diffusion" effect, called the *Soret effect*, which represents the effect of temperature gradients on diffusion. On the other hand, the reciprocal "diffusive-thermal" effect, called the *Dufour effect*, which indicates the effect of mass concentration gradients on the energy, also gives rise to the heat flux. These effects are given by

$$R^0 T\sum_{i=1}^{N}\sum_{j=1}^{N}\left(\frac{X_j \alpha_i}{W_i D_{ij}}\right)(\mathbf{V}_i - \mathbf{V}_j),$$

where D_{ij} is the mass diffusivity, and α_i is the thermal diffusivity. Thus, the total heat flux is of the form

$$\mathbf{q} = \mathbf{q}^{(c)} + \mathbf{q}^{(R)} + \rho\sum_{i=1}^{N} H_i Y_i \mathbf{V}_i + R^0 T\sum_{i=1}^{N}\sum_{j=1}^{N}\left(\frac{X_j \alpha_i}{W_i D_{ij}}\right)(\mathbf{V}_i - \mathbf{V}_j), \tag{5.12.24}$$

in which the heat flux contributions consist of conduction (the first term), radiation (the second term), chemical diffusion (the third term), and Soret and Dufour effects (the fourth term).

Let us now consider the energy equation (5.3.18),

$$\rho\frac{\partial \epsilon}{\partial t} + \rho(\mathbf{v}\cdot\nabla)\epsilon = \sigma_{ij}\frac{\partial v_j}{\partial x_i} - \nabla\cdot\mathbf{q}, \tag{5.12.25}$$

where the stress tensor σ_{ij} is as given by Eq. (5.2.29) and ϵ is the specific internal energy, defined as:

$$\epsilon = H - \frac{p}{\rho} = \sum_{i=1}^{N} H_i Y_i - \frac{p}{\rho}. \tag{5.12.26}$$

Here, the sign of $\nabla\cdot\mathbf{q}$ is the reverse of what it was in Eq. (5.3.18), because the definition for $\mathbf{q}$ has the opposite sign in this case from what it had in Eq. (4.4.36).

In general, the Soret and Dufour effects and the radiative heat flux are negligible. Thus, the energy equation becomes

$$\begin{aligned}\rho c_p\frac{\partial T}{\partial t} &+ \rho c_p(\mathbf{v}\cdot\nabla)T - \frac{\partial p}{\partial t} - (\mathbf{v}\cdot\nabla)p - p\nabla\cdot\mathbf{v}\\ &- \sigma_{ij}\frac{\partial v_j}{\partial x_i} - k\nabla^2 T + \nabla\cdot\rho\sum_{k=1}^{N} H_k Y_k \mathbf{V}_k = 0.\end{aligned} \tag{5.12.27}$$

A more popular form of the energy equation results from the so-called Shvab–Zel'dovich formulation, as presented in subsection 5.12.6.

5.12.5 Physical Derivation of Multicomponent Diffusion

In subsection 5.12.4 it was observed that the diffusion velocity is related to Fick's first law. The exact method of finding the diffusion velocity, however, can be introduced from the kinetic theory of gases (Hirschfelder, Curtis, and Bird 1954). The three-dimensional dynamical problem of binary collision between two particles with masses m_i and m_j is found to be mathematically equivalent to a one-body problem in a plane with the reduced mass

$$\mu_{ij} = \frac{m_i m_j}{m_i + m_j}, \tag{5.12.28}$$

which is related to the sum $\mathbf{S}_i$ of the contributions due to collision and body forces:

$$\mathbf{S}_i = \sum_{j=1}^{N} \mu_{ij} Z_{ij}(\mathbf{V}_j - \mathbf{V}_i) + \sum_{j=1}^{N} \rho Y_i Y_j \mathbf{f}_i \quad (i = 1, 2, \ldots, N), \tag{5.12.29}$$

where $\sum_{j=i}^{N} Y_j = 1$ and Z_{ij} is the total number of collisions per unit volume per second between the molecules of types i and j.

The partial pressure p_i of species i physically represents the momentum of molecules of type i transported per second across a surface of unit area, traveling with the mass-average velocity of the fluid. Thus, the quantity $\mathbf{S}_i$ may also be expressed by the relation

$$\mathbf{S}_i = \rho_i \frac{D\mathbf{v}}{Dt} + \nabla p_i = \rho Y_i \frac{D\mathbf{v}}{Dt} + \nabla p_i \quad (i = 1, 2, \ldots, N). \quad (5.12.30)$$

Let us now introduce Dalton's law of partial pressure,

$$p_i = X_i p \quad (i = 1, 2, \ldots, N) \quad (5.12.31)$$

from which

$$\nabla p_i = p \nabla X_i + X_i \nabla p. \quad (5.12.32)$$

If the viscous forces are neglected, the momentum equation is of the form

$$\rho \frac{D\mathbf{v}}{Dt} = -\nabla p + \rho \sum_{j=1}^{N} Y_i \mathbf{f}_j. \quad (5.12.33)$$

It follows from Eqs. (5.12.29)–(5.12.33) that

$$\nabla X_i = \sum_{j=1}^{N} \frac{\mu_{ij} Z_{ij}}{p} (\mathbf{V}_j - \mathbf{V}_i) + (Y_i - X_i) \frac{\nabla p}{p} + \frac{\rho}{p} \sum_{j=1}^{N} Y_i Y_j (\mathbf{f}_i - \mathbf{f}_j). \quad (5.12.34)$$

The multicomponent diffusion equation derived more rigorously from the kinetic theory of gases leads to

$$\nabla X_i = \sum_{j=1}^{N} \frac{X_i X_j}{D_{ij}} (\mathbf{V}_j - \mathbf{V}_i) + (Y_i - X_i) \frac{\nabla p}{p} + \frac{\rho}{p} \sum_{j=1}^{N} Y_i Y_j (\mathbf{f}_i - \mathbf{f}_j) + \sum_{j=1}^{N} \frac{X_i X_j}{\rho D_{ij}} \left(\frac{\alpha_j}{Y_j} - \frac{\alpha_i}{Y_i} \right) \frac{\nabla T}{T} \quad (i = 1, 2, \ldots, N), \quad (5.12.35)$$

where α_j is the thermal diffusion coefficient of species j and

$$D_{ij} = \frac{X_i X_j p}{\mu_{ij} Z_{ij}}. \quad (5.12.36)$$

If pressure-gradient diffusion, body forces, and thermal-gradient diffusion (the Soret and Dufour effects) are negligible, then

$$\nabla X_i = \sum_{j=1}^{N} \frac{X_i X_j}{D_{ij}} (\mathbf{V}_j - \mathbf{V}_i). \quad (5.12.37)$$

When the binary-diffusion coefficients of all pairs of species are equal, the diffusion equation, (5.12.37), reduces to

$$D\nabla X_i = X_i\sum_{j=1}^{N} X_j\mathbf{V}_j - X_i\mathbf{V}_i \quad (i = 1, 2, \ldots, N). \qquad (5.12.38)$$

Multiply Eq. (5.12.38) by Y_i/X_i and sum over i to obtain

$$\sum_{i=1}^{N}\frac{Y_i}{X_i}D\nabla X_i = \sum_{i=1}^{N}Y_i\sum_{j=1}^{N}X_j\mathbf{V}_j - \sum_{i=1}^{N}Y_i\mathbf{V}_i,$$

from which it follows that

$$\sum_{j=1}^{N}X_j\mathbf{V}_j = \sum_{i=1}^{N}Y_iD\nabla\ln X_i = \sum_{j=1}^{N}Y_jD\nabla\ln X_j, \qquad (5.12.39)$$

where we have used the relations

$$\sum_{i=1}^{N}Y_i = 1 \quad \text{and} \quad \sum_{i=1}^{N}Y_i\mathbf{V}_i = 0.$$

When we substitute Eq. (5.12.39) into Eq. (5.12.38) and divide by X_i, we obtain

$$D\left(\nabla\ln X_i - \sum_{j=1}^{N}Y_j\nabla\ln X_j\right) = -\mathbf{V}_i \quad (i = 1, 2, \ldots, N). \qquad (5.12.40)$$

By making use of the relation

$$X_i = \frac{Y_i/W_i}{\sum_{j=1}^{N}(Y_j/W_j)} \qquad (5.12.41)$$

we obtain Fick's first law from Eq. (5.12.40):

$$\mathbf{V}_i = -D\nabla\ln Y_i \quad (i = 1, 2, \ldots, N), \qquad (5.12.42a)$$

$$Y_i\mathbf{V}_i = -D\nabla Y_i \quad (i = 1, 2, \ldots, N), \qquad (5.12.42b)$$

or

$$\rho_iV_i = -\rho D\nabla Y_i \quad (i = 1, 2, \ldots, N), \qquad (5.12.42c)$$

which is in the same form as Eq. (5.12.7a). Thus, Fick's first law is the simplified version of the rigorous form given in Eq. (5.12.35).

5.12.6 Shvab–Zel'dovich Approximations

In the derivation of Fick's first law, a number of simplifications were made. We neglected (1) body forces, (2) Soret and Dufour effects (terms involving α_i), (3) pressure-gradient diffusion, and (4) bulk viscosities. The so-called Shvab–Zel'dovich formulation requires additional simplifications: (1) steady flow, (2) viscous effects, and (3) constant pressure. Thus, the energy equation, (5.12.27), reduces to

$$\rho(\mathbf{v}\cdot\nabla)H = \nabla\cdot(k\nabla T) - \nabla\cdot\left(\rho\sum_{i-1}^{N} H_i Y_i \mathbf{V}_i\right),$$

or

$$\nabla\cdot\left(\rho\sum_{i=1}^{N} H_i Y_i(\mathbf{v}+\mathbf{V}_i) - k\nabla T\right) = 0, \tag{5.12.43}$$

where we use the relations

$$H = \sum_{i=1}^{N} H_i Y_i \quad \text{and} \quad \sum_{i=1}^{N} H_i Y_i \nabla\cdot(\rho\mathbf{v}) = 0.$$

The conservation-of-species equation, (5.12.17), for steady flow becomes

$$\nabla\cdot[\rho Y_i(\mathbf{v}+\mathbf{V}_i)] = \omega_i \quad (i = 1, 2, \ldots, N). \tag{5.12.44}$$

By substituting Eqs. (5.12.23), (5.12.42c), and (5.12.44) into Eq. (5.12.43), we obtain

$$\nabla\cdot\left[\rho\mathbf{v}\int_{T^0}^{T}\left(\sum_{i=1}^{N} Y_i c_{p_i}\right)\mathrm{d}T - \rho D\sum_{i=1}^{N}\nabla Y_i\int_{T^0}^{T} c_{p_i}\,\mathrm{d}T - \rho D\frac{k}{\rho c_p D}c_p\nabla T\right] = -\sum_{i=1}^{N} H_i^0\omega_i. \tag{5.12.45}$$

Assuming that the Lewis number is unity, i.e., that

$$Le = \frac{k}{\rho c_p D} = 1$$

and using the relation

$$\sum_{i=1}^{N} Y_i c_{p_i} = c_p$$

from Eq. (5.12.45) we derive:

$$\nabla\cdot\left[\rho\mathbf{v}\int_{T^0}^{T} c_p\,\mathrm{d}T - \rho D\left(\sum_{i=1}^{N}\nabla Y_i\int_{T^0}^{T} c_{p_i}\,\mathrm{d}T + c_p\nabla T\right)\right] = -\sum_{i=1}^{N} H_i^0\omega_i. \tag{5.12.46}$$

The second term on the left-hand side in Eq. (5.12.46) may be recast in the form

$$\begin{aligned}\sum_{i=1}^{N}(\nabla Y_i)\int_{T^0}^{T} c_{p_i}\,\mathrm{d}T + c_p\nabla T &= \sum_{i=1}^{N}(\nabla Y_i)\int_{T^0}^{T} c_{p_i}(T)\,\mathrm{d}T + \sum_{i=1}^{N} Y_i c_{p_i}(T)\nabla T\\ &= \sum_{i=1}^{N}(\nabla Y_i)\int_{T^0}^{T} c_{p_i}(T)\,\mathrm{d}T\\ &\quad + \sum_{i=1}^{N} Y_i\nabla\left(\int_{T^0}^{T} c_{p_i}(T)\,\mathrm{d}T\right)\\ &= \nabla\left(\sum_{i=1}^{N} Y_i\right)\int_{T^0}^{T} c_{p_i}\,\mathrm{d}T = \nabla\left(\int_{T^0}^{T} c_p\,\mathrm{d}T\right).\end{aligned} \tag{5.12.47}$$

Substituting Eq. (5.12.47) into Eq. (5.12.46) yields

$$\nabla\cdot\left[\rho\mathbf{v}\int_{T^0}^{T} c_p\,\mathrm{d}T - \rho D\nabla\left(\int_{T^0}^{T} c_p\,\mathrm{d}T\right)\right] = -\sum_{i=1}^{N} H_i^0\omega_i. \tag{5.12.48}$$

This is known as the Shvab–Zel'dovich energy equation. Likewise, substituting Eq. (5.12.42a) into Eq. (5.12.44) leads to the Shvab–Zel'dovich species equation:

$$\nabla\cdot(\rho\mathbf{v}Y_i - \rho D\nabla Y_i) = \omega_i. \tag{5.12.49}$$

Note that the specific heat or any transport coefficient of the mixture is not assumed to be constant and the specific heats of all the species are not necessarily considered equal.

If the specific heats are constant for a single species, then, in view of Eqs. (5.12.27) and (5.12.45), the energy equation may be written as:

$$\begin{aligned}&\rho c_p\frac{\partial T}{\partial t} + \rho c_p(\mathbf{v}\cdot\nabla)T - \frac{\partial p}{\partial t} - (\mathbf{v}\cdot\nabla)p - p\nabla\cdot\mathbf{v} - \sigma_{ij}\frac{\partial v_j}{\partial x_i}\\ &\quad - \nabla\cdot(\rho c_p DT\nabla Y) - \nabla\cdot(\rho c_p DLe\nabla T) = -H^0\omega.\end{aligned} \tag{5.12.50}$$

Similarly, the species equation becomes

$$\rho\frac{\partial Y}{\partial t} + \rho(\mathbf{v}\cdot\nabla)Y - \rho\nabla\cdot(D\nabla Y) = \omega. \tag{5.12.51}$$

For a single irreversible reaction step, we have

$$\sum_{i=1}^{N}\nu_i' M_i \rightarrow \sum_{i=1}^{N}\nu_i'' M_i \tag{5.12.52}$$

and

$$\omega = \frac{\omega_i}{W_i(\nu_i'' - \nu_i')} \quad (i = 1, 2, \ldots, N). \tag{5.12.53}$$

If we denote

$$\xi_T = \frac{\displaystyle\int_{T^0}^{T} c_p \, \mathrm{d}T}{\displaystyle\sum_{i=1}^{N} H_i^0 W_i(\nu_i' - \nu_i'')} \tag{5.12.54}$$

$$\xi_i = \frac{Y_i}{W_i(\nu_i'' - \nu_i')} \quad (i = 1, 2, \ldots, N) \tag{5.12.55}$$

the energy equation, (5.12.48), becomes

$$\nabla \cdot (\rho \mathbf{v} \xi_T - \rho D \nabla \xi_T) = \omega, \tag{5.12.56}$$

and the species equation, (5.12.49), is given by

$$\nabla \cdot (\rho \mathbf{v} \xi_i - \rho D \nabla \xi_i) = \omega. \tag{5.12.57}$$

It follows that the nonlinear term ω can be eliminated from N of the $N + 1$ equations corresponding to

$$L(\xi) = \omega, \tag{5.12.58}$$

where L is the linear operator (if ρD is independent of ξ), such that

$$L(\xi) = \nabla \cdot (\rho \mathbf{v} \xi - \rho D \nabla \xi) \tag{5.12.59}$$

and where ξ can be $\xi_T, \xi_1, \xi_2, \ldots, \xi_N$. For $\xi = \xi_1$, we have

$$L(\xi_1) = \omega, \tag{5.12.60}$$

from which other variables are determined by the linear function

$$L(\beta) = 0, \tag{5.12.61}$$

where β and be $\beta_T, \beta_2, \beta_3, \ldots, \beta_N$:

$$\begin{aligned} \beta_T &= \xi_T - \xi_1 \\ \beta_2 &= \xi_2 - \xi_1 \\ &\vdots \\ \beta_N &= \xi_N - \xi_1. \end{aligned}$$

It is obvious that the above approach (Shvab–Zel'dovich formulation) is efficient in a reacting combustion process for reactants that are unmixed initially. By solving the linear equations between flow variables, burning rates can be determined without solving the nonlinear equation.

Remarks

The governing equations for reacting fluids differ from those for nonreacting fluids mainly in the form of the respective continuity equations. There are $N - 1$ species equations for the N species in addition to the continuity equation for the mixture. Therefore, the variables to be solved are

$$\rho, \quad Y_k \quad (k = 1, 2, \ldots, N), \quad \mathbf{v} \quad (i = 1, 2, 3), \quad T, \quad p.$$

The $N + 6$ equations consist of

1 overall mass continuity	Eq. (5.12.12)
$N - 1$ species equations	Eq. (5.12.49) or (5.12.51)
1 equation relating all Y_i	$Y_1 + Y_2 + \ldots + Y_N = 1$
3 momentum equations	Eq. (5.12.22)
1 energy equation	Eq. (5.12.48) or (5.12.50)
1 equation of state	Eq. (5.12.2)

It is possible to solve the earlier energy and species equations, (5.12.18) and (5.12.27), with the diffusion velocities as unknown variables. In this case, however, there are other equations to be solved, including (5.12.35) or (5.12.37) and (5.12.41).

Many of the complicated problems of reacting flows remain unresolved because of the lack of understanding about such physical phenomena as chemical kinetics, turbulence, and phase change. Furthermore, there are computational difficulties, such as nonlinearity caused by convection and "stiffness" of equations due to the widely disparate reaction rates that various species may exhibit. If reacting flows are involved in shock wave discontinuities and turbulence, solutions of appropriate governing equations constitute perhaps the most challenging tasks in continuum mechanics. Some initial attempts are described in Chung (1993).

Problems

5.1 Use the first law of thermodynamics to derive the FLT equation (5.3.12) and the most general form of the governing equations (continuity, momentum, and energy) for a nonisothermal compressible viscous flow in Cartesian coordinates.

5.2 Differentiate the CNS equation or the conservation form of the Navier–Stokes system of equations, and show that the conservation of mass is prerequisite to the conservation of momentum and the conservation of both mass and momentum is the prerequisite to the conservation of energy. Show also that the results of the first law of thermodynamics are recovered in this process.

5.3 Integrate the conservation form of the Navier–Stokes system of equations, obtain the CVS equations, show all components of mass, momentum, and energy, and identify them on the two-dimensional free-body sketches, contributing to both the upstream and downstream sides of the square faces as shown in Fig. 5.3.1.

5.4 Use the results obtained in Problem 5.1 to show all governing equations for three-dimensional, compressible viscous flows completely expanded in terms of the primitive variables (u, v, w, p, ρ, T) and the independent variables (x, y, z).

5.5 Use the second law of thermodynamics to derive the energy equation with the thermal expansion coefficient.

5.6 Repeat Problem 5.1 for curvilinear coordinates. Expand the results to cylindrical coordinates to prove Eqs. (5.4.22)–(5.4.25).

5.7 Repeat Problem 5.1 for spherical coordinates and verify the results given in Eqs. (5.4.26)–(5.4.28).

5.8 (a) Consider two functions, A and B, that are continuous in the spatial coordinates $z_i (i = 1, 2, 3)$. Prove that the divergence of the cross product of gradients A and B vanishes

$$\nabla \cdot (\nabla A \times \nabla B) = 0.$$

(b) Use the relation in (a) to develop the relations between the velocity components and the stream function in two-dimensional, axisymmetric cylindrical and axisymmetric spherical flows.

5.9 Derive the Bernoulli equation, (5.5.23). Hint: First consider a one-dimensional domain with a length of $\mathrm{d}L$ and a height of $\mathrm{d}h$ inclined an angle θ from the horizontal axis. A component of ρF_i in this direction is $-\rho g \sin\theta = -\rho g \,\mathrm{d}h/\mathrm{d}L$ (the negative sign implies the gravity acting in the direction opposite to the positive Cartesian coordinates) in balance with pressure. Apply this process to all three directions (x, y, z), one at a time, and collect the resulting equations.

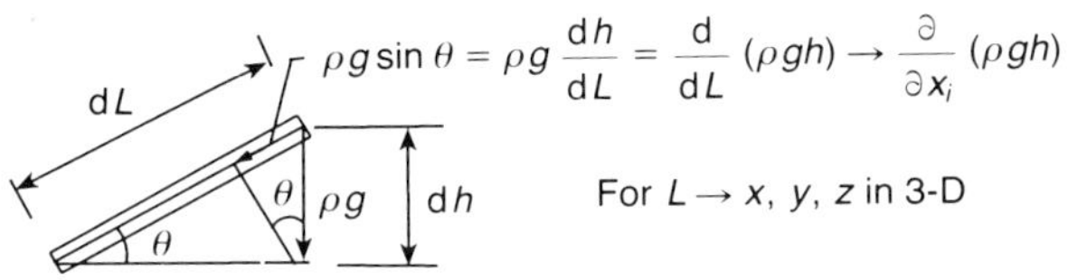

Figure P5.9.

5.10 Derive the Bernoulli equation equivalent to the energy equation, including the head losses due to internal energy, heat transfer, viscous dissipation (friction), and shaft work corresponding to the constant of integration. Hint: Use the third part of Eq. (5.3.27) or (5.3.26), integrate over the domain to obtain the surface integral, and adjust the result for appropriate physical quantities.

5.11 Derive the vorticity transport equation for two-dimensional incompressible flows.

5.12 Use tensorial derivations to show that the steady-state momentum equation may be written in the form

$$\nabla^4\psi = \frac{1}{\nu}\left[\frac{\partial^3\psi}{\partial x\partial y^2}\frac{\partial\psi}{\partial y} - \frac{\partial^3\psi}{\partial x^2\partial y}\frac{\partial\psi}{\partial x} + \frac{\partial^3\psi}{\partial x^3}\frac{\partial\psi}{\partial y} - \frac{\partial^3\psi}{\partial y^3}\frac{\partial\psi}{\partial x}\right].$$

5.13 Derive the boundary conditions required in Problem 5.12. Provide comments on each boundary term.

5.14 Prove Eq. (5.6.9).

5.15 Derive the transport equations for turbulent kinetic energy and turbulent dissipation energy.

5.16 Carry out an order-of-magnitude analysis for the Navier–Stokes system and arrive at the laminar and turbulent boundary layer equations, (5.8.12)–(5.8.17).

5.17 Derive the full potential equation for compressible inviscid flows.

5.18 Show mathematically that an incompressible flow can be defined by the flow speed with $M \ll 1$. Hint: Use the continuity equation, the Bernoulli equation, and the definition of the speed of sound.

5.19 Show that the Crocco equation is of the form in two dimensions,

$$u\omega = T\frac{\partial\eta}{\partial n} - \frac{\partial H}{\partial n},$$

where $\omega = \omega_3$ is the component of a vorticity vector normal to the two-dimensional flow domain.

5.20 Prove Equation (5.10.24).

5.21 Derive the acoustic wave equations in terms of pressure, velocity, and velocity potential. State the assumptions made in your derivation.

5.22 Derive all governing equations required for reacting fluids. State the assumptions made in your derivation.

References

Ackerman, C. C., and B. Berman. 1966. Second Sound in Solid Helium. *Phys. Rev. Lett.* **16**, 789–791.

Aris, R. 1962. *Vectors, Tensors, and the Basic Equations of Fluid Mechanics*. Prentice-Hall, Englewood Cliffs, N.J.

Blasius, H. 1908. Grenzschichten in Flüssigkeiten mit Kleiner Reibung. *Z. Math u. Phys.* **56**, 1–37. English translation in NACA TM 1256.

Boussinesq, J. 1877. Theorie de L'ecoulement Tourbillant. *Mem. Pre. Par. Div.*, Sa. 23, Paris, France.

Chung, T. J. 1988. *Continuum Mechanics*. Prentice-Hall, New York.

Chung, T. J. (ed.) 1993. *Numerical Modeling in Combustion*. Taylor & Francis, New York.

Chung, T. J., and J. Y. Kim. 1984. Two-Dimensional, Combined-Mode Heat Transfer by Conduction, Convection and Radiation in Emitting, Absorbing, and Scattering Media. *ASME Trans. J. Heat Transfer* **106**, 448–452.

Coleman, B. D., and W. Noll. 1959. On Certain Steady Flows of General Fluid. *Archive Rational Mech. Anal.* **3** (4), 289–303.

Dutt, P. 1988. Stable Boundary Conditions and Difference Schemes for Navier–Stokes Equations. *SIAM Num. Anal.* **25**, 245–267.

Eringen, A. C. 1962. *Nonlinear Theory of Continuous Media*. Academic Press, New York.

Ferziger, J. H. 1977. Numerical Simulation of Turbulent Flow. *AIAA Journal* **15**, 1261–1271.

Ferziger, J. H. 1983. High Level Simulations of Turbulent Flows. In *Computational Methods for Turbulent, Transonic, and Viscous Flows*, J. A. Essers (ed.), 93–182.

Frederick, D., and T. S. Chang. 1965. *Continuum Mechanics*. Allyn & Bacon, Boston.

Green, A. E., and R. S. Rivlin. 1957. The Mechanics of Nonlinear Materials with Memory, I. *Archive Rational Mech. Anal.* **1**, 1–21.

Green, A. E., and W. Zerna. 1954. *Theoretical Elasticity*. Oxford University Press, London.

Gurtin, M. E. 1981. *An Introduction to Continuum Mechanics*. Academic Press, New York.

Gustafsson, B., and A. Sundstrom. 1978. Incompletely Parabolic Problems in Fluid Dynamics. *SIAM J. Appl. Math.* **35**, 343–357.

Hanjalic, K., and B. E. Launder. 1976. Contribution Towards a Reynolds-Stress Closure for Low-Reynolds Number Turbulence. *J. Fluid Mech.* **74**, 593–610.
Hinze, J. O. 1975. *Turbulence*. McGraw-Hill, New York.
Hirschfelder, C. F., C. F. Curtis, and R. B. Bird. 1954. *Molecular Theory of Gases and Liquids*. John Wiley, New York.
Hooke, R. 1678. Lectures de Potentia Restitutiva, or Early Science in Oxford 8, 331–356. 1931 in R. T. Gunther.
Huser, A., and Biringen, S. 1993. Direct Numerical Simulation of Turbulent Flow in a Square Duct. *J. Fluid Mech.*, **257**, 65–95.
Lamb, H. 1879. *A Treatise on the Mathematical Theory of the Motion of Fluids*. Cambridge University Press, Cambridge, England.
Launder, B. E., and D. B. Spalding 1972. *Lectures in Mathematical Models of Turbulence*. Academic Press, London.
Liepmann, H. W., and A. Roshko. 1957. *Elements of Gas Dynamics*. John Wiley, New York.
Malvern, L. E. 1969. *Introduction to the Mechanics of a Continuous Medium*. Prentice-Hall, Englewood Cliffs, N.J.
Mooney, M. 1940. A Theory of Large Elastic Deformations. *J. Appl. Phys.* **11**, 582–592.
Noll, W. 1965. A Mathematical Theory of the Mechanical Behavior of Continuous Media. *J. Rational Mech. Anal.* **4**, 3–81.
Oden, J. T. 1972. *Finite Elements in Nonlinear Continua*. McGraw-Hill, New York.
Oden, J. T., and J. N. Reddy. 1976. *Variational Methods in Theoretical Mechanics*. Springer-Verlag, Berlin.
Pierce, A. D. 1981. *Acoustics*. McGraw-Hill, New York.
Prandtl, L. 1904. Über Flüssigkeitbewegung bei sehr Kleiner Reibung. *Verhandlungen IIIrd Intern. Math Kongress Heidelberg 1904*, 484–491.
Reiner, M. 1945. A Mathematical Theory of Dilatancy. *Am. J. Math.* **67**, 305–362.
Ritz, W. 1909. Uber Eine Neue Methods Sur Losung Gewisser Variations-Probleme der Mathematishen Physik. *J. Reine Angew, Math.* **135**, 1.
Rivlin, R. S. 1948. The Hydrodynamics of Non-Newtonian Fluids: 1. *Proc. Roy. Soc. (London)* **A193**, 260–281.
Rivlin, R. S., and J. L. Ericksen. 1955. Stress-Deformation Relations for Isotropic Materials. *J. Rational Mech. Anal.* **4**, 323–425.
Sneddon, I. N. 1957. *Elements of Partial Differential Equations*. McGraw-Hill, New York.
Sokolnikoff, I. S. 1958. *Tensor Analysis: Theory and Applications*. John Wiley, New York.
Sparrow, E. M., and R. D. Cess. 1966. *Radiation Heat Transfer*. Brooks/Cole, Monterey, Calif.
Strikwerda, J. C. 1976. Initial Boundary Value Problems for Incompletely Parabolic Systems. *Comm. Pure Appl. Math.* **30**, 797–822.
Temkin, S. 1981. *Elements of Acoustics*. John Wiley, New York.
Truesdell, C., and W. Noll. 1965. The Non-linear Theories of Mechanics. In S. Flügge (ed.), *Encyclopedia of Physics*, vol. **3-3**. Springer-Verlag, Berlin.

Truesdell, C., and R. A. Toupin. 1960. The Classical Field Theories. In S. Flügge (ed.), *Encyclopedia of Physics*, Vol. **3-1**, 226–793. Springer-Verlag, Berlin.

Von Mises, R. 1913. Göttinger Nachrichten. *Math. Phys.* **K1**, 582.

Index